A

CONCISE BUILDING
ENCYCLOPAEDIA

A
CONCISE BUILDING ENCYCLOPAEDIA

ILLUSTRATED

COMPILED BY

T. CORKHILL

M.I.Struct.E.

*Examiner, 16 years, Building Construction and Woodwork (National
Union of Teachers), Masonry, Brickwork and Concreting
(Union of Lancashire and Cheshire Institutes)
Edi'or, Pitman's "Brickwork, Concrete, and Masonry"*

THIRD EDITION

LONDON
SIR ISAAC PITMAN AND SONS LTD.

Third edition 1951
Reprinted 1955
Reprinted 1957
Reprinted 1960
Reprinted 1963
Reprinted 1965
Reprinted 1968

SIR ISAAC PITMAN AND SONS Ltd.
PITMAN HOUSE, PARKER STREET, KINGSWAY, LONDON, W.C.2
THE PITMAN PRESS, BATH
PITMAN HOUSE, BOUVERIE STREET, CARLTON, VICTORIA 3053
P.O. BOX 7721, JOHANNESBURG, TRANSVAAL
P.O. BOX 6038, PORTAL STREET, NAIROBI, KENYA

ASSOCIATED COMPANIES

PITMAN MEDICAL PUBLISHING COMPANY Ltd
46 CHARLOTTE STREET, LONDON, W.1

PITMAN PUBLISHING CORPORATION
20 EAST 46TH STREET, NEW YORK, N.Y. 10017

SIR ISAAC PITMAN & SONS (CANADA) Ltd.
PITMAN HOUSE, 381–383 CHURCH STREET, TORONTO

SBN: 273   40827   5

MADE IN GREAT BRITAIN AT THE PITMAN PRESS, BATH
F8

PREFACE

THIS encyclopaedia has been compiled to assist the younger members of the various professions and crafts, and as a supplement to the books dealing with the respective subjects. It provides an easy source of reference for an explanation of both common and unusual words and phrases. The difficulty with unfamiliar words is to find their association, and it is hoped that sufficient guidance is given in this respect; although in nearly all cases the explanations are complete in themselves and require no further research. In a few cases where a useful explanation was too lengthy to be included, only sufficient information is given to suggest an appropriate textbook.

The large number of proprietary and patent materials on the market made it impossible to include more than a representative number in a book of this character, and there has been no attempt to differentiate between the respective merits of those included. The further references, at the end of a definition, suggest either an illustration or supplementary information, and the reader should continue as directed to obtain a thorough understanding of the subject. The object has been to keep the book within the means of all engaged in building; and for this reason the well-known tools and appliances, common to the various trades, have been omitted. Some of the words are local words and have no significance outside the particular area in which they are used; this applies especially to some of the terms used in Scotland. About 14,000 words are explained, and 1200 illustrated.

T. C.

A
CONCISE BUILDING
ENCYCLOPAEDIA

A

a.b. Abbreviation for *as before*.

Abaciscus. 1. A square compartment in mosaic work. 2. Any flat member. 3. Small pieces, or tesserae, used in tessellated pavements.

Abacus. The crowning member of the capital to a column, forming a seating for the horizontal members, or a springing for arched members. In the Tuscan, Doric, and Ionic orders it is rectangular; but it is curved inwards in the Corinthian and Composite, and has a central rose.

Doric

Abatement. Waste of material to obtain the required sizes.

Abattoir. A public slaughter-house.

Abat-vent. A roof to a tower, with small inclination.

Abbey. A monastery or convent of the highest rank.

Abele. The white poplar, or *Populus alba*.

Abel Test. An apparatus for determining the flash point of paraffin.

Corinthian

ABACUS

Abies. The botanical name for fir. Now called whitewood.

Abrasives. Natural or artificial materials for shaping, cleaning, or smoothing surfaces. Artificial abrasives are usually either silicon carbide (carborundum) or aluminous (alundum and aloxite). Natural abrasives include emery, corundum, and diamond. Grinding wheels are generally aluminous abrasives but vary according to their use. The abrasives used in the building trades are usually carborundum, emery, flint, garnet, sand and glass-paper. *Wood* is made smooth by glass-paper; *masonry* by rubbing with a similar piece of stone; *brickwork* by a piece of similar kind of brick; paintwork is cleaned (cutting down) by pumice stone or re-cast pulverised stone and glass-paper; ironwork by sand-blasting, file, wire brush, emery cloth. Fine abrasive powders such as powdered pumice stone, Tripoli powder, and rouge are used for varnish, polish, and special purposes. See *Glass-paper*, *Flour-paper*, *Sand-paper*, and *Sander*.

Abrasive Tools. Files, rasps, and glass-paper.

Abscissae. Measurements parallel to the x co-ordinate, in the plotting of points on squared paper. See *Graphs*.

Absorption. This property demands careful consideration in the selection of building materials for damp situations, and in acoustical problems. Porous materials have a higher coefficient of absorption than dense materials.

Abstracting. The operation of collecting the squared dimensions under the various headings, preparatory to billing for quantities.

Abura. A Nigerian light brown hardwood, with mahogany texture but plain figure. Used for general purposes, vats, interior joinery, etc. Very resistant to acid. The wood is rather light and soft, shrinks considerably in seasoning, and is not strong or durable but is easily seasoned and wrought and polishes well. Weight about 30 lb per c. foot.

Abut, To. To adjoin at the end.

Abutment. The pier from which an arch springs.

Abutment Cheeks. The surfaces on each side of a mortise receiving the thrust from the shoulders of the tenon.

Acacia. A large number of species and a big variety of woods. They vary considerably but they are strong, heavy, and mostly difficult to work. The best known in this country are Babul, Blackwood, Cutch, Gidgee, Kanti, Knobthorne, Koa, Myall, Raspberry Jam Wood, Talane, Thorn, Wattle. Also see *Robinia* and *Queensland Acacia*.

Academy. A centre of instruction of a particular art or science. A school for secondary education. A place for the display of works of art.

Acajou. A yellowish hardwood with red markings and beautiful figure from C. America. It is not durable but resists insects and is used chiefly as veneers. The name implies mahogany. Weight about 30 lb. per c. ft.

Acanthus Leaf. The leaf of the bear's breech plant, which is imitated in the Corinthian and Composite capitals (*q.v.*).

Acapú. A Central American hardwood, chocolate brown in colour with lighter stripes. It is hard, heavy, durable, and strong, polishes well but is difficult to work. It is used for superior joinery, flooring, etc. Weight about 58 lb. per c. ft.

Accelerators. 1. The various chemicals used for speeding up the setting of cement and concrete. They are usually products of potash and soda silicates, or calcium chloride. Setting may also be accelerated by using a rocking frame, when casting. 2. Electrically driven accelerators are appliances, or pumps, for forcing the circulation in low pressure hot-water heating, and for removing the steam from brick kilns.

Access Eye. An opening in a pipe, usually with screw cap, for cleaning purposes.

Access Gulleys. Inspection gulleys having a special cover for the easy access of rods when cleaning branch drains. A drain junction with removable cover plate.

Access Plate. A plate bolted on the side of a tank for a hot water system, etc., that can be removed when required.

Accrington Bricks. Hard red bricks made from Accrington shale, which contains iron oxide. See *Lancashire Bricks*.

Acer. The botanical name for the maple tree.

Acetas. A proprietary acid resisting asphalt for floors, tanks, etc.

Acetate of Lead. *Sugar of lead.* When ground in oil, it is used as a drier for the lighter tints of paints.

Acetylene Gas. A colourless gas yielding a white light. The gas is usually generated by allowing water to come in contact with calcium carbide.

Acid-proof Bricks. Salt glazed fire bricks; or vitrified bricks containing a high percentage of true clay.

Acid Treatment. 1. The application of muriatic acid to concrete to remove the cement face to expose the aggregate. See *Retarders*. 2. Forming a patterned or obscured face on glass by means of acid.

Acoustele. A Gyproc acoustic ceiling.

Acoustic Plaster. A special form of plaster with a porous texture and rough surface for absorbing sound.

Acoustics. The science of sound. The phenomena and laws governing the reception of sound in public halls.

Acratex. A proprietary hard-gloss bituminous mastic paint.

Acre. An area of 4840 sq. yd. or 10 sq. chn.

Acropolis. The principal part of an ancient Grecian city.

Acroter. A small pedestal to a pediment, to carry an ornament or statue. Also applied to pinnacles, or ornaments, on parapets or horizontal copings.

Acrow. A registered name for numerous building appliances : travelling cradles, steel scaffolding, road forms, etc.

Action. A term used in mechanics to denote the effect of forces, or loads. The forces acting in the opposite direction, to produce equilibrium, are the reactions. Action and reaction are equal and opposite, for equilibrium.

Activated Sludge. Aerated sewage used for the purification of untreated sewage, by combination.

Acts. The Public Health Acts, 1875, 1888, 1890, 1907, 1925, 1936, that control the by-laws of Local Authorities. These Acts, together with the Factories Acts, 1937 and 1948, are continually being amended. The London Building Acts are specially applicable to London, q.v.

a.d. Abbreviation for *average depth,* and *air dried.*

Adamant Clinkers. Paving bricks, usually with chamfered edges. They are about 6 in. $\times$ 1$\frac{3}{4}$ in. $\times$ 2$\frac{5}{8}$ in.

Adamant Plaster. A patent plaster with plaster of Paris as the base, and suitable for two-coat work. The mixture for undercoats contains sand, glue, sawdust, with a little borax and washing soda. The glue acts as a retarder.

Adam Architecture. A style of architecture of a classical and refined character prevalent in the latter part of the eighteenth century, and introduced by the four brothers Adam.

Adam's Formula. A practical formula for brick piers. Safe load per sq. ft. $= \dfrac{24W - Wr}{18}$, where $W =$ safe crushing strength of brickwork in tons per sq. ft., and $r =$ ratio of height to least thickness of pier.

Adam's Water Bar. A patent bar for the bottom of a door or french casement opening inwards, to exclude water and draughts.

Addling. A term applied to rendering that fails through lack of adhesion producing hollowness and hair cracks. The failure is due to weak undercoat and relatively strong finishing coat together with insufficient key in the brickwork.

Adhesion. The act of sticking together, as with glue, cement, etc. The holding power of nails. The adhesion of nails across the grain is twice that with the grain in softwoods.

Adit. The approach to a building.

Admixtures. The various chemicals, etc., added to mortars and concretes to produce special properties in the concrete, such as water-proof, acid-proof, etc.

Adobe. Unburnt sun-dried bricks.

Advice Notes. The advice by post that certain goods have been dispatched for delivery.

Adze. A carpenter's tool having a thin arched blade at right angles to the handle, and used chiefly for squaring logs.

Aegricanes. Ornamentation in the form of ram's heads.

Aegrit. A patent plaster for facing or skimming wall-boards.

Aerobic. Organisms in filter beds for the purification of sewage.

Aerocrete. A proprietary material for making cellular concrete. Aluminium powder mixed with the cement acts like yeast in dough.

Aerodrome. The hangars, repair sheds, landing ground, etc., constituting a base for aviation.

Aerograph. A patent spraying equipment.

Aeron. A patent spraying machine.

Aerostyle. A patent portable spray equipment.

Aetoma. The tympanum of a pediment.

Afara. A yellow lustrous hardwood from Nigeria. It is light but strong and fairly hard, and easily wrought. It is used for match-boarding, shelving, etc. Weight about 30 lb. per c. ft. Also called Eghoin and Limba.

After Flush Cistern. A W.C. cistern having a small chamber which is closed when the cistern is full. The chamber fills during the flush of water and gradually fills the basin.

Agba. A Nigerian pinkish-brown hardwood somewhat like mahogany. Used for interior joinery, flooring, etc. The wood is also called Moboron, Nigerian Cedar, etc., and is fairly hard, heavy, strong, and stable, and may be obtained in large sizes. Weight about 35 lb. per c. ft.

Agg. Abbreviation for *aggregate*.

Aggregate or Agglomerate. The broken material and sand used with cement, or lime, to form concrete.

Agora. The market places of ancient Greece.

Ah-set. A patent rack and anchor for embedding in concrete and brick-work, to provide a fixing.

Air Bricks. Perforated bricks made specially to provide ventilation.

Air Brush. See *Aerograph*.

Air Cock. A tap to allow air to escape from hot-water fittings in the event of air lock.

Air Conditioning. Plant for purifying the air in public buildings and for retaining the required temperature.

Air Doors. Used to regulate air currents in mines.

Air Drain. An open space, or narrow *area*, round buildings to keep a basement, below ground level, damp-proof.

Air Ducts. An enclosed passage for conveying air from one point to another.

Airey. A prefabricated house, especially suitable for rural areas.

Air Flue. A tube for conveying air to or from a room for ventilating purposes.

Air Grates. Serve the same purpose as air bricks but they are larger and made of iron or terracotta or stone.

Air Inlet. A valve with mica flap to allow the fresh air free access to the drains. See *F. A. I.*

Air Lock. 1. The chambers for allowing men and materials to be admitted to caissons without releasing the high-pressure air. 2. Air admitted into water pipes and which causes *knocking*.

Air Pipe. A by-pass to prevent the unsealing of traps.

Air Seal. An acoustical material consisting of air-blown glue.

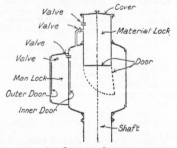

AIR LOCK TO CAISSON

Air Seasoned. Applied to wood that has been seasoned in the open air in stacks or steers, to distinguish from kiln seasoned wood. See *Seasoning*.

Air Siphon. An inclined tube built into a wall to prevent damp.

Air Slaked. The slaking of cement and lime by absorption of moisture from the air.

Aisle. Literally meaning a wing, but generally applied to the lateral divisions in a church. A longitudinal division in the plan of a building.

Akle. A dark brown hardwood from the Philippines very similar to black walnut. It is hard, heavy, strong, stable, durable, and with beautiful figure. It is not difficult to season or work. The wood forms a lather when rubbed in water and is odorous. Used for superior polished work. Weight about 45 lb. per c. ft.

Akousticos. A proprietary insulating and acoustical material.

Akoustolith Tiles. Made to the specification of Professor Sabine for walls and ceilings for the rapid absorption of sound, to prevent reverberation.

Ala. A space opening on to the atrium of old Roman buildings.

Alabascote. A proprietary plastic material for textured surfaces.

Alabaster. A semi-transparent sulphate of lime, or gypsum, used for decorative purposes, electric light bowls, etc.

Alabastine. A proprietary filling for cracks in plaster.

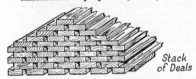

Stack of Deals

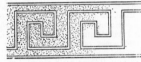

AIR SEASONING   A LA GRECQUE

A la Grecque. The Greek fret ornament.

Albagloss. A registered name for a white enamel. Albamatte is the same firm's flat oil paint.

Albite. Soda felspar, $NaAlSi_3O_8$, found in granites.

Alburnum. The outer part, or sapwood, of a tree trunk.

Alclad. A registered name for an alloy coated with aluminium to prevent corrosion.

Alcove. A deep recess in a room, usually containing a seat.

Alcrete. A registered construction for a prefabricated house.

Alder. *Alnus.* A European hardwood. White when first cut, then red, then fading to reddish yellow. Durable under water. Seldom used in buildings, but used for piles and foundations. Weight about 38 lb. per c. ft.

Alframe. A registered construction for prefabricated houses.

Alidade. A surveying instrument used in topography for obtaining heights and distances.

Alignment. Arranging in a straight line.

Alignment Chart. A chart in which several related values are shown. It serves the same purpose as an experimental graph, and is used in structural design to avoid the repetition of calculations.

Alkalis. Potash, soda, lime. They neutralize acids. Small quantities increase the fusibility of fire-clays.

Alkathene. A proprietary plastic tubing for plumbing services, etc.

Alley. A long narrow passage between buildings; in parks and gardens, a walk bordered by trees.

Alligator. See *Timber Joint Connectors.*

Alloy. A mixture of two or more metals. For example, bronze is an alloy of copper and tin.

All-Rowlock. Alternate courses of headers and stretchers on edge.

Almery. See *Ambry.*

Almonry. A building attached to a monastery from which alms are distributed.

Aloring. The parapet wall protecting the alur, *q.v.*

Alpax. An alloy of aluminium and silicon.

Altar. An elevated place or structure for sacrificial purposes. The communion table in a Christian church.

Altar Courses. The steps forming the side walls of graving docks.

Altar-piece. A decorative screen, or reredos, placed behind the altar.

Altazimuth. A telescopic surveying instrument for measuring altitudes.

Alternating Current. When the electric current generated by a dynamo is reversed at every half revolution it is alternating.

Alto relievo. High relief. Sculptured figures projecting at least half of their thickness from the background.

Altro. A registered name for non-slip treads and tiles.

Alum. A mineral salt, used to control the setting of plaster.

Alumina. The characteristic ingredient of ordinary clay.

Alumina Cement. Rapid-hardening cement, prepared by burning a mixture of lime and bauxite at a high temperature. It is mixed and used like Portland cement, but the two *must not* be mixed together. The constituents are approx. 10 per cent silica, 40 per cent alumina, 40 per cent lime, 10 per cent iron oxide. See *Ciment Fondu.*

Aluminex. A registered roof-glazing material.

Aluminium Alloys. These are used extensively in Building for many purposes, because of their light weight : structural sections, patent glazing, metal windows, gutters, etc. For structural work the metal consists of about 92 per cent aluminium with other elements (copper, nickel, manganese, magnesium, silicon, zinc, iron) according to the requirements. The sections are obtained by extrusion, and plates by rolling. The tensile strength is up to 40 tons per sq. in.

Aluminium Metal. A very light metal, S.G. 2·6. It is only used in Building as an alloy.

Aluminium Paint. Silver bronze composition. It consists of aluminium powder suspended in a medium, which may be slow-drying oil varnish or quick-drying spirit varnish. Used chiefly on metalwork. A good preservative and very durable.

Alundum Aggregates. Manufactured from very hard aluminous ore and used to make non-slip cement surfaces.

Alur. A passage or gallery round a parapet.

Amalgaline Plumbing. A method of joining pipes by using a metal ribbon having a low melting point which, when the blow-lamp is applied, causes the surfaces of the pipes to fuse together.

Amalgam. A compound consisting of mercury and some other metal.

Amaranthus. A decorative wood from Brazil. Turns purple when seasoned.

Amazon. A registered figured rolled obscured glass.

Amazonite. A semi-precious stone used with marble for decorative purposes.

Amazu. A proprietary, plastic, damp-proof material.

Amber. Fossilized resin.

Ambetti. An ornamental glass used in leaded lights and stained glass work.

Ambit. The perimeter of a figure. A circuit. A space round a house.

Ambo. A reading desk or pulpit used in the early Christian churches.

Amboyna. A decorative timber from the East Indies. Orange brown with small curls and burrs.

Ambry. A cupboard or closet near the altar to hold the sacred utensils.

Ambulatory. The aisles of a church, or the cloisters of a monastery. A corridor. The continuation of the aisles of the choir round the apse.

American Whitewood. *Liriodendron tulipifera* (Canary wood) and *Tilia americana* (Basswood). Used as substitutes for superior hardwoods. Stain and polish well. Warp freely, and used for inside work only. S.G. ·5.

Ammeter. An instrument for measuring the force of an electric current.

Ammonia. One of the constituents of town atmospheres, specially noticeable where there is stagnant organic matter, as in public lavatories.

Ampere. The unit of electric current.

Amphiprostyle. Structures in the form of Ancient Greek or Roman rectangular temples, with a portico at each end but with no side columns. See *Temple*.

Amphitheatre. An oval or circular open space, or arena, surrounded by tiers of seats.

Amugis. A reddish brown, lustrous, hardwood from the Philippines. It is hard, heavy, strong, and resists insects, but is not durable. The wood is easy to work and season and polishes well. Used for all kinds of superior interior work. Weight about 48 lb. per c. ft.

Anaerobic. Organisms that break down the solid matter in sewage purification.

Anaglyph. An ornament in relief; may be chased, embossed, or engraved.

Analyptic Wallpaper. Embossed wall coverings. Liquid paper pulp is deposited on to a mould, and retains the shape when dry.

Anamorphosis. A perspective drawing showing the object distorted, but when viewed from a particular position the true proportions are shown.

Ancaster. Limestones from Lincolnshire. Two varieties: free bed and weather bed. The former is cream colour with fine grain and used for interior dressings. The weight is 135 lb per c. ft. with crushing strength about 210 tons per sq. ft. It is rather absorbent and does not weather well. The latter is light brown with coarser grain and is hard and compact with good weathering properties. Weight about 156 lb. per c. ft. Crushing strength about 550 tons per sq. ft.

Anchor. 1. Used with the egg in ornamentation, and called *egg and dart* moulding. 2. Any form of metal fixing, one end of which is built in, or anchored, in the brickwork, stonework, or concrete. See *Door Frame*.

Anchor Bolt. A bolt anchored into the masonry and holding down some part of the structure or a machine. See *Masonry Fixings* and *Balcony*.

Anchor Box. A dovetailed cast-iron box, designed by Goetz, and built into the wall. The wooden beam rests in the box and is notched over a projecting lug on the base plate, thus anchoring the beam to the wall.

Anchoring. Fixing masonry to steel-work.

Anchor Plates. The plates, or washers, used with an anchor bolt.

Anchor Towers. The towers for a derrick crane, to which the stays are anchored. See *Derrick Towers*.

Ancient Architecture. Usually applied to Egyptian, Assyrian, and Persian architecture, typical examples of which are the Pyramids, 3700 B.C., Nineveh, 1290 B.C., and Persepolis, 520 B.C.

Ancient Lights. The Act of 1833, by which if a tenement has had continuous access of light for 20 years it is legally entitled to the continuance of the light, which must not be obstructed by other buildings.

Ancon. 1. An angle or elbow. See *Ancones* and *Crosette*. 2. A quoin or corner stone. 3. A carved ornamental bracket under a cornice to a window or door opening.

Ancona. A decorative walnut, used as a veneer.

ANCON

Ancones. 1. Ornaments carved on keystones or pilasters. Also called trusses or consoles. 2. Projections on stones for fixing purposes. They are cleaned off after fixing.

Anderite. Bitumen on a hessian base, for damp-proof courses.

Anderson. A registered name for waterproof feltings, damp courses, wall boards, etc.

Anderson's Formula. A formula used to obtain the weight of the girder, in the design of plate girders, as an approximate allowance.

Andirons. See *Fire-dogs.*

Andrew's Formula. A formula used in the design of struts for frame buildings.

Anemometer. An instrument for recording the velocity of the wind, from which the pressure p is obtained. $p = \cdot 003\,V^2$.

Angel Beam. A hammer beam carved at the end into human form.

Angel Choir. A room enclosed by screens of adjustable louvres to regulate the volume of sound from a hidden choir. See *Dimmers.*

Angica. A hard heavy decorative timber from Brazil. Yellowish brown with black markings. It is used chiefly for constructional work because of its hardness, although it is suitable for cabinet work and polishes well. Weight about 56 lb. per c. ft.

Angiosperms. The hardwoods of commerce. Deciduous trees.

Ang-Kary. A patent clip for shuttering, or formwork, to eliminate nails, loose packings, etc.

Angle Bar. The vertical bars at the angles of windows.

Angle Bead. A vertical bead fixed to the salient angle of a wall. It serves as a screed for the plastering and prevents injury to the corner. It may be of wood or metal.

Angle Bond. The arrangement at the quoins in footings. The heading bond is broken in alternate courses on one face, to interlace the courses.

Angle Brace. A tie to strengthen the angles of framing. A *diagonal brace* goes from corner to corner of rectangular framing.

Angle Bracket. The corner bracket at the mitre of a plaster cornice or coved ceiling. See *Angle Rib.*

Angle Capital. An Ionic capital on the flank column of a portico. The volutes are on three sides.

Angle Iron. 1. A standard steel section used in structural work. See *Rolled Steel Section.* 2. Any metal bracket to strengthen two pieces of timber forming an angle.

Angle Joints. The jointing of timbers, etc., not in the same straight line or in the same plane. There is a large variety, such as halving, mortise and tenon, dovetailing, housing, bridle, cogging, keying, etc. See *Joints.*

Angle of Friction. The tangent of the angle of friction = coefficient of friction, and is denoted by μ (mu). Also called angle of repose.

Angle of Repose. The inclination at which one body will just rest on another without slipping. The natural slope of a bank of earth under the action of the weather.

ANGLE JOINTS

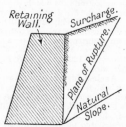

ANGLE OF REPOSE AND PLANE OF RUPTURE

Angle Pier. An attached pier at the salient angle of two intersecting walls.

Angle Post. The large corner post in half-timber structures.

Angle Rafter. Alternative name for hip rafter.

Angle Rib. A curved piece of timber at the mitre of a coved ceiling.

Angle Staff. See *Angle Bead*, and *Metal Trims*.

Angle Tie. The timber tying together the wall-plates and carrying one end of the dragon beam. Also called *dragon tie*.

Anglo-Classic. English Renaissance architecture of the seventeenth to eighteenth centuries. Best known examples are St. Paul's; Banqueting House, Whitehall; Royal Naval College, Greenwich; Colonnade, Hampton Court. Architects of the period included Inigo Jones and Sir Christopher Wren.

Anglo-Norman. English style of architecture A.D. 1066–1189. Also called English Romanesque or twelfth century Gothic. The style was bold and massive, with short massive columns, having bowl or cushion capitals, and richly treated round arches. *Examples*: Winchester, Ely, and Peterborough Cathedrals, etc.

Anglo-Saxon (A.D. 500–1066). The first of the English styles of architecture. *Characteristics*: Walls of rough rubble with ashlar quoins of rude construction; narrow openings with round or triangular heads; square towers; stumpy and massive columns; little ornament. *Examples*: Dover Castle and Church, Earl's Barton Church, etc.

Angular Capital. A modern variety of the Ionic capital with four sides alike and the volutes placed octagonally with the frieze.

Angus Smith's Solution. A preservative for iron pipes consisting of tar, pitch, and linseed oil. The pipes are heated and dipped in the solution.

Anhydrous. Applied to materials from which the moisture has been extracted. The term is especially applied to certain plasters, *q.v.*

Anime, Gum. A fossil resin used in varnish. It dries quickly and is durable and hard, but tends to crack.

Anjan. An Indian wood brick red to dark brown in colour with black streaks. It is hard and very strong and durable and used for superior joinery and where strength and durability are of first importance. Weight about 55 lb. per c. ft.

Ankar. A proprietary wood-fibre insulating wallboard.

Annealing. 1. Changing the physical properties of metals and glass by heating and controlled cooling. The process refines the crystalline structure, removes stresses, and modifies the hardness, ductility, etc. 2. The continuation of the vitrification of bricks until the necessary hardness is obtained.

Annex. An addition to a building.

Annual Ring. The concentric rings of wood fibre added annually to the growing tree. Growth rings. See *Timber*.

Annular Vault. A vault springing from two walls circular in plan and forming concentric circles.

Annulated. Ring-like. Formed of rings or marked with rings.

Annulated Column. A column consisting of a cluster of small columns, much used in early English architecture.

Annulet. A small moulding, usually semicircular in section, circumscribing a column.

Annulus. The area enclosed by two concentric circles. $A = \pi (R^2 - r^2)$.

Anodium. Aluminium alloy used for metal windows.

Anorthite. A lime felspar.

Anston. A dolomite stone from Yorkshire. It is cream in colour, and weighs about 134 lb. per c. ft., with a crushing strength of 833 tons per sq. ft. Selected stone is very good for general purposes.

Antae. The projecting ends of the side walls, beyond the end wall, to form piers for the portico, in Greek and Roman temples. A pilaster complementary to a column but with varied details.

Antaefixae. Ornamented upright blocks placed at regular intervals on the crowning member of a cornice.

Ante-chapel. The outer part at the west end of a collegiate chapel.

Ante-pagmenta. The decorative stone dressings to a doorway.

Ante-room. A room or apartment leading to a more important room. A waiting room.

Anthemion. The honeysuckle ornament.

Anthracite Stoves. Closed stoves burning Welsh anthracite coal, which is smokeless and slow-burning.

Anti-actinic Glass. A patent glass that transmits light but excludes the heat of the sun.

Anti-clockwise. Applied to forces having a turning effect, or moment, in the opposite direction to the movement of the hands of a clock.

Antimony Oxide. A white pigment used in paint. It is obtained by burning *stibnite*, or grey antimony, in air. Also called Antimony white, and sold as Timonex. See *Pigments*.

Antique Bricks. Dark red bricks made with rough and sandy faces to appear old for architectural purposes.

Antique Glass. Made as medieval glass. Large range of colours and used for stained glass windows.

Anti-siphonage Pipes. Air pipes used to prevent the unsealing of W.C. traps, etc., by siphonage. The sizes vary from 2 in. to $3\frac{1}{2}$ in. for soil pipes and from 1 in. to $1\frac{1}{2}$ in. for waste pipes. See *Syphonage*.

Antium. A porch to a southern door, in ancient architecture.

Anti-Vac Trap. A patent anti-siphonage trap.

Anti-vibration. Devices are used to absorb, or damp, the vibrations from machinery, consisting of cork, compressed fibre, or springs.

Ant Queen. A mobile extensible hoist. It operates up to 100 ft.

Ants. Wood-destroying insects of which there are many varieties. The most destructive are the white, black, and yellow ants. Arsenic, paraffin, and tar are protections.

Antwerp Blue. See *Prussian Blue*.

Anvil Washer. See *Concealed Fastener*.

Apa. A brownish-red hardwood with variegated streaks, from Nigeria. There are several species, but they are chiefly hard, heavy, strong and durable, with attractive appearance, and can be used for nearly all purposes. They are easy to season but rather difficult to work and need care in polishing. Weight about 52 lb. per c. ft.

Apartment. A room or chamber. In the plural it denotes a suite of rooms in the same sense as a self-contained flat.

Apco. A proprietary wallboard. Some brands are faced with bakelite etc.

Aperture. An opening in the wall of a building, as a door opening, etc.

Apex Stones. The closing stone at the top, or apex, of a dome or the intersection of vaults or of copings to a gable. See *Gable*.

Apitong. A reddish-brown hardwood from the Philippines. It is heavy, fairly strong and hard, but it should be treated for exterior work in contact with the ground. The ribbon grain and fine texture make it suitable for superior joinery. It is rather difficult to work and requires careful seasoning to prevent warp. Weight about 48 lb. per c. ft.

Apodyterium. 1. See *Atrium*. 2. A dressing room to a bath in ancient Roman villas.

Apophyge. The top and bottom members of the shaft of a column. It is usually in the form of a cavetto. Also called spring, scape, or congé. See *Mouldings*.

Applique. Ornamentation planted, i.e. not worked in the solid.

Approximate, or False, Ellipse. See *Arches*.

Apron. 1. A plank flooring on which lock gates close. 2. *Apron Flashing.* 3. A projecting ornamentation on the lock rail of a door. 4. The stone paving to prevent the toe of a sea-wall from being undermined. 5. A lining at the bottom edge of a stair string to hide the carriages, etc. 6. Deep plates at the front and back of a circular saw and under the saw bench as a protection. 7. The wall between floor level and window sill.

Apron Feeder. A conveyer used in brick making.

Apron Flashing. Sheet lead, or zinc, used to prevent the percolation of water where a vertical surface penetrates a roof. A horizontal flashing. See *Flashing, Stepped,* and *Soaker.*

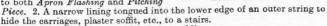

Apron Lining. Vertical linings round the well of a staircase.

Apron Moulding. The raised ornamentation on the lock rail, or apron rail, of a door.

Apron Piece. 1. Sometimes applied to both *Apron Flashing* and *Pitching*

APRON MOULDING

Piece. 2. A narrow lining tongued into the lower edge of an outer string to hide the carriages, plaster soffit, etc., to a stairs.

Apron Rail. A lock rail with apron moulding.

Apse, or Apsis. 1. A semicircular or polygonal end to a building, usually with a vaulted roof. 2. A term used in quarries for rock enclosed in "faults."

Apteral. Applied to a temple having no columns along its sides or flanks.

Aquacrete. A proprietary water repellant Portland cement.

Aquadag. A graphite lubricant, with water as the vehicle.

Aqualithic. A proprietary, liquid, concrete floor-hardener.

AQUEDUCT

Aquaseal. A proprietary cold-applied asphaltic roofing.

Aqueduct. A bridge for carrying water across a valley. An artificial channel.

Aqueous Rocks. Rocks formed by deposits in water or air. Most sandstones and limestones used for building are aqueous rocks.

Aquex. A waterproofing powder for cement renderings and concrete.

Arabesque. Ornamentation representing flowers, fruit, and other objects interwoven together, and characteristic of Mohammedan architecture.

Araeostyle. The columnar arrangement in classic temples where the spacing is $3\frac{1}{2}$ to 4 diameters apart. See *Intercolumniation.*

Aranga. A very hard, heavy, strong and durable hardwood from the Philippines. It is yellowish to pale chocolate in colour, and lustrous. Difficult to work and requires slow seasoning. Used for structural work, superior joinery and where durability and strength are essential. Weight about 55 lb. per c. ft.

Arbour. A bower, or place of shelter or retreat, in a garden, usually of lattice work.

Arcade. A series of arches supported by pillars and carrying a roof to form a covered walk. The arches and pillars are replaced by shops in a modern arcade.

Arcanal. A proprietary anti-corrosive paint.

Arc Boutant. A flying buttress.

Arch Bar. 1. A semicircular bar in a sash. A cot bar. 2. A chimney bar, *q.v.*

Arch Block. A voussoir, *q.v.*

Arch Braces. Additional braces, near the supports, in framed timber bridges. They are used to supplement the ordinary braces where the vertical shear is greatest.

Arch Bricks. Wire-cut bricks made to the required shape for arches, chimneys, etc.

Arches. An arrangement of wedge-shaped blocks mutually supporting each other over an opening and designed to carry the wall and load above. An arch is named according to (a) the curve of its intrados, (b) the purpose it serves in the structure, (c) its resemblance to some familiar object, (d) the number of centres used in describing the outline, (e) some particular style of architecture. The names are as follows: camber, catenary, cycloidal, drop, elliptical, equilateral, flat, Florentine, French or Dutch, gauged, Gothic, horseshoe, hyperbolic, inverted, lancet, Moorish, ogeval, parabolic, rampant, relieving, rere, rough, Saracenic, segmental, semicircular, soldier, squinch, stilted, three (or more) centred (for an

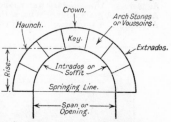

SEMICIRCULAR ARCH: TERMS USED

approximate, or false, ellipse), trimmer, Tudor, Venetian. The parts of an arch and its supports are as follows: abacus, abutment, capital, column, crown, extrados, haunch, impost, intrados, key, label course, lacing course, orders, pier, respond, rise, skewback, soffit, span, spandril, springer, springing, voussoirs. See page 13.

Architectural Order. An entablature supported by columns, with the relative parts carefully proportioned. The three main orders are Doric, Ionic, and Corinthian, to which are usually added the Tuscan and Composite. They form the distinguishing feature of classic architecture. See *Columns and Mouldings*, and page 15.

Architecture. The art of designing buildings, structurally sound, with appropriate materials, suitable for their purpose, and pleasing to the eye.

Architrave. 1. The ornamental mouldings round a door or window opening; usually covering the joint between the plaster and framing. See *Finishings*. 2. The lowest part of the entablature, or epistylium, that rests on the capital of the column. See *Mouldings* and *Pediment.*

Architrave Cornice. An entablature without frieze.

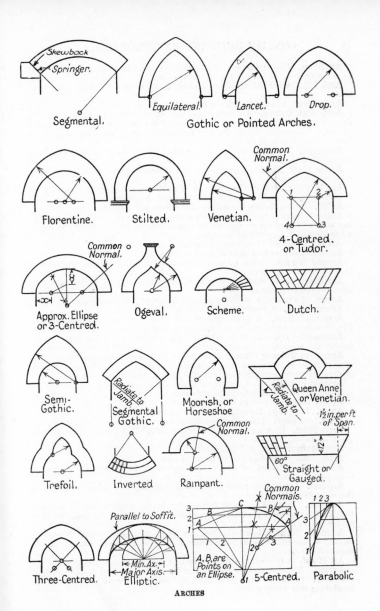

Skewback
Springer.

Segmental.

Equilateral. Lancet. Drop.

Gothic or Pointed Arches.

Florentine. Stilted. Venetian.

Common Normal.

1 2
4 3

4-Centred, or Tudor.

Common Normal.

Approx. Ellipse or 3-Centred.

Ogeval. Scheme. Dutch.

Semi-Gothic.

Radiate to Jamb.
Segmental Gothic.

Moorish, or Horseshoe.

Radiate to Jamb.
Queen Anne or Venetian.

1½ in. per ft of Span.

Trefoil. Inverted.

Common Normal.
Rampant.

½
60°
Straight or Gauged.

Three-Centred.

Parallel to Soffit.
Min. Ax.
Major Axis.
Elliptic.

Common Normals.
3 2 1
C
B B
A A
1 2 1 2 3

A, B, are Points on an Ellipse. 6 1 5-Centred.

1 2 3
3 2 1
Parabolic

ARCHES

Archivolt. The architrave or mouldings on the face of an arch parallel to the intrados. A soffit with moulded edges.

Arch Rib. A curved member of a roof truss, as in a hammer-beam roof truss.

Arch Ring. A ring of bricks or stones forming an arch.

Arch with Orders. An arch consisting of several concentric rings, in thick walls. The opening is in the middle of the thickness of the wall and the rings increase in size to the two faces of the wall, thus allowing for a better distribution of the light.

Arcosolia. Semicircular recesses in the outer walls of ancient churches intended as tombs.

Arctic Glass. Rough patterned rolled obscured glass. Also made with wire reinforcement.

Arc Welding. One of the methods of welding by fusing the metal by an electric arc, struck from a carbon or metallic electrode.

Ardor. A proprietary insulating aluminium-foil sheeting.

Area. 1. A space below ground level, between the main wall and a retaining wall, as a precaution against dampness, and to provide light and air to a basement. 2. Superficial contents.

Arena. The floor of an amphitheatre.

Arènes. Natural mixtures of sand and clay.

Argand Burner. A gas burner in the form of a hollow ring having fine holes at regular intervals.

Argillaceous. A chemical classification of stones and shales in which alumina predominates.

Argon. An inert gas used for electric lamps of low candle-power.

Armar Wood. A proprietary veneered sheet metal.

Armoured Door. A solid fire-resisting door built up of thicknesses of ⅞ in. boards at right angles to each other, and covered with tin-plate.

Armoured Plywood. Plywood faced on one or both sides with metal: galvanized steel, monel metal, aluminium, stainless steel, bronze, or copper.

Armoured Tubular Floors. A fire-resisting self-centring floor consisting of reinforced concrete inverted T-beams, or webs, resting on the flanges of girders and supporting hollow concrete blocks, or tubes. See page 15.

Armour Plate Glass. A registered tough shock-resistant glass.

Arnolite. A proprietary paint remover.

Arnott Valve. A patent ventilating valve.

Arras. Tapestry. Hangings woven with figures.

Arris. An edge. The salient corner where two plane surfaces meet.

Arris Fillet. 1. A local term for tilting fillet. 2. A square fillet used in place of an angle bead.

Arris Gutter. V-shaped gutters.

Arris Rail. A triangular rail used for fences.

Arris-wise. Bricks or tiles laid diagonally. Materials cut or placed diagonally.

Arterial. Applied to main roads connecting up a number of towns or the suburbs of large cities.

Artesian. Applied to wells formed by boring until water is reached that rises by natural pressure. If necessary it is assisted by centrifugal or air-lift pumps.

Artificial Stone. Reconstructed, pre-cast, or patent, stone. Concrete having a face mix of crushed natural stone and Portland cement of the required colour.

Art Metal. Any metal worked to form a decorative feature.

Artoco. A proprietary plastic sheeting in the form of a thin veneer. See *Ioco*. There are over 30 finishes to imitate different materials. It is insulating and damp proof.

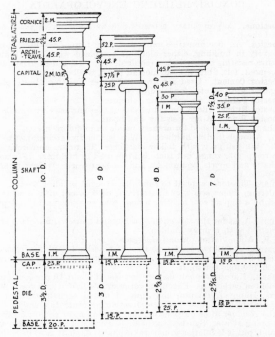

ARCHITECTURAL ORDERS

D = diam. of column. M = module = $\frac{1}{2}D$. P = parts = $\frac{1}{30}M$

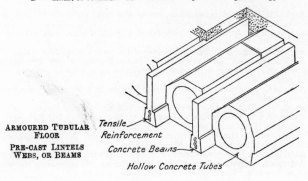

ARMOURED TUBULAR FLOOR

PRE-CAST LINTELS WEBS, OR BEAMS

Tensile Reinforcement

Concrete Beams

Hollow Concrete Tubes

Asbestone. A proprietary asbestos-cement material, in plain or corrugated sheets.

Asbestos. An incombustible mineral of a fine fibrous texture. It is used extensively in building, either alone or combined with other materials, where fire-resisting properties are required. The term is used in conjunction with cement, felt, fibre, mastic, mortar, paint, wallboard, etc.

Asbestos Cement. Used for fire- and water-resisting sheets, plain or corrugated, gutters, etc., for building purposes.

Asbestos Paint. A fire-retarding solution.

Asbestos Rubber Tiles. Asbestos-cement tiles surfaced with rubber.

Asbestos Wood. A fire-resisting wall-board.

Ascham. A store shed for archery tackle and equipment.

Ash. A tough flexible hardwood. Weathers badly, and subject to insect attack. Whitish grey with yellow markings. There are numerous species in the temperate zones, but there is little used in building. It is used for tool shafts and sports' equipment, and for bent work. Weight from 38 to 48 lb. per c. ft.

Ashlar. The term applied to carefully wrought building stones.

Ashlering. Short studs, or quarterings, used to cut off the acute angle formed by a sloping roof meeting a floor, and carrying laths and plaster. See *Foot Plate*.

Ashpit. A pit for the collection of ashes and house refuse. It is often in conjunction with a privy midden.

A.S.P. Abbreviation for *anti-siphonage pipe*.

Aspatria. A dull red sandstone from Cumberland, with fine close grain. It is used for building purposes and weighs 125 lb. per c. ft. with crushing strength 760 tons per sq. ft.

Asphalt. Natural asphalt usually consists of about 10 per cent bitumen and calcium carbonate. European varieties are Limmer, Seyssell, Val-de-Travers, Hanover, Ragusa, and Jura. Trinidad asphalt contains about 40 per cent bitumen, and Guanoco about 60 per cent. It is suggested that it is the residue, after evaporation and oxidation, of petroleum from an outcropping strata. Synthetic asphalt is composed of tar, pitch, tallow, and sand, but it is inferior to natural asphalt. Asphalt used in building is a mixture of asphaltic bitumen with a large proportion of crushed stone. Sand and lime are also used in its manufacture. See *Asphaltic Rock* and *Mastic Asphalt*.

Asphaltex. A proprietary bituminous plastic roofing cement in several colours.

Asphaltic Rock. A bituminous, calcareous rock, pulverized and mixed with melted bitumen, and run into moulds. This is heated and mixed with grit and is applied, in plastic state, to form a tough and waterproof surface.

Asphaltic Slag. A proprietary material for tarred macadam roads.

Asphaltum. An anti-corrosive bitumastic paint made by melting asphalt and thinning with coal-tar naphtha as a solvent. Also resists acid gases.

Assignee. A person to whom the sole interest in a lease has been transferred by the original tenant.

Astel. A plank used for overhead partitioning in tunnelling.

Astos. A proprietary asbestos-asphalt roofing, lead-lined D.P.C., etc.

Astragal. 1. A small semicircular moulding, like a raised bead. It is often carved into the form of berries. See *Mouldings*. 2. T-shaped glazing bars in metal casements. Moulded sash bars.

Astrolabe. A surveying instrument for the determination of latitude.

Astroplax. A patent cement used in plastering. It is used in the same way as Keene's. The basis is white hydraulic cement.

Astylar. Having no columns.

Asymmetrical Façade. The front of a building out of symmetry.

Asymptotes. The asymptote to a curve is a line always approaching the curve but never coinciding with it.

Atala. A reddish-brown to purple red hardwood from Nigeria. It is very hard, heavy, and strong, smooth, and retains sharp arrises. Used for hard-wearing floors. Weight about 62 lb. per c. ft.

Atelier. The workshop of a sculptor or painter. A studio.

Athena Floor. A patent composition for jointless floor coverings.

Atkinson's Cement. See *Cement*.

Atlantes. Male figures used as columns. Also see *Caryatides*.

Atlas. Proprietary asbestos-cement products.

Atlasite. A proprietary asbestos sheeting patterned to represent marbles, timbers, etc.

Atomic Draught Strip. A patent sealing strip of phosphor-bronze alloy for draughty doors and windows.

Atrium. A courtyard in ancient Roman architecture.

Attached Column. One that is attached to a wall, and usually projecting three-quarters of its diameter.

Attached Pier. A projection bonded with a brick wall to provide stability, or support for a beam, roof truss, etc.

Attic. A room in the roof of a building. The ceiling is square with the walls as distinct from a garret. A low storey above an entablature.

Attic Base. A base used in the Ionic and Doric orders consisting of torus, scotia, and lower torus, with fillets between.

Attic Order. An "Order" of small square pillars above the main cornice, in classic architecture.

Attoc. Proprietary floor and wall tiles and partition blocks of hard burnt clay.

Audiograph. An instrument to denote the rate of decay of sound, to detect echo defects.

Auditorium. That part of a public hall allotted to the audience.

Auger. A long twist bit for boring in timber; it is turned with the hands.

Augite. A mineral closely allied to hornblende, and found in many igneous rocks.

Aumbry. See *Ambry*.

Austral Windows. A patent window in which the sashes are balanced on arms to eliminate weights, pulleys, and cords.

Australian Walnut. See *Queensland Walnut*.

Autaram. A patent sewage lift.

Autoclaves. Hardening chambers, or cylinders, for sand-lime bricks, etc. They are long steel cylinders from 30 to 70 ft. long by 6 ft. or more in diameter and hold up to 20,000 bricks. When the chamber is closed the steam is admitted to the required pressure. They are used for curing resin-bonded woodwork, etc.

Autocrat. A proprietary water-paint or distemper.

Auxiliary Planes. Other planes than the *planes of projection* used in geometry.

Auxiliary Projection. The projection of lines, planes, or solids on to a plane other than the ordinary planes of projection.

Auxiliary Rafter. An additional principal rafter often used in large queen-post roof trusses.

Av. Abbreviation for *Average*.

Avodire. A Gold Coast hardwood, cream-coloured with fine interlocked grain. It is somewhat like sycamore in appearance and suitable for superior joinery. Weight about 35 lb. per c. ft.

Axed Arch. One in which the voussoirs are roughly cut to shape, from common bricks, with the bricklayer's axe.

Axial Load. A centric load. A load or force acting through the C.G. of the section of a structural member.

Axis. An imaginary straight line through the centre of gravity of a surface or solid, so that the surface or solid is symmetrical about the axis.

Axle Pulley. Used for sliding sashes. The sash cords, carrying the balance (or sash) weights, pass over the pulleys which are fixed near the top of the pulley stiles. See *Sash and Frame.*

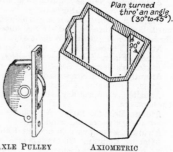

Plan turned thro' an angle (30° to 45°).

AXLE PULLEY   AXIOMETRIC

Axonometric, or **Axiometric.** A form of pictorial drawing in which the plan of the object is turned through an angle of 30° or 45°. Vertical lines are then drawn to show the front and side views. Also see *Planometric.*

B

Bachelor. See *Stone Slates.*

Back. 1. The top surface of an inclined or horizontal timber, or a slate. 2. Same as extrados. 3. The upper, or convex, side of a saw tooth. 4. The hind part furthest removed from the face of anything.

Back Boxing. See *Back Lining.*

Back Drop. A branch drain at a higher level than the main drain, at the manhole. The object is to save deep trenching for the branch drain.

Backer. A narrow slate on the back of a broad square-headed slate where a diminishing of the width commences.

Back Fillet. When door or window jambs project over the face of the wall the *return* of the projection is a back fillet.

Backfilling. Replacing excavated earth in trenches or against a building.

Back Flap. 1. A hinge with large flaps for screwing on to the face of the door and framing. 2. The back leaves for folding window shutters.

Back Flow. The term applied to sewage that *heads up* in the drainage system and flows back into the rain-water connections.

Back Gauge. The distance of the centre line of a bolt or rivet hole from the back edge of an angle cleat.

Back Hearth. The back part of the hearth, between the jambs and under the fire.

Backing. 1. The shaping of a hip rafter or arch rib to the dihedral angle. 2. The inferior materials used behind facings, as in walls.

Backings. 1. Levelling fillets fixed on the top of uneven joists. 2. The horizontal grounds dovetailed into the vertical grounds to carry the linings of internal door jambs. See *Finishings.* 3. Stone suitable for random rubble.

Back-inlet Gulley A drainage fitment, of cast iron or stoneware, consisting of trap, water seal, and open grating. It is arranged so that the waste and rain water pipes discharge below the grating and above the water seal level. See *Gulley.*

Back Linings. 1. Rough thin material fixed to the edges of the linings, in a sash and frame, to enclose the balance weights and to keep the box free from mortar. See *Sash and Frame.* 2. The linings at the back of a recess for folding shutters.

Back Nut. A screwed nut for gas or water pipes.

Back Putty. The layer of putty placed in the rebate, into which the glass is bedded.

Backs. 1. Large settling tanks used in brickmaking, for the washed clay. 2. Principal rafters.

Back Shore. The outer member at the foot of a system of raking shores.

Back Sight. Directing the sight of the instrument back towards the starting point, when surveying.

Back Stairs. A staircase in a large house to the servants quarters.

Back-to-back Grate. See *Combination Grate.*

Bacteria Beds. Used on sewage farms for the purification of sewage.

Badger. 1. A rebate plane similar to a jack plane but with the irons flush with the sides of the plane. 2. An appliance for cleaning the cement from the inside of drains.

Badger Softener. A painter's soft hair brush for fine-grain stippling and for blending, or softening, one colour with another adjacent colour. It is not used for the application of paint.

Badigeon. A mixture of plaster and finely-ground stone dust for internal repair work to stone.

Baffles. 1. Sheet-iron plates fixed above gas fires to catch mortar droppings from the flue. 2. A cage to a geyser flue, near the ceiling, to disperse the foul air in the event of down-draughts. 3. A plate to check any objectionable feature, as noise, vibrations, down-draughts, etc.

Bagac. A bastard teak from the Philippines. See *Teak.*

Bagging. Making good the pinholes in cement surfaces. The small holes are due to trapped air.

Baguette. An astragal or bead.

Bail, or Bale. A wooden bar for dividing the stalls in stables.

Bailey. 1. The spaces between the defence walls of ancient castles, forming courts, or wards. A courtyard. 2. A prison.

Bait. A long rod with hooks, or other arrangement, that *gathers* the glass from the furnace for drawn sheet glass.

Baked Bricks. Underburned bricks such as rubbers, ordinary firebricks, etc.

Baked Plasters. Gypsum calcined at very high temperatures. They are slow setting but give a very hard surface.

Bakelite. A thermo-setting synthetic resin from phenol and formaldehyde. It is made with a filler, fine sawdust, etc., and is hard, and infusible. It is used extensively for decorative articles, veneering plywood, etc. There are many variations sold under trade names. See *Plastics.*

Baker's Rules. Sir Benjamin Baker's rules for the design of retaining walls.

Balance Beam. A long wooden lever fixed to a lock gate to open and close the gate against the pressure of the water.

Balanced Sashes. See *Sash and Frame.*

Balanced Shutters. Shutters sliding vertically and balanced with weights like a "sash and frame" window.

Balanced Steps. Dancing steps, *q.v.*

Balanced Winders. Winders made wider than the usual quarter space winders to give more foothold at the narrow end. They do not radiate to a common centre.

Balance Weights. The weights used to balance vertically sliding sashes

There are two weights to each sash. The top sash weights should be slightly heavier than the sash, and the bottom ones slightly less than the weight of the sash. They are of cast-iron or lead.

Balconette. A balustrade of a balcony on the façade of a building as a decorative feature only.

Balcony. 1. A platform projecting from the face of a building and supported by columns or brackets and enclosed by a balustrade. 2. Tiers of theatre seats between circle and gallery.

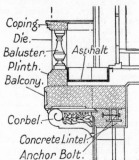

Baldachino. A canopy to an altar.

Baldaquin. See *Ciborium*.

Balection Moulding. See *Bolection*.

Bale Tacks. See *Tingle*.

Balk, or Baulk. A squared log.

Ballast. 1. The excavated material from sea and river deposits in the form of sand and shingle used as an aggregate in concrete. 2. The broken stone or slag used on railways for drainage purposes. See *Metal*.

BALCONY

Ball Bearings. Hardened steel balls used in bearings to lessen the friction.

Ball Catch. A plate carrying a steel ball controlled by a spring. It is let into the edge of a swing or cupboard door to keep the door closed.

Ball Flower. A carved ornament resembling the petals of a flower enclosing a ball. One of the characteristics of the decorated style of architecture.

Ballistraria. A cross-shaped opening in the walls of a turret.

Balloon. A ball surmounting a pier as an ornament. It may be of stone, terra-cotta, concrete, or wood.

Balloon Framing. Framing for timber houses in which the vertical posts run through and the horizontal members are nailed to them.

Ball Taps, or Valves. Taps actuated by a spherical copper float at the end of a cranked lever. There are various types for low pressure, high pressure, and automatic flushing cisterns.

Balsa. A very light wood from South America. Non-conducting, very elastic and difficult to split. It is used where a very light wood is required, for buoyancy, and for insulating purposes. Weight about 9 lb. per c. ft.

Balsam Wool. A proprietary acoustic blanket.

Baltic Timber. Pines and firs from Baltic ports.

Baltimore Truss. A modification of the Pratt truss, used for long spans. The panels are subdivided.

Balusters. The small vertical pillars supporting handrails. See *Balcony*.

Baluster Seatings. The dies worked on masonry plinths and cappings between which the balusters are fixed.

Balustrade. A row of balusters with base and rail, or cap, forming a protective enclosure. See *Balcony*.

Bamboo. Endogenous trees, such as the palm, etc. They are used in the countries in which they grow for structural purposes, but they are not *timber* trees.

Bancal. A decorative covering for a seat. A hardwood from the Philippines, yellow to orange with darker streaks in colour. There are several species but little is imported to this country. Weight about 34 lb. per c. ft.

Band. 1. Any flat member with small projection. 2. A continuous ornamental feature along a wall or around a building. 3. A moulding round the shaft of a column.

Band and Gudgeon. A long strap hinge consisting of a wrought iron band rotating on a pin fixed to the frame. Strictly, the term gudgeon is used only when the pin is fixed in brick or stonework, otherwise it is called *band and hook.*

Bandelet. A narrow band.

B. & J. Abbreviation for *bed and joint.*

Band Moulding. A simple form of architrave.

B. & P. Abbreviation for *bed and point.*

Band Saw. An endless ribbon saw running round two pulley wheels, like a driving belt. They are used for curved work and are usually about 12 ft. long and from ¼ in. wide. Log band saws used for converting logs may be up to 40 ft. long and over 16 in. wide.

Band Stand. A partially enclosed and covered platform on which entertainers give public performances in the open air.

Banister. A local name for a baluster.

Banjo. An adjustable flexible rod for copying curves.

Bank. 1. A bench or seat. 2. A mound, a long sloping mass of earth.

Banker. 1. A block of stone used by the mason as a bench on which to work. 2. A small platform, fenced on three sides, on which concrete is mixed.

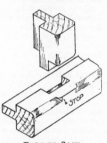

BANKER

Banqueting Hall. A large sumptuous apartment used for important functions and feasts.

Banquette. 1. A footway, on a bridge, raised above the roadway. 2. A seat in a window recess.

B.A.P. Abbreviation for *brass axle pulley.*

Baptistery. A portion of a church in which baptism is performed by immersion.

Bar. 1. The intermediate members of a sash, dividing the glass into smaller squares. 2. A local name for bolts used for securing doors. They are made from flat wrought iron. 3. See *Ledged Door.* 4. A rail or barrier

BARS TO SASH

used as a protection. 5. A counter at which liquid refreshments are served 6. See *Crowntrees.*

Barbican. An advanced work before the gate of a fortified town or castle.

Bareau. A proprietary waterproofing powder for cement and concrete.

Barefaced Tenon. A tenon with one side only shouldered, as used for the lower rails of framed and batten doors.

Barff's Process. A preservative process, in which a film of magnetic oxide of iron is formed on the surface of the metal, by heating the metal to red heat and applying superheated steam.

Barge. The projection of a chimney beyond a gable wall.

Barge Boards. Verge or gable boards. The inclined timbers, on the gable of a building, used to cover the ends of the roof timbers, and for ornamental purposes.

Barge Couple. The outside couples of a roof when they project over the gable.

Barge Course. 1. A brick coping for a gable wall; usually brick on edge with tile creasing. 2. The row of roof tiles against the gable.

Barium Carbonate. $BaCO_3$. A scum preventer used in brickmaking.

Barium Hydrate. Used, in solution, as a preservative for stone.

Barium Plaster. Used for plastering the walls of X-ray rooms in hospitals.

Barium Sulphate. $BaSO_4$. Found native as *barytes, heavy spar,* or *cawk*. A white pigment used in paint and in glazes for bricks, and known as blanc fixe.

Bark Pocket. A timber defect in which the bark is partially or wholly enclosed in the wood.

Barlux Lamps. Tubular vacuum lamps with low surface temperature. The filament is on a flexible metal spine that allows for curvature.

Barmac. A wheelbarrow with flat top for conveying bricks.

Barmesh. A patent road reinforcement constructed on the site.

Barn. A covered farm building for storing hay, grain, etc.

Barn Doors. Large double doors to admit loaded wagons.

Baroque. Grotesque or whimsical ornamentation; an architectural style embodying this.

Bar Posts. Posts mortised for rails and forming gate posts for a field gate.

Barrack. A collection of huts and buildings for the housing of soldiers, or workmen.

Barred Door. A northern term for a ledged door, *q.v.*

Barred Gate. A field gate consisting of framing only.

Barrel Bolt. A bolt sliding in a cylindrical case, for securing doors. See *Tower Bolt.*

Barrel Locks. Mortise locks in which the case is cylindrical. The edge of the stile is prepared by boring with a twist bit.

Barrel Roof. A roof with a semi-cylindrical ceiling.

Barrel Vaults. The simplest form of vaulting, usually semicircular in shape. They are actually continuous arches. Also called tunnel or wagon vaults. See *Vaults.*

Barricade. A temporary barrier. A fence or railing to prevent approach.

Barrol System. A patent extractor ventilator.

Barrow. 1. A small hand-cart; a vehicle with one wheel for transporting small loads. 2. A hillock, or funeral mound of earth.

Barr's Method. The method of *substitution* used for drawing reciprocal diagrams, where there are too many unknowns at a point for the ordinary method to apply. The method consists of replacing two members by a single member.

Bartisan. A small overhanging turret at the angle of a building.

Mortise for Dowel. *Groove for Glazing.*

BAR TRACERY

Barton. A farm-yard.

Bar Tracery. See *Tracery.*

Barwood. An African timber, dark red in colour, used for decorative purposes.

Barytes. See *Barium sulphate.*

Basalt. A compact black igneous rock containing porphyritic crystals of plagioclase felspar, augite, olivine, and magnetite. The constituents vary according to the volcanic origin.

Bascule. Any appliance operated as a lever so that as one end is raised the other end is lowered.

Bascule Bridge. A draw-bridge. The Tower Bridge, London, is an example.

Base. 1. That part of a column between the pedestal and the shaft. See *Mouldings.* 2. Usually applied to the lowest member of anything. 3. A moulded skirting. 4. One of the constituents of paint, usually white or red lead, zinc white, or oxide of iron.

Base Bed. A fine-grained oolitic limestone obtained from the lowest of the Portland beds. It is softer than whitbed.

Base Board. See *Skirting*.

Base Line. An imaginary fixed line for reference purposes.

Basement. 1. A story below street level. 2. The ground floor on which the *order* or columns, which decorate the principal story, are placed.

Base Mouldings. The mouldings immediately above the plinth of a pillar or wall.

Base of Wall. The underside of the course immediately above the footings.

Base Plate. A plain skirting without mouldings.

Basic Bricks. Made from bauxite or magnesia. The bricks are weak but fire-resisting.

Basil. See *Bezel*.

Basilica. The public hall and Court of Justice of the Romans.

Basin. See *Lavatory Basin* and *Lock* (2).

Basite. A damp-proof course made of pure bitumen on a hessian base.

Basket. See *Bell*.

Basket Grate. A loose iron receptacle, or brazier used in an open fireplace.

Basket Handle Arch. An arch with rise less than half span, as the semi-elliptical arch.

Basket Weave. Ornamental brick panels, in 9 in. squares, consisting of alternate squares of three bricks on end and three on flat.

Basket Work. Interlaced work.

Basoolah. An adze with short haft peculiar to Eastern countries.

Bas Relief. Carving projecting slightly over the face of the material of which it forms a part. In low relief. See *Alto-relievo*.

Bassett. The edge of an outcrop of a geological stratum.

BASKET WEAVE OR DIAPER BOND

Basswood. See *American Whitewood*.

Bastard Masonry. Masonry that only serves as a facing, and does not serve any structural purpose.

Bastard Pointing. Applied to tuck pointing in which the ridge, or projecting fillet, is of the same material as the pointing. See *Pointing*.

Bastard Stucco. Used for painted internal work and containing a little hair.

Bastille. A fortress in Paris used as a state prison, but applied to any prison in which the prisoners are treated severely.

Bastion. 1. A bulwark or tower projecting from the face of a fortification. 2. The supports for a heavy dome.

Baston. Alternative name for the torus moulding.

Bat. 1. A piece of a brick larger than a closer. A part of a brick cut crosswise. 2. A lead wedge. 3. An iron cramp to secure frames to masonry. 4. See *Dresser* (2). 5. A square toothed plate timber connector, *q.v.*

Batch. A mixing of concrete.

Batching Plant. Large concrete mixers, *q.v.*

Batete. A reddish-brown hardwood with darker streaks caused by oil secretion; from the Philippines. It is fairly hard, heavy, and strong, and resists insects, but not durable. Works readily and polishes well, but requires careful seasoning. Used for constructional work and superior joinery. Weight about 37 lb. per c. ft.

Bath Stone. A soft easily-worked limestone used extensively for interior dressings and carvings. The best known varieties are Monk's Park, Combe

Down, Farleigh Down, Corsham Down, and Box Ground. Cream to light brown in colour. S.G. 1·95 to 2·2.

Baton. A glass-bead.

Batswing. A gas burner with a slit to give a fan-shaped flame.

Batted Surface. Regular vertical chisel marks, on ashlar, formed with a "broad" or "batting" tool.

Battenboard. A variation of laminated board, *q.v.* The core consists of narrow battens, up to 3 in. wide, running at right angles to the grain of the outer veneers.

Batten Door. A ledged or barred door, consisting of a number of narrow boards to build up the required width and held together by nailing them to three or more ledges on the back.

Battening. Narrow grounds fixed to a wall to carry laths for plastering.

Battens. 1. The boards used to build up ledged doors. 2. Timber of small section, as slate battens, etc. See *Pantiling.* 3. Softwoods from 2 in. to 4 in. thick and 5 in. to 8 in. wide. The sizes vary with different ports.

Batter. Inclined from the vertical, as the face of a retaining wall that gradually decreases in thickness.

Batter Rule. A board with one edge cut to the batter of the wall and with plumbline and bob.

Battery. Either wet or dry Leclanché cells in series, for supplying the current for electric bells.

Batting Tool. A "mallet head" mason's chisel, about 4 in. wide, for *tooling* surfaces.

Battle Axe. A registered name for a patent plaster used extensively for covering plasterboard.

Battlemented. An indented parapet or capping to a screen. Indentations are called crenelles, or embrasures, and projecting parts merlons.

Baulk. See *Balk.*

Baumé Gravity Instrument. Used for testing the S.G. of chemicals in solution.

Bauxite. Impure aluminium hydroxide. A natural earthy mineral used in the manufacture of refractory materials and cement.

Bawk. A tie beam.

Bax Ventilator. The principle is the same as that of the Tobin ventilator, but it is bracketed from the wall.

Bay. 1. A part of a room in the form of a recess. 2. The floor, ceiling, or roof between two main beams or trusses. 3. An opening between two pillars. 4. A principal compartment or division in the arrangement of a building.

Bay Window. A window projecting over the face of the wall and built up from the ground. It may be square or polygonal in plan. If circular it is a bow-window.

Baywood. Originally implied Honduras mahogany, but the name is now seldom used. See *Mahogany.*

B.B. Abbreviation for *bead butt.*

bdg. Abbreviation for *boarding.*

Beacon. A registered welded steel floor.

Bead. A round moulding with a small quirk, used to remove a sharp arris, and to break the joint between boards. See also *Sunk Bead, Return Bead, Reeds, Cock Bead,* and *Mouldings.*

Bead and Reel. An ornament used in Jacobean and Grecian mouldings. It consists of a bead carved into alternate semi-cylinders and hemispheres.

Bead Butt. As *bead flush* panels, but the horizontal beads are omitted.

Bead Cuts. The preparation of the beads for a pivoted sash.

Beaded Drip. A small cylindrical turn under at the end of the zinc covering to a flat roof.

Bead Flush. Thick panels, flush with the framing on one side and beaded all round. The horizontal beads, across the grain, are planted and the vertical beads stuck on the solid.

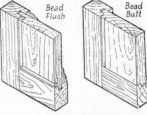

BEADED PANELS

Beak. 1. See *Bird's Mouth.* 2. The bottom edge of the corona, forming the drip.

Beak Head. A Norman enrichment over doorways. It resembles a head with beak.

Beam Blocking. Packing pieces nailed on joists for carrying sham beams in ceilings formed of building board.

Beam Box. A cast iron socket for the end of a wood beam.

Beam Compass. A trammel. A wooden rod, or beam, and two pointed legs, one of which carries a pencil. It is used for describing large circular arcs.

Beam Filling. The brickwork filling between timbers. Usually applied to the brickwork between the rafters and to the pitch of the roof.

Beam Hangers. Stirrups hanging from main beams to carry secondary beams. See *Flitch Beam.*

Beams. Strong horizontal timbers, reinforced concrete, or B.S.B.'s, supported at intervals by walls or columns. They are used to carry transverse loads and have transverse reactions. The term includes girders, joists, lintels, cantilevers, bressummers, etc.

Beam Seat. A steel bracket to a continuous column to carry a beam, in framed timber construction.

Beanstalk. A portable hydraulic lift with platform for workmen, rising to 17 ft. to obviate ladders.

Bearers. The joists supporting winders or landings.

Bearing. 1. The portions of a beam resting on the supports. 2. The resistance of a rivet against crushing. 3. The mechanism supporting a rotating shaft.

Bearing Pile. A pile resting on a solid stratum and supporting a load.

Bearing Plate. Any wood, stone or metal plate, or template, distributing the weight of a structural member on a wall.

Bearing Wall. One supporting other structural members.

Beat. Figure. The ornamental grain in timber cut tangentially to the annual rings.

Beating Hair. Preparing the hair for plastering by beating it with laths to remove dust and to open out the hair.

Beatl. See *Urea.*

Beaulate. A registered name for natural-slate interior decorative work: wall linings, bath panels, etc.

Beauvais Tiles. Interlocking tiles, deep chocolate red in colour, from Northern France. They are about 10½ in. long and laid to about 9 in. gauge. They resemble flat pantiles and require about 210 per square.

Beaver Board. A proprietary wood-fibre insulating wallboard.

Beck. A term used in stone slating. See *Stone Slates.*

Becket. A contrivance for securing the loose end of a rope.

Bed. 1. The bearing surface of anything, normal to the pressure. 2. A thin layer of mortar between two surfaces.

Bedding. The setting of bricks and stones on mortar beds. Puttying rebates to receive glass. See *Back Putty.* Placing plastic material between a wood frame and the brickwork to seal the joint against the weather.

Bedding Dots. Fixing short pieces of laths, 3 to 4 in. long, in a small bed of plaster on walls and ceilings. The floating rule working on the dots rules off the screeds, preparatory to filling in the bays, when plastering.

Bedding Frames. The placing of plastic material between wood frames and the brick or stonework, when fixing. Either oil putty or hair plaster is used.

Bedding Stone. See *Rubbing Bed.*

Bed Joints. The joint between two bricks or stones pressing against each other, as the voussoirs in arches, dome stones, etc. The horizontal joints in walls.

Bed Mould. A templet cut to the true shape of the bearing surface of a voussoir, etc. See *Mould.*

Bed Moulding. 1. A moulding in the angle between a vertical surface and an overhanging horizontal surface. 2. The cornice mouldings below the corona.

Beech. *Fagus sylvatica.* Hard heavy timber with close texture. Wears evenly. Light to dark brown. Used for planes, bentwork, beetles, etc. It is seldom used in building. Weight about 48 lb. per c. ft. Australian beech is used for joinery.

Beefwood. An Australian timber, rich red in colour. Used for decorative work.

Beer. A white to pale cream limestone from Devonshire. It is compact with fine grain and used for interior work. Weight is 129 lb. per c. ft. and crushing strength 107 tons per sq. ft.

Beetle, or Beedle. 1. A heavy wooden maul, or mallet, used by the pavior. 2. Pounders for beating and pounding materials. 3. There are several beetles destructive to wood : bark, black carpenter, cockchafer, death watch, elephant, furniture, longhorn, pinhole borer, powderpost, sipalus, weevil, wharf borer. Also see *Death watch beetle.* 4. A proprietary adhesive.

Bel. The unit of sound intensity.

Belaying Pin. A hook for securing the free end of a cord to a ventilator etc.

Belco. A proprietary cellulose lacquer.

Belfast Truss. A shed roof truss for large spans, consisting of a lower horizontal member and a curved top member connected together by thin lattices. The outline of the truss is drawn on the floor and a bow is bent to the outline and secured between blocks fixed to the floor. The lattices are nailed in position radiating to a point, equal to half the span, down the

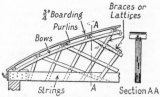

BELFAST ROOF TRUSS

support. The top bow is then fixed in position. The lattices are nailed where they cross each other and between the bows and strings, which should be cambered. See *Roofs* and *Bowstring.*

Belfry. That part of a steeple containing the timbers from which the bells are hung. A bell tower or campanile.

Belgium Truss. A truss suitable for a suspended ceiling. The number of panels in the tie beam is two less than in the principal rafters. The struts are at right angles to the principal rafter. See *Roofs.*

Bell. The body of a Composite or Corinthian capital with the foliage removed. Also called *vase* or *basket.*

Bell Capital. Characteristic of early English architecture.

Bell Cistern. A siphonic flushing cistern. When the bell is lowered, it raises sufficient water to start siphonage.

Bell Gable. A bell-cote. A small turret on the apex of a gable of a church, containing one or two bells.

Bell Hanging. The term applied to the running of wires and fixing of bells, whether mechanical, electric, or pneumatic.

Bellied. Applied to panels that have buckled through swelling. 2. A convex curve. Anything that curves outwards.

Bell Roof. A roof with a section of contrary flexure like a bell.

Bell Trap. A trap, used in sanitary appliances, consisting of a metal bell suspended in the water forming the trap.

BELL CAPITAL

Belt. An endless strip of balata or leather for transmitting motion from one pulley to another. The speed may be up to 6000 ft. per minute.

Belt Course. A plain string course with little projection.

Belt Rail. A lock rail.

Belt Stresses. Horizontal stresses in a dome.

Belvedere. A projecting room above a roof, for observation purposes.

Bema. A transept, or the raised portion at the end, in early churches.

Bench. 1. Any rigid framing with a top on which the workman prepares the work. 2. Also see *Berm.* 3. A church seat, or pew.

Bench End. The ornamental end of a church seat, or pew.

Bench Holdfast. A clamp for fixing the material to the bench whilst being wrought.

Bench Hook. A bench appliance for steadying small stuff whilst sawing.

Benching. 1. Forming horizontal steps in foundations on sloping ground. 2. Rounding the corners of manholes, etc. with cement.

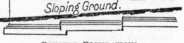

BENCHED FOUNDATIONS

Bench Knife. A short piece of broken knife for steadying the free end of a piece of timber when being worked on the bench.

Bench Mark. A fixed point, on a line of survey, left for future reference.

Bench Planes. The joiner's planes that usually rest on the bench; jack, smoothing, and try planes.

Bench Screw. The screw operating the vice to a joiner's bench.

Bench Stop. An adjustable projection on a bench, against which the timber presses when being planed.

Bench Table. A stone seat fixed to a wall or round a pillar.

Bender. A machine for bending steel, especially for reinforced concrete.

Bending Bolt. A curved steel bolt from 9 in. to 2 ft. long tapered at one end and used by the plumber in lead working.

Bending Moment. At any cross section of a beam the B.M. is the "algebraic sum of the moments of the external forces acting on one side of the section," and is equal to the moment of resistance of the beam, i.e. $BM = MR$. See pages 28 and 366.

Bends. Curved pipes for changing the direction of drains, etc.

Benfix. A registered prefabricated floor joist, to economize in wood.

Benieflex. A registered wire-reinforced waterproof felt.

Benin Walnut. Nigerian walnut.

Bénitier. A stoup, *q.v.*

Bentee. A fitting for flues to gas-heated boilers, to prevent down-draught.

Bentonite. A natural clay (U.S.A.). It is a hydrous silicate of alumina and in the form of a light creamy powder, 60 per cent silica, 20 per cent alumina. It swells, with water, to 15 times the dry volume and forms a stiff gel. It is the most plastic clay known and reinforces other clays greatly. Used for refractory materials, rubber compounds, etc. Added to hydrated lime to increase plasticity.

Berangan. Also called Kata and Malayan chesnut. It is a yellowish grey to light brown hardwood from the E. Indies. It is fairly hard, heavy, and stable, and resists insects but is not durable. Better qualities are used for interior fillings. Weight about 45 lb. per c. ft.

Berge. A Scotch term for an apron flashing.

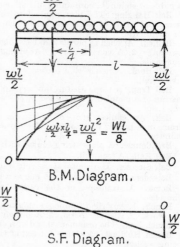

B.M. Diagram.

S.F. Diagram.

BENDING MOMENT AND SHEAR FORCE
w = load per ft. W = total load

Berlin Blue. See *Prussian Blue*.

Berm. A horizontal step or terrace in a trench.

Berner's Window. A large type of Hope's metal casement. It may be subdivided into panes by metal bars or left in one large sheet.

Bernoulli's Theorem. A theorem in hydraulics: "the total energy per pound of flow at all sections is constant."

Berries Graining. See *Cut Graining*.

Berth, Pullman. A sleeping bunk, on board ship, that folds up against the bulk-head when not in use. Also a railway sleeping carriage.

Bescot. A patent locking latch for doors.

Bessemer Steel. Comparatively pure iron produced by forcing dense air through molten cast-iron to which spiegeleisen has been added.

Bestac. A composition for bedding and pointing corrugated asbestos sheeting.

Bethel's Process. A method of preserving timber by placing it in iron cylinders, exhausting the air, and forcing creosote into the timber under pressure.

Béton. French for concrete, especially when consisting of puzzolano, lime, sand, and broken stone. Hydraulic concrete.

Betonac. A hardening material for cement surfaces, consisting of metallic particles which are added to the cement and sand.

Bettaskaf. A patent steel scaffolding.

Bevel. 1. An adjustable tool that can be set to any angle. 2. The term is used for any angle not a right-angle. See *Mouldings*.

Bevel Cut. The term applied to a method of preparing the wreath for a handrail in which the edges are cut bevelled from the plank instead of "square cut."

Bevel Halving. A halving joint with a slight bevel to prevent the two pieces from pulling apart when fixed.

Bevelled Glass. See *Bevelling*.

Bevelled Siding. See *Sidings*.

Bevelling. Grinding and polishing the edges of a glass surface into bevelled, or splayed, edges.

Bevels. A term specially applied to the cuts of roof timbers and to wreaths for handrails.

Bexoid. A proprietary urea-formaldehyde plastic.

Bezel. The bevel of a tool to form the cutting edge.

B.F. Abbreviation for *bead flush*.

Bhotan Pine. A wood similar to Baltic Redwood from the Himalayas It is excellent timber and also called Blue Pine. Weight about 28 lb. per c. ft.

Biancola. A synthetic marble.

Bib and Spanner. A bib tap operated by a removable key.

Bib Tap, or Bib Cock. A tap designed to discharge water vertically, as to a bath or sink. They are named according to their special characteristics, as screw down, spring, lever, etc.

Bicoca. A turret or watch tower.

Bicuiba. A pale to reddish-brown hardwood, with purplish tinge from C. America. It is fairly hard, heavy, and is stable and durable, and is easily wrought and polishes well. The stripes and lustre make it suitable for superior joinery. Also used for construction. Weight about 39 lb. per c. ft.

Bidet. A small pedestal hip bath with a rising spray and supply for mixing hot and cold water.

Big Ben. A registered self-locking tubular-steel trestle.

Big-span Floor. A Diespeker fire-resisting floor.

Bijou House. A very small house.

Bilge Blocks. The blocks on which a vessel rests when in a dry dock for repairs.

Billet. 1. An ornament in Norman architecture consisting of rows of short cylinders, placed at intervals, usually in a hollow moulding. 2. A balk

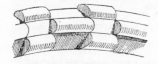

ROLL BILLET

SQUARE BILLET

with three sawn sides only. 3. A blob of metal for rolling.

Billian. See *Ironwood*.

Billing. Totalling the separate trades when preparing the Bill of Quantities.

Bill of Quantities. The total quantities and descriptions of work set out conveniently for the estimator to fill in the unit prices, to prepare the estimate. The preparation includes taking off, abstracting and squaring, billing or casting up.

Bills. 1. Knee timbers. 2. The points of compasses.

Bin. A receptacle for storage purposes. A wooden box for washing sand. A compartment in a wine cellar.

Binburra. The white beech of Australia. Strong and durable, easy to work. Weight about 37 lb. per c. ft.

Bind, To. The term applied to any hung frame that fits tightly on the stop to the hanging stile and prevents easy closing. See *Hinge-bound.*

Binder. 1. A beam supporting the common joists in a double floor. They may be of steel or timber or a combination of both. 2. Any material used for binding and consolidating the metalling in macadam.

Binding Rafters. A term sometimes applied to *purlins.*

Bintangor. A reddish hardwood from the East Indies. There are numerous species and several names. It is fairly hard, strong, and heavy but not durable. It is used for structural work and interior joinery. Weight about 36 lb. per c. ft.

Biotite. Ferro-magnesium mica. It gives a glistening appearance to granite and is dark or black in colour.

Birch. *Betula spp.* Fairly hard tough timber. Pinkish-brown. Stains and polishes well. Retains sharp arrises. There are numerous species and Canadian figured birch is marketed with trade names: Birnut, Flame, Ice, Karelian and Masur birch. Weight about 44 lb. per c. ft.

Bird's Beak. A Gothic moulding resembling a bird's beak in outline. See *Mouldings.*

Bird's-eye. An irregular growth in maple forming a decorative feature resembling birds' eyes.

Bird's Mouth. A re-entrant angle at the end of a piece of timber, as the forked joint where a rafter sits on the wall plate.

Birmingham Gauge. A standard, legal, gauge for iron and steel plates and hoops, from 15/0 to 52. See *B.W.G.*

Biscuiting. The second burning of glazed bricks after the slip, or glaze, is applied.

Bishop. A rammer consisting of a heavy cylindrical piece of wood with spindles at the top and side for handling.

Bishop's Mitre. A *mason's mitre* in which a prominent member of the moulding is produced beyond the intersection. See *Mitres.*

Bison. A patent *hollow-beam* fire-resisting floor. Self-centering.

Bit, Copper. A soldering iron.

Bitch. A dog with the tangs at right angles to each other, and used for fixing temporary timbering.

Bite. Applied to the cutting edges of tools, especially boring tools, as a measure of the efficiency.

Bitrol. A proprietary bituminous quick-drying paint.

Bits. The small interchangeable tools used in a brace, lathe, drilling machine, etc.

Bitubond. A proprietary composition for pouring in cavities as a damp-proofer.

Bitumarin. A proprietary bituminous paint and enamel for iron, etc.

Bitumastic. A proprietary waterproofing bituminous paint, which may be obtained in several colours.

Bitumen. Obtained from inflammable mineral products of organic origin containing a high percentage of carbon and hydrogen, as petroleum, naphtha, pitch, asphaltum, etc.

Bitumen Emulsion. Mixtures of boiling water, melted bitumen, and other finely dispersed materials, which solidify and produce a waterproof surface.

Bituminous Felt. Any fibrous material treated with asphalt, or bituminous material coated with sand. This produces a flexible sheet which is used for damp-proof courses, etc.

Bituminous Stone. Granites impregnated with refined tar to provide a dustless and durable wearing surface.

Bituplastic. A proprietary waterproofing compound.

Bituseal. A proprietary cold process bituminous compound for roofs, floors, etc. It may be obtained either black or grey.

Bitusol. A proprietary bituminous paint for ironwork.

Bi-vault. A double vault.

Black Bean. A dark brown hardwood from Australia. It has lighter and darker streaks, and is hard, heavy, strong, durable. It shrinks and warps freely but is not difficult to work and polishes well. Its greasy nature sometimes makes it difficult to glue. It is used for all kinds of superior joinery, electrical appliances, etc. Weight about 48 lb. per c. ft.

Black Bolts. A term used in structures for cheap unfinished bolts, for which a larger factor of safety must be used in design, than for the accurately finished bolts.

Black Bricks. See *Staffordshire Blue Bricks*.

Blackbutt. A species of Australian Eucalyptus. It is pale yellow in colour, very hard, durable, dense and tough. It is too hard for ordinary work but is used for structural work, piling, paving, etc. Weight about 54 lb. per c. ft.

Black Chuglum. See *Chuglum*.

Blackjack. A term used for the various black bituminous mixtures used for roof repairs.

Black Pigments. Used for paints, and include lamp-black, vegetable black, ivory black, carbon black, and drop black.

Black Walnut. *Juglans nigra.* Hard close-grained timber. Beautiful figure when cut tangentially to the annual rings. Weight about 41 lb. per c. ft. See *Walnut*.

Blackwood. Numerous woods are marketed under this trade name. They are all hard and heavy, and highly decorative dark brown or purplish hardwoods. The best known is Bombay blackwood or Indian rosewood. Several others are *Dalbergia spp.*, or rosewood. Australian blackwood, used for superior joinery and fittings, is *Acacia sp.*

Blades. 1. A term sometimes applied to principal rafters. 2. The cutting part of a tool, as the blade of a chisel.

Blaes. Clays that are mined from a considerable depth. Also the partly calcined clays and shales, from collieries, used in brick-making.

Blanc Fixe. A white pigment obtained by precipitating solutions of barium salts in sulphuric acid.

Blank Door, or **Window.** A recess in a wall, having the appearance of a door or window opening bricked up.

Blast-furnace Slag. See *Slag*.

Blazoned Glass. Decorative rolled glass for special lighting effects, surrounds, etc. Several registered designs.

Blebs. Air pockets causing defects in gelatine moulds for fibrous plaster.

Bleeding. 1. The rise of water to the surface when pouring concrete. 2. Applied to any under-surface penetrating a covering surface, as in paint. 3. Preservatives exuding, or leaching, from the surface of treated timber.

Blender. A bath or lavatory tap for combining hot and cold water.

Blind. 1. A screen to a window, either inside or outside, for privacy or as a protection from the sun. There are many varieties such as Venetian, Florentine, Spanish, Canaletti, Helioscene, Pinoleum, etc. 2. Applied to numerous things that do not function in the usual way: arches, screws, openings, etc.

Blind Alley. A cul-de-sac. A narrow street or passage closed at one end.

Blind Area. Same as *dry area, q.v.*

Blind Boxes. Frames into which blinds fold when not in use.

Blind Header. A false header, or half brick.

Blind Story. See *Triforium*.

Blind Wall. One without an opening.

Blind Window. A blank window.

Blister. Ornamental figure, resembling large bird's eye, in walnut.

Blistering. This is caused in painted work by the sun acting on trapped moisture, or by an excess of oil, or due to unseasoned timber. In plastering it is due to slaking of particles of lime after completion. In veneering it is due to trapped air.

Blister Steel. Steel produced by the cementation process.

Blob Foundations. Isolated shallow footings and bases.

Block, or Blocking. To secure two pieces of wood together by gluing triangular pieces in the angle. The triangular fillets are called blocks or blockings. See *Angle Joints*. Also see *Plinth Block, Cutter Block, Mitre Block*, and *Pulley Tackle*.

Block and Start. See *Long and Short Work*.

Blockboard. A variation of laminated board *q.v.* The core is formed of square wood strips with alternating grain, between two veneers.

Block Floor. See *Wood Blocks*.

Block-in-Course. Used chiefly in engineering masonry. It consists of large blocks of hard stone, or granite, with worked beds and hammer-dressed faces.

Blocking Course. A course of stones or bricks placed on the top of a cornice or coping. See *Oblique Projection*.

Blockings. The timbers in an underground excavation, supporting the roof.

Block Joint. A joint for fixing long lengths of large vertical lead pipes. A wood or stone projecting block is built in the wall and the pipe is passed through a hole in the block and flanged over so that it is hanging from the block. The wall is chased to receive the pipe.

Block Plan. A line drawing of the plan of a building and its surroundings drawn to a small scale. It usually includes the plan of the drainage system.

Block Quoins. Blocks of bricks three courses in height and showing alternately 1 and 1½ bricks on the face. Sometimes they project a little over the face of the wall.

Blocks and Tackle. See *Pulley Tackle*.

Blondins. Aerial cableways for conveying quarried stone to the surface.

Bloom. 1. A mass of molten metal collected for rolling; or glass for blowing. See *Glass*. 2. A bluish smoky cast on varnish due to excessive moisture in the atmosphere before the varnish had hardened. 3. A large block of faced steel used as a base for a stanchion.

Blow Lamp. A lamp used chiefly by painters and plumbers that provides a very hot, clean, and concentrated flame. The usual type burns paraffin, but petrol lamps are also used. It is dangerous to use petrol in the paraffin lamp.

Blown Joint. A joint for lead pipes in which one end is opened out and the end of the other pipe is rasped to fit. The joint is then soldered by means of a blow-pipe flame.

Blow Pipe. A bent tube to direct a jet of air from the mouth through a flame to concentrate the heat on to any particular position. The device mixes oxygen with the combustible gas which gives a mixture with a high-temperature flame.

Blows, or Blowing. 1. See *Blistering* in plasterwork. 2. The disintegration of bricks through the presence of uncombined lime. 3. The admission of water into the working chamber of a caisson, in under-water work. 4. See *Glass*.

Blubs. Air pockets in the jelly, when gelatine moulding, caused by using an excess of boiled oil when preparing the model.

Blue. A primary colour from pigments produced from mineral, vegetable, or artificial substances. There are several kinds: Chinese, Monastral and Prussian blue, and Ultramarine.

Blue Bricks. See *Staffordshire Blue Bricks*.

Blue Circle. A registered name for a Portland cement.

Blue Goods. Applied to softwoods in which the sapwood has a bluish stain through lack of seasoning before shipment. The stain is due to a fungal agency but it does not cause decay.

Blue John. Fluor spar in which the crystals have a bluish tinge.

Blue Lias Lime. Hydraulic lime from limestone of the Lias formation. Its properties approach those of cement and it is suitable for damp situations. It can be obtained ready ground.

Blue Pigments. Used for colouring paints, and including ultramarine blue, Prussian blue, cobalt blue, lime blue, and celestial blue.

Blue Plaster. A coloured thickness of plaster used in waste moulding as a guide for chipping the sections of the plaster casts.

Blue Prints. A reproduction of an Indian ink tracing on ferro-prussite paper. The paper is originally white but turns blue on exposure to light. Ferro-gallic paper produces black lines on a white ground. Coralin and Ozalid papers give reddish-brown lines on white paper. Modern apparatus can produce the prints from pencil drawings. See *True to Scale, Dialine, Dyeline*, and *Printing*.

Blundell's. A transparent waterproofing liquid.

Blythe's Process. A preserving process in which the moisture is extracted from the timber and replaced by carbolic or tar acids.

B.M. Abbreviation for *Bird's Mouth* and *Bending Moment*.

B.N.R. A proprietary material for damp courses, obtainable with a lead, hessian or fibre base.

Board. Applied to hardwoods of any width and up to $1\frac{1}{4}$ in. thick and to softwoods over 4 in. wide and less than 2 in. thick. See *Floor-, Match-* and *Weather-boards*, and *Market Sizes*.

Board and Batten. Applied to the covering of timber houses with alternate thick and thin boards to give a recessed, or panelled, effect. The thick boards are grooved for the thin boards.

Boarding Joists. Common joists carrying the floor boards.

Board Measure. The unit of measure for timber in N. America. The unit is 1 ft. sq. by 1 in. thick, or 144 c. in.

Boaster. A mason's chisel, about $1\frac{1}{2}$ in. wide, with a *mallet head*.

Boasting. 1. Knocking off superfluous stone preparatory to carving. 2. Preparing ashlar with a boaster. The chisel leaves a regular series of chisel marks at an angle to the edge of the stone.

Boat. A scaffold cradle.

Bobbin. A boxwood semi-oval tool with a hole through its length and used for truing up bends in lead pipes.

Body. 1. Usually applied to paint to denote the covering power. 2. The undercoat for the glaze in brickmaking. 3. The main part of anything to which auxiliary parts are attached.

Body of Niche. The vertical surface of a niche.

Bogey. A low truck on four wheels. The body is pivoted at the middle of the two axles. See *Derrick Crane* and *Trolley*.

Boiled Oil. Linseed oil that has been heated to over 400 degrees F, during which driers are added. It is used for paints, etc.

Boiler. See *Fire-back Boiler* and *Low-pressure System*.

Bolding's Gulley. A reversible gulley. It is made in two parts so that the top can be turned through any required angle.

Bole. 1. A recess or opening in a wall. 2. The stem of a tree. 3. A mineral pigment used in paints.

Bolection. A rebated panel moulding, the face of which stands above the face of the framing. See *Mouldings*.

Bollard. Strong posts, usually cast-iron, anchored to quay walls, for mooring vessels by means of cables.

Bolman Truss. A trussed beam used for timber deck bridges, and carrying the load on the top chord. See *Fink Truss* and *Roofs*.

Coach Screw.

Bolomey's Curves. Curves showing the proportion and grading of aggregate for concrete with low permeability.

Bolsover Moor. A yellow-brown limestone from Derbyshire. It is very good for general building and weighs 152 lb. per c. ft. with a crushing strength of 484 tons per sq. ft.

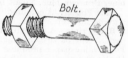

Bolt.

Bolster. 1. A short horizontal timber on top of a pile or post to provide a bearing surface. 2. The lateral part joining two volutes on the Ionic capital. 3. As *balusters*. 4. A bricklayer's cutting chisel, about 5 in. wide.

Bolt Carton. A cardboard bolt-sleeve (*q.v.*).

Boltel, or **Boultin.** A quarter-round moulding.

Handrail Bolt.

Bolting Iron. A drawer-lock chisel.

Bolts. 1. Strong metal fastenings consisting of a cylindrical shank with a head,

BOLTS

and threaded for a nut. They are named according to the shape of the head, as square, hexagonal, etc. 2. A bar or pin used to secure a door or window. See *Tower Bolt*. 3. See *Dead Bolt*. 4. See *Bit, Copper*. 5. Blocks of wood cut to special sizes for a particular purpose, as lath-bolt.

Bolt Sleeve. A wood or cardboard casing round a bolt in shuttering to prevent the concrete from adhering to the bolt.

Bombway. See *White Bombway*.

Bon-accord. See *Granite*

Bond. The arrangement in brickwork and stonework so that the units are tied together to form a solid mass, to give stability. The principal brick bonds are English, single and double Flemish, stretcher, heading, Dutch, garden wall, diagonal, herring-bone, longitudinal, Sussex, rat-trap, flying, Monk, Quetta, and St. Andrew's.

Bonded. See Resin Bonded and Glue.

Bonder. A stone usually from front to back of a wall to tie the wall transversely. Headers or *through* stones.

Bonderising. A registered phosphate coating process. The method consists of immersing iron or steel articles in a boiling solution of an acid metal phosphate. The process is somewhat like Parkerising. Spray bonderising is adapted for mass produced articles. The name Bonderised is registered for numerous types of metal kitchen equipment. The coating increases the adhesion of paint and lacquer, and so prevents corrosion.

Bonding Bricks. Used for tying together the two leaves of a cavity wall. The bricks are bent so that one end is built into a lower course in the front wall. They are used instead of tie irons. See *Wall Ties*.

Bonding Pockets. Recesses formed, in vertical damp proofing, to provide bond for a 4½ in. inner wall to protect the asphalt. The pockets are usually 18 in. square and 4½ in. deep, so that the inner wall is 9 in. thick at the pockets.

Bonding Timbers. Long timbers built in thin walls to tie together the brickwork.

Bond Stone. See *Bonder* and *Kneeler*.

Bongor. A dark grey, brownish, or pale red hardwood from the E. Indies. It is fairly hard and heavy and is a good substitute for teak for construction, piling, etc. Weight about 32 lb. per c. ft.

Boning. 1. Determining the amount of twist, or winding, of a surface. 2. Obtaining the levels for excavating by means of boning rods. See *Sight Rails*.

Boning Falls. Obtaining the falls for trenches by means of boning rods.

Boning Pegs. Small hardwood cubes placed at the corners when preparing the face of large or hard stones. When the corners are out of twist they are connected up with the straightedge.

Boning Rod. A T-shaped appliance for obtaining levels and falls. They

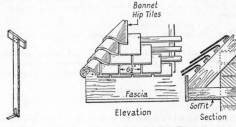

BONING ROD   BONNET HIP TILES

are made from 3 in. × ¾ in. deal. The upright is 3 ft. or more in length, and the crosshead about 8 in. long. Also see *Sight Rails*.

Bonnet. The roof of a bay window. A wire cover to a vent pipe.

Bonnet Hips. Tiles for covering hips, and shaped like a sunbonnet at the tail.

Boogie Pump. A pressure pump for consolidating underground strata, etc. It consists of a perforated steel tube and a hermetically sealed pan for the grout which is connected to a compressor.

Boom. The horizontal members of a trussed girder; or the flanges of a beam.

Booster. An electric heater for producing a continuous flow of hot water.

Booster Pump. One for increasing the pressure in a pipe system.

Boot. A registered name for a prefabricated house.

Boot, or Shoe. A bend for discharging water at the foot of a pipe.

Boot Boiler. See *Fire-back Boiler*.

Booth. A temporary shelter, usually of canvas.

Border. An edging used for strength or ornamentation.

Bordex. A proprietary building board. It is water- and fire-resisting.

Bordorflex. A registered fire-resisting light-weight wallboard. It is very flexible and suitable for curved surfaces.

Bore. The internal diameter of a pipe.

Boring. 1. Obtaining a knowledge of the strata of earth at various depths by boring holes about 4 in. diameter. The object is to obtain the bearing strength before structural operations begin. 2. Making holes in wood with brace and bit.

Borning. See *Boning*.

Borrowed Light. An interior window obtaining light from another window.

Boss. 1. A carved block covering the intersection of two or more cross-timbers or stones in ceilings and vaulting, or at the end of a cantilever. 2. The conical *keystone* to a niche or dome. See *Niche*. 3. A plumber's turning pin.

Bossage. The boasting of stone preparatory to carving.

Bossing. Working sheet lead to the required shape with a bossing stick.

Bossing Stick. A piece of boxwood, about 14 in. long, shaped for handling, and used to beat lead to any required shape.

Bottle-nosed Curb. A round nosing to a drip on a lead flat, or to a step.

Bottle Trap. A sanitary trap in the shape of a bottle; a vertical pipe dips into the water for the inlet, and the outlet is at the side.

Bottom Rail. The lowest rail in timber framing.

Boucherie Process. A process for preserving timber. The moisture is expelled by fluid pressure, and copper sulphate in solution is forced into the timber.

Boudoir. A lady's private room.

Boulder Clay. Stony clay. It is stiff and tenacious, and the remains of glaciation.

Boulder Wall. Walls formed of round flints and strong mortar.

Boulevard. A broad walk, or promenade, bordered by trees.

Boule Work. Inlays of ivory, tortoise shell, mother of pearl, or metal.

Boultine. See *Boltel*.

Boulton. A registered tubular scaffolding.

Bow. 1. See *Belfast Truss*. 2. An arch or arched gateway. 3. A curved member of a frame.

Bow Compasses. Small compasses for drawing. They are usually fitted with a spring and a screw for fine adjustment. See *Spring Bows*.

Bower Barff. A process of heating iron water pipes and exposing them to superheated steam to give them a protective coat of black oxide to resist corrosion.

Bowl Capital. A cushion cap to a column in the late Anglo-Saxon and Anglo-Norman styles of architecture.

Bowranite. A registered name for a protective coating against corrosion. Also for several other special paints, etc.

Bow Saw. A saw for cutting circular work. It consists of two hardwood ends, with a stretcher between the blade and a twisted string which is used to give the required tension to the blade.

BOWL CAPITAL

Bowser. A short piece of hardwood with two holes for fixing the return end of a rope in the required position. It is also called a toggle.

Bow's Notation. A system of lettering, or numbering, the spaces between the forces in graphical problems in mechanics, to give a circuital reading. See *Graphic Statics*.

Bowstring Floor. A patent floor built up of "Ridley" Lewis dovetail sheeting units.

Bowstring Girder. A built-up steel girder with a segmental top boom, and horizontal bottom boom.

Bowstring Roof. See *Belfast Truss*.

Bowtell. 1. The shaft of a clustered pillar or jamb. 2. A plain round moulding. See *Mouldings*.

Bow Window. A segmental or semicircular bay window.

Box. 1. A compartment in a theatre to accommodate several people. 2. See *Boxwood*. 3. The bottom removable handle of a pit saw. 4. See *Sash and Frame*. 5. "To Box" usually means "to enclose."

Box Culvert. A culvert with vertical sides, and roofed with flag stones.

Box Dam. A coffer dam, rectangular in shape, for under-water work.

Boxed. Generally applied to anything built-up into rectangular shape and hollow.

Boxed Eaves. Projecting eaves with closed soffit and fascia.

Boxed Frames. See *Cased Frames.*

Boxed Mullions. The built-up mullions, in which the weights slide, for sliding sashes.

Boxed Plane. A plane that has had the mouth closed by an inlay.

Boxed, or Box, Tenon. A tenon in the form of a right angle, used on corner posts.

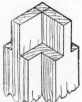

Boxform Lintel. A patent pressed steel lintel for use with Hope's metal door frames.

Box Girder. A built-up, or plate, girder with a double web.

Box Ground. A cream coloured limestone from Wiltshire. It has coarse grain, weathers well, and is good for exterior dressings. Weight 129 lb. per c. ft. Crushing strength 107 tons per sq. ft. See *Bath Stone.*

BOX TENON

Box Gutter. A rectangular gutter, the sides of which are formed by a parapet wall and a pole plate or by two pole plates. A parallel gutter. The depth varies from 3 in. to 9 in. according to the fall and number of drips.

Box Horse. A box-shaped support for a barrow-run.

Boxing. 1. The upper ballast on railways. 2. A recess to receive folding window shutters, or *boxing shutters.* 3. A casting, or stool, bolted to the top of a cast-iron column to carry the girders and to form a seating for the column above.

Box Key. A T-shaped appliance for turning a nut to a bolt. It has a square recess in the end to fit over the nut.

Box Scarf. See *Lap Scarf.*

Box's Rule. A formula for calculating the discharge of drains or the diameter of the pipes. Used in water supply problems. The formula is $H = \dfrac{G^2 \times L}{(3d)^5}$ where H is in feet, G = galls. per min., L = length in yd., and d = diameter in inches.

Box Staple. The metal receiver for the bolt of a dead lock or rim lock.

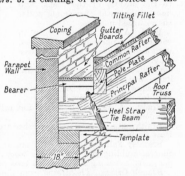

BOX GUTTER

Boxwood. A very hard and heavy timber from Asia Minor and S. Europe. Yellow colour; uniform, close and even texture. Sold by weight. Used for tools, screws, etc. Weight about 62 lb. per c. ft. The name Box or Boxwood is applied to many other woods having similar qualities and characteristics.

Boyle's Law. "The volume of a given mass of gas is inversely proportional to its pressure." *P V = Constant.*

Boyle's Mica Flap. Inlets to ventilating flues in chimneys, in which a mica flap closes the ventilator during down-draughts.

B.P. Board. A registered wood-fibre wallboard.

B.P.P. Abbreviation for *British polished plate*.

Brace. 1. A member of a framed structure crossing a space diagonally and able to resist tension or compression. See *Centering* and *Partition*. 2. A cranked tool for controlling boring bits.

Brace Bits. The small interchangeable tools used in the brace for boring holes, etc.

Brace Blocks. Keys used in beams, built up in depth, to prevent sliding due to horizontal shear.

Bracket. 1. A right-angled support, either solid or with angle brace, for a projecting horizontal surface, such as a shelf. See *Mantel*. 2. The ornamental return of a riser for a cut string, in stairs. 3. Rough supports to steps, nailed to the carriage of a flight of stairs.

Bracket Scaffold. A support for workmen whilst repairing walls. A wedge-shaped spike, with an eye, is driven firmly into a joint of the brickwork and the bracket is hooked into the eye. Scaffold boards rest on the bracket.

Bracketed Cornice. A lath and plaster cornice carried on brackets fixed to wall and ceiling.

Bracketed Stairs. A flight of stairs in which the string is shaped to the outline of the steps, and has ornamental brackets under the returned nosings.

Brad. A cut nail tapering in width, but parallel in thickness, with head projecting on one edge only. They are used chiefly for flooring, and are usually $2\frac{1}{2}$ in. long. See *Nails*.

Bradawl. A small hand-boring tool, for nail and screw-holes.

BRACKET SCAFFOLD

Brady. A registered name for rolling shutters and other building components.

Braithwaite. Registered prefabricated houses also called Telford and Unit Frame Construction.

Bramley Fall. A light brown sandstone from Yorkshire. It is used for general building and engineering work. Difficult to work. Weight 162 lb. per c. ft. Crushing strength 552 tons per sq. ft.

Branch Drains. The drains from the various parts of a building leading to the main drain. They are usually collected at a manhole.

Branched Work. Carved ornaments in the form of leaves and branches, for panels and friezes.

Branches. Any pipes branching off from a main pipe.

Brandering. 1. Counter lathing. 2. Small fillets nailed on wide timbers before lathing, to provide a key for the plaster.

Branding. Stencilling or stamping marks on timber to denote the quality.

Brandishing. See *Brattishing*.

Brandreth. A rail, or fence, round the opening of a well.

Brash. Small pieces of disintegrated rock.

Brashy. Applied to short-grained timber. Also to timber from over-matured or dead trees.

Brass. An alloy of copper and zinc having more than 50 per cent copper. A yellow metal that tarnishes on exposure to the atmosphere.

Brass Thimble or Sleeve. See *Ferrule.*

Bratisheen. A proprietary tough sarking felt.

Brattice. A fence round machinery. A partition in a coal mine.

Brattishing. The term applied to an embattlemented or perforated parapet.

Brazier. A receptacle for a portable fire; used for drying out new buildings, etc.

Brazing. Hard soldering. Joining metals together by means of spelter, moistened with borax water, which fuses on the application of heat.

B.R.C. Fabric. A steel wire mesh reinforcement for concrete, consisting of mild-steel main wires crossed by thinner wires, and electrically welded at the points of contact. Obtained in rolls 240 ft., or longer, and 7 ft. wide.

Break. A change in direction on a plane surface. A recess or projection on a wall.

Breaker. A pneumatic drill.

Breaking Down. Sawing logs into small stuff.

Breaking Joint. 1. Crossing the heading joints in built-up stuff. The heading joints in floor- and match-boarding on different joists or studs. Avoiding vertical joints in consecutive courses in brick and stonework. 2. Chamfering or beading the arrises of the joint between two boards so that the joint is not so obvious after shrinkage.

Breaking Stress. The amount of force or load that will just overcome the resistance of the material.

Breaking Weight. The amount of load that will *just* break a beam or member of a framed structure.

Break Iron. A dog with sharpened edge on which the slater cuts the slates to size.

Break Jambs. Stop jambs. Recesses in masonry jambs.

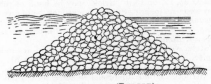

BREAKWATER (RUBBLE)

Breakwater. A wall to protect a harbour, or part of the coast, against the sea.

Breast. A narrow vertical wall surface.

Breast, Chimney. The projections on a wall containing the flues and enclosing the fireplace.

Breasting. Levelling the points of saw teeth preparatory to sharpening.

Breast Lining. The panelled framing between a window sill and skirting.

Breast Wall. A retaining wall.

Breeze. Finely broken coke, or cinders.

Breeze Blocks. Building blocks made from breeze and cement. They are used for internal walls.

Bressummer. A strong wooden beam serving as a lintel.

Bricanion Lathing. A square mesh wire netting, with the intersections cased in burnt clay. It is used as a reinforcement for thin solid plaster walls.

Brick Backing. A term applied to a wall having an inner lining of bricks. The outer face may be stone, concrete, terra-cotta, etc.

Bricklayer's Scaffold. A scaffold used in the erection of brick buildings. It consists of standards, ledgers, braces, putlogs, and scaffold boards.

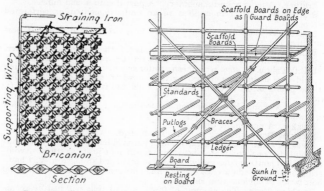

BRICANION LATHING   BRICKLAYER'S SCAFFOLD

Bricknogged. A lath and plastered partition having the lower part (to shoulder height) filled with bricks or pre-cast slabs between the studs.

Bricks. Rectangular blocks of burnt clay. Large variation in quality, colour, and texture owing to the variation in brick earth and care in manufacture. Standard bricks are $8\frac{3}{4} \pm \frac{1}{8}$ in. by $4\frac{3}{16} \pm \frac{1}{16}$ in. by 2, $2\frac{5}{8}$, or $2\frac{7}{8}$ in., rising 4 courses to $9\frac{1}{2}$, 12, or 13 in. Bricks may be hand or machine made, and may be pressed or wire cut. Special bricks may be of any size or shape but they are usually within the above maximum dimensions. See *Accrington, Engineering, Facing, Fire, Fixing, Fletton, Purpose-made, Enamelled, Rubber, Salt-glazed, Lime-sand, Staffords, Stocks.*

Bricktor. Patent wire mesh used to assist the bond in brickwork.

Brick Trimmer. A brick arch under the hearth of a fireplace to give support to the hearth and to protect the ceiling from the fire. See *Hearth.*

Brickwork. Structures built of bricks to some recognized bond. It is valued by the rod or by the cubic yard. See *Bond.*

Bridge. 1. A structure for carrying traffic, supported at intervals with open spaces below. See *Wing Wall.* 2. A cover piece for the strut to fix the bricks in the cutting box, when sawing the voussoirs for an arch to the required bevel.

Bridge Board. See *Notch Board.*

Bridge Gutter. Boards supported by bearers and covered with lead to form a gutter.

Bridge Stone. A stone over an area before a doorway.

Bridging, or Strutting. Stiffening joists by nailing short pieces of timber between them. The strutting may be solid or herring-bone. See *Herringbone, Keyed* and *Solid Strutting.*

Bridging Joists. Wooden joists carrying floor boards and *bridging* a space.

Bridging Over. Laying timbers across a space, as bridging joists.

Bridging Piece. A strong piece fixed between joists to carry a partition.

Bridging Run. A scaffold, bridging a gap, for the passage of workmen. It should not be less than 18 in. wide if over 6 ft. above a floor, and the boards should be tied to prevent unequal sagging.

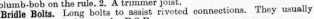

BRIDGING PIECE

Bridgwater Tiles. English-made interlocking tiles. Several patterns may be obtained but they are chiefly of the Double Roman or Marseilles type. The former only require about 65, and weigh about 6 to 7 cwts., to a square.

Bridle. 1. A guard to control the string of a plumb-bob on the rule. 2. A trimmer joist.

Bridle Bolts. Long bolts to assist riveted connections. They usually pass through both flanges of a B.S.B.

Bridle Joint. A joint similar to the stub mortise and tenon but with the positions of the mortise and tenon reversed. It is used chiefly for angles less than right angles as for the joint between the tie beam and principal rafter in a roof truss.

Bright. Applied to newly sawn wood that is not discoloured.

Bright Fronts. Better quality bricks from washed earth.

Brindles. Bricks not uniform in colour due to contact with other bricks in the kiln. The defect is only in the appearance. Second quality blue bricks with irregular reddish markings, and not so vitrified as the best quality.

Bringing, or Carrying, Up. The erection of brick or stone walls to a specified level.

Briquettes. 1. Bricks of compressed graphite or coke, used to withstand high temperatures. 2. Small specimens made to test the quality of cement.

Bristol Pennant. A blue-grey sandstone from Gloucester. It is very strong and durable and used for general building. Weight 172 lb. per c. ft. Crushing strength 1000 tons per sq. ft.

British Standard Sections. Rolled steel structural sections made to the dimensions recognized by the British Engineering Standards Committee. They include I- beams, channels, angles, T- and Z-sections. See *Rolled Steel Sections.*

British Standard Specification. A statement of the necessary qualities and sizes, etc., for materials used in building, such as Portland cement, steel, etc., as agreed upon by committees of the British Standards Institution, appointed for the purpose. A large number of standards have already been published for the different industries and new ones are continually appearing.

British Thermal Unit. See *B.Th.U.*

Broach. 1. A local term for a spire. 2. A tapered bit such as a rimer.

Broached. Ashlar that is tooled clean after it is droved.

Broaches. The inclined corners from a square base to intersect an octagonal spire.

Broach Spire. A spire rising from a tower without parapets. See *Splayed Foot.*

Broad. A cutting tool for turning soft wood.

Broad Flange Beams (B.F.B.'s). Rolled steel beam sections with specially broad flanges. Generally used as stanchions.

Broad-leaved Trees. Deciduous trees, which are classed as hardwoods, as distinct from conifers which are softwoods. They are dicotyledons and the wood is usually hard and has a more complex structure than needle-leaved trees, or softwoods.

Broadmead Floor. An improved *Armoured Tubular Floor.*

Broad Reeded. Reeded glass, *q.v.*, with specially wide reeds.

Brob. A spike with a right-angled head, for withdrawal by leverage.

Broche. See *Broach*.

Broduit. A registered copper-conduit system for electric wiring.

Broken Bond. Irregularity of the bond in brickwork as often occurs when filling between piers.

Broken Pediment. A pediment with the raking mouldings not continued to the apex, for ornamental purposes. See *Pediment*.

Broken Soffit. The soffit of a stone staircase that is not flush.

Bromak. A patent pneumatic brush, fed with paint under pressure. It combines both spray and brush methods of painting.

Bronze. An alloy of copper and 5 to 20 per cent tin. Rich brown colour and malleable.

Bronze Sheathed. Thin drawn bronze on a hardwood core used extensively in shop fronts and fittings.

Brooming. 1. The bruising of a pile head with the pile driver. 2. Scratching the floating coat with a broom before applying the setting coat, in plastering. Devilling.

Brown Glaze. Obtained by adding iron and manganese oxides to the body or glaze, or by using coloured clays and umber.

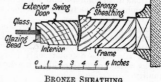

BRONZE SHEATHING

Brown Haematite. Hydrated ferric oxide. One of the ores from which iron is extracted.

Browning Coat. The floating coat in plastering, after rendering. Also called topping coat.

Brummer. A proprietary waterproof wood stopping.

Brunswick Black. A cheap, black varnish for ironwork. It consists of refined tar and resins soluble in turpentine.

Brunswick Blue. See *Prussian Blue*.

Brunswick Green. See *Pigments*.

Brushing Belco. A proprietary hard-setting lacquer.

B.S. Abbreviation for *both sides*.

B.S.I. Abbreviation for *British Standards Institution*.

B.S.B. Abbreviation for *British Standard Beam*.

b.s.m. Abbreviation for *both sides measured*.

B.S.S. Abbreviation for *British Standards Specification*.

B.Th.U. British Thermal Unit. The amount of heat required to raise the temperature of 1 lb. of pure water at $39 \cdot 1°$ through one degree Fahrenheit. See *Calorie* and *Heat*.

Bubblestone. Patent light-weight concrete for partitions. Air cells are pumped into the mix.

Bubinga. A West African dark brown hardwood, resembling Padauk. Also called Okweni, Kevazingo, etc. It is hard, heavy, stable, and strong and difficult to work. Used for interior fittings, and decorative wood for sliced veneers. Weighs about 52 lb. per c. ft.

Buck. A 3 in. post recessed to receive a breeze block partition, and to provide a fixing for the door linings.

Bucket. 1. A compartment of a water-wheel. 2. A scoop to a dredging machine.

Buckles. Large wooden pins used in thatching.

Buckling. Bending out of the straight. A saw is buckled when it is permanently bent and requires re-hammering. Deformation due to compression.

Buckling Factor. Used in the design of columns and dependent upon the fixing of the ends of the column. See *Slenderness Ratio*.

Bucrania. An ox head decorated with a wreath as employed in the frieze of the Ionic and Corinthian orders.

Budger. A painter's graining tool.

Budget. A pocket to carry pins, used by tilers.

Buffer. Any arrangement for reducing the effect of concussion.

Buffet. 1. A stool. 2. A sideboard or recessed cupboard. 3. A refreshment bar.

Buhl. See *Boule Work*.

Builders. A term applied to wire cut blue bricks, and to bricks of standard size to work with a ⅜ in. joint.

Builder's Scaffold. A strong scaffold built of square timbers, and used where heavy blocks of stone and girders have to be hoisted and handled.

Building Acts. See *Acts*.

Building Blocks. The various types of slabs and blocks used for partitions and walls. They may be made from concrete, clay, or moler, and may be solid, hollow, or reinforced.

Building Boards. The numerous types of wall-boards used for lining walls and ceilings. They are made from such varied materials as gypsum, asbestos, cane, straw, bamboo, sawdust, wood pulp, etc., with various finishes to the faces, and in many sizes and thicknesses. They may be veneered with ornamental woods, bakelite, celluloid or metal. See *Wallboards* and *Plywood*.

Building Jig. A steel profile used in brick-laying.

Building Lime. Greystone lime used for ordinary constructional work.

Building Paper. Bitumen with fibre reinforcement between sheets of tough kraft paper. It forms a thin flexible, waterproof lining and is used for many purposes. Special fixing glue is provided. See *Sisalcraft*.

Building Stones. See *Stones*.

Built-in. Applied to fittings that are fixed securely to the structure and so become part of a building; a term applied to cupboards, wardrobes, etc.

Built Rib. A laminated, or built-up, member. The layers are bolted together, and the rib is usually curved.

Built-up. A term used by woodworkers when the width or thickness is comprised of several pieces.

Bulb Angle. A special shape of rolled steel angle iron in which one edge is thickened into a small cylinder to increase the strength.

Bulbous. Pillars or columns having a large swelling like a bulb and characteristic of the Jacobean style of architecture.

Bulk Concrete. Ordinary concrete without reinforcement and not rendered as in foundations, etc.

Bulkhead. 1. A partition, in a ship, forming separate compartments. 2. A sloping top, or ceiling, to a flight of stairs leading to a basement-or under a shop window. 3. Vertical partitions erected when tunnelling.

Bulk Modulus. *K*. Corresponding to the volumetric strain produced when a body is stressed in three directions, perpendicular to each other, as when immersed in liquid under pressure.

Bull Dog Clips. A patent clip for securing battens or joists to concrete.

Bull Dog Plates. Patent friction plates, with projecting prongs stamped on both sides. They are used between overlapping timbers to prevent lateral movement. See *Timber Connectors*.

Buller Nails. Nails with round lacquered heads and short shanks.

Bullet Wood. A hard close-grained greenish-brown timber, resembling greenheart, from the West Indies.

Bull Heads. Bricks tapering in width for circular work.

Bullion. See *Crown Glass* and *Glass*.

Bullnose. 1. A small metal plane for rebates. 2. A brick with a rounded corner.

Bullnose Step. A step at the bottom of a flight of stairs with one or both ends rounded.

Bullnose Winder. Used when four winders are used, instead of three, in quarter-space winders. A riser intersects the angle between the strings and so weakens it, and leaves an acute angle for the dirt to accumulate. To avoid this the riser is rounded near the string in the form of a small bullnose.

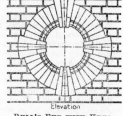

Bull's Eye. A small circular or elliptical window. 2. See *Crown Glass* and *Turret.*

Bulwark. Sea defence walls. Ramparts.

Bumping. The sharp second pressure applied to the press in brick moulding.

Bumping Blocks. Buffers constructed of timber only.

Bungalow. A domestic building of one story.

Bunker. A large hopper-like structure for storing coal, etc. The bottom slopes to a central opening so that the contained material may be loaded into carts as required.

Elevation

BULL'S EYE WITH KEYS

Bunsen. A particular type of gas burner with a tube for mixing the gas with in-drawn air.

Bureau. 1. A proprietary waterproofer. 2. A chest of drawers or writing table. 3. A room, office, or department for secretarial business or inquiries.

Burlap. Coarse canvas.

Burlington Slates. Dark blue slates from North Lancashire.

Burnettizing. Preservation of timber by immersing, under pressure, in a solution of zinc chloride, $ZnCl_2$.

Burning. Lead burning without solder. The pieces are melted together.

Burning In. Securing lead in holes and grooves by pouring in molten lead and caulking.

Burning Off. Removing old paint with a blow-lamp and knife.

Burnishing. Finishing off the surface of wood with a lustrous polish.

Burn's. A registered name for a variety of C.I. sanitary goods.

Burnt Sienna. See *Pigments.*

Burrs. 1. Partially melted bricks that have run together through being near the *live* holes in the kiln. 2. Irregular growth of timber due to injury to the cambium layer. In some cases the value of the timber is increased because of the beautiful figure, as in burr walnut, and the "injuries" are caused by skilled men. 3. A small washer to a rivet.

Burr Woods. Woods in which the burr is important for figured, or decorative, purposes. This applies to Amboyna, Ash, Elm, Oak, Thuya, Walnut, Yew, etc. In most woods burrs have no decorative value, and in all cases they lessen the strength.

Burtons. A registered name for several patents applied to steel scaffolding.

Bush. 1. A metal lining to a cylindrical hole. 2. A perforated plug.

Bushel. A measure of 8 gallons.

Bush Hammering. Tooling the surface of concrete with an electrically driven hammer, to expose the aggregate.

Butments. Abutments.

Butt. 1. The base of a tree trunk, which provides the

BUTT HINGE

strongest timber. 2. Hinges intended to be sunk into the edge of the door or casement.

Butt and Break. The term used in plastering for the laths when the ends butt together; and when the ends, of about every yard super, meet on different joists or studs.

Butted. A heading joint, *q.v.*

Buttered. A brick bedded on mortar round the edges only.

Butterfly Nut. See *Wing Nut.*

Butterfly Roof. A V-roof with outer vertical lights.

Buttery. A butler's pantry.

Butt Joint. Two pieces of timber joined together with a plain square joint.

Button. A small piece of w od or metal secured by one screw so that it is free to revolve. It is used to secure a cupboard door; or one piece of framing to another, as a counter top to the pedestals and battens. See *Turnbuckle.*

Button Rivet. One with a snap head.

Buttonwood. American Plane or Sycamore, *q.v.*

Buttress. A projection from a wall to give strength and to resist a thrust. It is usually of an ornamental character. See *Flying Buttress.*

Buttressing Pier or Wall. One giving lateral support to a wall.

Butt Weld. A weld when the throat lies in a plane at 90° to the surface of at least one of the joined pieces.

Butt Work. See *Butt and Break.*

B.W. Abbreviation for *bath waste.*

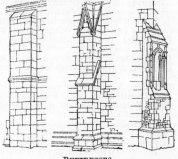

BUTTRESSES

B. W. G. Birmingham wire gauge. A standard gauge for wire, from 4/0 to 36. See *Standard Wire Gauge.*

By-laws, or **Bye-laws.** Regulations of Local Authorities controlling the erection of buildings. London Building (Constructional) By-laws, 1952. See *Acts.*

By-pass. Applied to a road specially built for fast through traffic, to relieve a busy thoroughfare or the main streets of a town.

By-products. The by-products from wood and trees include acetic acid, bark (tanning, etc.), cellulose, charcoal (gunpowder, etc.), dyes, excelsior, gutta-percha, medicines, oils, pitch, plastics, potash, pulp (paper, etc.), rosin, rubber, scents, turpentine, wood meal, wood flour, wood spirit, etc.

Byre. A cow-house.

Byzantine. Asiatic architecture with Grecian characteristics. The best example is the Church of S. Sophia (A.D. 532), Constantinople (Istanbul).

C

C. An abbreviation for *centigrade* and for *constant.* It is used in formulae to denote some invariable quantity. A symbol for *compressive stress.*

c.a. Abbreviation for *cart away.*

C.A. Abbreviation for *cleaning arm.*

Cabin. A hut, or temporary wooden building, as used for the convenience of the workmen during the erection of buildings. See *Derrick Crane.* A building from which signals and points are controlled for a railway.

Cabinet. 1. A small room or private apartment. 2. A nest of drawers : or a small ornamental cover or frame.

Cabin Hook A long hook for engaging in an eye for fixing open casements or doors in position.

Cable. The protected wires used in electric lighting and heating, telephone and telegraph systems. Also any strong wire or chain used for anchoring structural work.

Cable Moulding. A cylindrical moulding carved to imitate a rope.

Cabling. Slender cylindrical mouldings in the flutes of columns, for about one-third of the height.

Cabot's Quilt. A sound absorber used in walls and floors. It consists of layers of kraft paper between which are leaves of eel-grass. It is supplied in rolls about 3 ft. wide and 80 ft. long.

Cabreuva. A walnut-brown hardwood from tropical America. It is fragrant, lustrous, with darker streaks, hard, heavy, durable, and strong. It is fairly difficult to work because of its hardness and roey grain. Used for superior joinery, etc. Weight about 56 lb. per c. ft.

Caen Stone. A fine oolitic limestone quarried in Normandy. Excellent for interior carving.

Café. A place for light refreshments. A coffee house. A restaurant.

Cafeteria. A restaurant in which customers serve themselves from the counter.

Cafferata. A proprietary partition slab.

Cage Construction. Steel frame construction. See *Framed Structure*.

Cain's Formula. Used in structural work to obtain the working stress for variable loads.

Cairn's Lodge. One of the best Irish limestones. Grey in colour and durable.

Caisson, or Coffer. 1. A deeply sunk panel in a ceiling or soffit. 2. A watertight structure used when laying foundations below water.

Caisson Disease. Sickness due to working in compressed air, in underwater work. The body becomes partly saturated with nitrogen and it is necessary to *decompress* slowly before entering the ordinary atmosphere.

Caithness. A dark-grey sandstone from Caithness. It is used for landings, steps, paving, etc. Weight 157 lb. per c. ft.

Calacata. An Italian white marble with grey veinings.

Calamander, or Coramandel. A hard brown timber with black stripes from Ceylon. Difficult to work. A decorative timber used for veneers.

Calamansanay. A yellowish-red to rose-coloured hardwood from the Philippines. It is hard, dense, durable, tough, and difficult to work. Used for furniture, structural work, etc. Weight about 38 lb. per c. ft.

Calcareous. Applied to stones in which calcium carbonate predominates.

Calcareous Bricks. Pumice bricks with dolomitic lime as the binding agent.

Calcareous Clays. Those having a proportion of calcium carbonate.

Calcarium. A washable water-paint.

Calcination. Burning with great heat to cause chemical changes, as limestone to lime. The *roasting* of iron ores to expel moisture before smelting.

Calcium Carbide. CaC_2. Formed by heating lime and coke in an electric furnace, and used for acetylene lamps.

Calcium Carbonate. $CaCo_3$. The basis of limestone, and the chief calcium mineral. Chalk.

Calcium Chloride. $CaCl_2$. Used in macadam roads as a dust preventer. The chemical absorbs the moisture from the atmosphere and keeps the road surface damp in dry weather. Also sometimes used in concrete as a protection against frost; and in refrigeration plants.

Calcium Fluoride. CaF_2. Fluorspar. Used as a flux in glass and steel industries.

Calderwood Cement. See *Cement*.

Calf. The wedge-shaped part of a scarf used when lengthening timbers

Calibre. The bore of a pipe.

Califont. A registered geyser. A self-contained hot-water system.

Californian Pine. Sequoia, or Redwood. The largest softwood (up to 20 ft. diameter). Dull brown colour with reddish markings. Raggy and coarse texture, short grained and brittle. Weight about 28 lb. per c. ft.

Calime. A proprietary hydrated lime. It is ready slaked and finely sifted.

Calipers. Used for measuring internal and external diameters of cylindrical and spherical objects. Similar to a compass in shape, but with arched legs.

Callender System. A process of damp proofing specially applicable to basements in water-logged soil. See p. 85.

Callendrite. Pure bitumen in roll form for damp-proof courses.

Callow. The overlying stratum of soil in brickfields.

Callow Rock. A proprietary lime, either quick or hydrated.

Calorex Glass. Anti-actinic or heat-excluding glass. It excludes 75 per cent of the heat from the sun, and is rough-cast or in sheet form.

Calorie. The amount of heat required to raise the temperature of 1 gram of water through 1° C. The large calorie is 1 kilogram through 1° C.

Calorifiers. Appliances for heating water by means of steam tubes. They are used for the hot water supply in hospitals, public buildings, etc.

Calp. A dark grey limestone.

Cam. An irregular projection to a revolving shaft to give intermittent or alternating motion to another part of the mechanism.

Camagon or **Camagoon.** A species like ebony from the E. Indies.

Camber. The rising at the middle of a horizontal structural member to counteract sagging, or for appearance. The rise of a very flat segmental arch. A camber bar is a chimney bar. The convexity of a road surface.

Camber Arch. See *Straight Arch* and *Arches*.

Camber Beam. A tie beam of a roof truss.

Cambering Machine. An appliance dropped down the cylinder, when pile-driving, to form an extended base for the pile.

Camber Slip. 1. A *turning piece*. A single piece of timber used as a centre when the curvature is small. 2. A triangular templet to describe a very flat segmental arch.

Cambium Layer. The wood-forming layer in exogenous trees immediately under the bark, which protects the layer. An injury to the cambium layer prevents the growth of new cells.

Cameoid Relief. An embossed wall-paper. The raised design is formed as the paper is pressed between rollers.

Camera. A suite of lodgings for an abbot or prior in a monastery.

Cames. Lengths of "fret lead." The leads in leaded lights, or stained glass work, for securing the small pieces of glass, and to build up the required size of square.

Camloc. A patent coupling for tubular scaffolding.

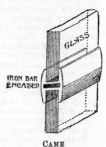

CAME

Campana. See *Bell.*

Campanile. A detached clock or bell tower.

Camphorwood. A tough, durable hardwood from Borneo. It is light red in colour and similar to mahogany, and obtained in large sizes. True camphor, *Cinnamonum sp.*, is not used for building. Kenya Camphorwood, *Ocotea sp.*, is also an excellent wood for superior joinery. There are numerous other species, as they are widespread, but little is used in this country

They are fragrant, smooth, decorative, and polish well, and very good for cabinet work, joinery, etc. Weight from 36 to 46 lb. per c. ft.

Camp, or Coom, Ceiling. A sloping or convex ceiling. An attic ceiling in the form of a truncated pyramid. Part of ceiling follows slope of roof.

Campshedding. Sheet piling round a barge bed.

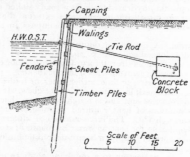

CAMP SHEETING FOR QUAY WALL

Camp Sheeting. Sheet piling in loose soil, or to a wharf wall. The cap, or sill, of a wharf wall.

Campus. The grounds of an American school or college.

Canada Balsam. A turpentine from balsam fir. It is used for a jointing compound for glass, as it has the same refractive index, airtight show-cases, mounting of slides, etc.

Canadian Red Pine. Very similar to yellow pine.

Canal. 1. An artificial waterway to connect natural waterways. 2. A flute, especially when it is in the soffit of a cornice.

Canary Ash. A lemon-yellow hardwood from Queensland. It is fairly hard, heavy, stable, and strong. Silica content dulls saws and knives. Used for flooring, superior joinery, etc. Weight about 42 lb. per c. ft.

Canary Sassafras. A yellow, lustrous, hardwood from Queensland. It is fairly light and soft and is easily wrought and seasoned. Used for joinery, flooring, etc. Weight about 34 lb. per c. ft.

Canary Whitewood. A variegated hardwood from the U.S.A. Also called Yellow Poplar. The colour varies from greyish white to yellow and green. It is soft, light, and easy to work. Weight about 27 lb. per c. ft. See *American Whitewood.*

Candelabra. A branched stand to carry lights. A chandelier.

Candle Power. The illuminating power of a standard sperm candle which is the unit of lighting.

C. & P. Abbreviation for *cut* and *pin.*

C. & W. Abbreviation for *claircolle and whiten.*

c. & w. Abbreviation for *cutting and waste.*

Canec. A registered wood-fibre wallboard.

Canelle. A brown timber, light in weight and not durable, from Brazil. Similar to laurel. Large sizes.

Canister Elbow. A T-shaped junction for iron flue pipes. One end of the cross piece has a movable stopper for cleaning purposes.

Canker. An infectious disease in trees.

Cannon End. The knuckle end of the arm of a sunblind. It is usually of cast brass.

Canopy. A projecting hood, supported on brackets, to a door, window, or niche. Also see *Sounding Board*.

Cant. To turn over at an angle not a right angle. An external splayed angle. An alternative name for *wane*, *q.v.* To remove slabs from the sides of a log.

Cant Bay. A bay window with hexagonal or octagonal sides.

Cant Board, or **Lear Board.** The sloping side of a gutter.

Cant Column. A column with a polygonal section.

Canted. Splayed, or on the bevel, or not square.

Canteen. A building for refreshments. It is specially applied to one in barracks or to a temporary one.

Cantharus. A vase in the atrium of the early churches.

Cantilever. A beam supported at one end only. See *Reinforced Concrete Cornice*, and page 366.

Cantilever Bracket. One built into the wall to withstand the leverage of the load, and implying a large projection, as for a cistern.

Cantilever Foundations. To keep the grillage, or base, inside the building line, wall columns often rest on cantilever girders.

Cantilever Steps, or **Hanging Steps.** Those built into the wall at one end only, and free at the other. It is better to support the free ends by a rolled steel section to prevent collapse in case of fire.

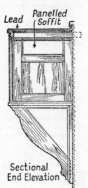

Lead / Panelled Soffit

Sectional End Elevation

CANOPY

Cant Moulding. A moulding with a bevelled face.

Cantoned. Quoins that project over the face of the wall.

Cantonment. A military depot. Lodgings for soldiers.

Cap. 1. A planted piece on the top of a post or column, for weathering or ornamentation. See *Knee*. 2. A rotating box-like structure housing the shaft that carries the sails of a windmill. 3. A horizontal roof timber in a mine. 4. See *Capping*.

Capillarity, or **Capillary Attraction.** The climbing of a liquid above its own level, in confined spaces, due to the surface tension of the liquid. See *Throating*.

Capital. The head of a column or pilaster immediately under the entablature. Each "Order" has its characteristic capital, but in other styles of architecture there is considerable variation. See *Mouldings, Bell, Bowl, Carved,* and *Egyptian*.

Capped Pipe. A drain pipe with cover-plate for cleaning purposes.

Capping. A long length, serving as a cap, for gates, framing, etc. A capping serves the same purpose as a coping. Also see *Dwarf Screen*.

Capstan. A machine based on the principle of the wheel and axle. The wheel is replaced by long spokes, or levers, which are turned by men or power, to haul heavy loads.

Caracole. A spiral staircase.

Carbide of Calcium. Manufactured by fusion, in an electric furnace, of calcium and carbon, and used for generating acetylene gas for lighting. Carbide must be kept dry in storage.

Carbolineum, or **Peterlineum.** A preservative for timber, similar to creosote.

Carbonate, Magnesium. $MgCO_3$. Contained in many limestones, and characteristic of dolomites.

Carbon Black. See *Black Pigments*.

Carborundum. A carbide of silicon, SiC, obtained by heating coke with sand in an electric furnace. Extremely hard, and used as an abrasive and for oilstones, etc.

Carbunk. A trolley for carrying timber in seasoning kilns.

Carcass or **Carcase.** The main parts of framing or structures before the coverings or decorative features are applied.

Carcassing. 1. The arrangement of the gas pipes in a building. 2. Preparing the carpentry in a building.

Card Process. A wood preservation treatment, which consists of pressure penetration of an oil and salt-solution mixture, usually creosote and zinc chloride.

Carling Sole, or **Runner.** A grooved timber, or head, for securing the top edges of framing on ship decks.

Carlisle. A patent retarded hemi-hydrate plaster.

Carnation. See *Granite.*

Carol. An enclosure, recess, or small closet, for quiet study.

Caroline. A decorative cream-coloured marble.

Carpenter's Boast. The joint between the collar tie and rafter in a collar tie roof truss. The joint is a half dovetail. See *Collar Tie Roof.*

Carpentry. The structural part of the woodwork of a building. The woodwork in the carcass. Also the craft of preparing structural and constructional woodwork.

Carpentry Joints. Constructional joints which should be designed to resist tension, compression, bearing and shear. To obtain the maximum strength of the timber the resistance of the various parts should be equal. The various kinds include: bird's mouth, bridle, cogged, simple dovetail, fished, halving, joggle, keyed, mitred, saddle, scarfed, step, tabled, tenon, trenched, tusk tenon. There are many variations and combinations with metal components. See *Joinery Joints*, and *Timber Connectors.*

Carrara Marble. One of the best known Italian marbles, often called Sicilian marble. It is uniform in texture, white with a bluish cast but free from dark veins. Weathers well. Other kinds are also quarried in the Carrara area.

Carriage. 1. The timber framing carrying a large bell and its accessories. 2. A small frame with a pivoted lever carrying a house bell. 3. A movable support or conveyance for anything.

Carriage Gates. Wide ornamental gates at the entrance of a drive to a house.

Carriage Piece. A rough bearer to strengthen wood steps at the centre, in a flight of stairs.

Carryall. A patent mechanical excavator.

Carrying Up. See *Bringing Up.*

Carter's Cradle. See *Travelling Cradle.*

Carton Pierre. Used by the plasterer for casts and reproductions, but little used in recent years. It consists of paper pulp with glue and whiting, with a little plaster of Paris added.

CARVED CAPITAL

Cartouche. 1. A Modillion. 2. A tablet, in the shape of a scroll or roll, for an inscription.

Carved Capital. The capital to a column carved to represent foliage, and characteristic of Early English architecture.

Caryatides. Columns in the form of female figures.

Cascade. 1. The difference in level between inlet and seal in an intercepting trap. 2. A registered decorative blazoned glass.

Casco. A registered name for several building products, and a moisture-resisting cold water glue.

Case Bays. The bays of a roof or floor except the tail, or end, bays. Joists framed between two girders.

Cased Columns. Columns surrounded by fire-resisting materials : bricks terra-cotta, patent and plaster slabs, etc.

Cased Frames. Hollow built-up frames, to receive the weights for sliding sashes. See *Sash and Frame.*

Cased Glass. Flashed glass, *q.v.*

Case Hardening. 1. Surface drying of timber which sets up internal strains and causes warping when converted. 2. Converting the face of W.I. into steel or hardening the surface of steel. 3. Stone in which the quarry sap has hardened on the face. Seasoned stone.

Casein. A protein of milk used in glues and plastics, such as Erinoid, Lactoid, Galalith, etc. See *Glues* and *Plastics.*

Casemate. A strong building able to withstand attack in warfare, a bomb-proof vault.

Casement. 1. A small hollow moulding. See *Mouldings.* 2. A hinged or pivoted sash.

Casement Lights. Windows with hinged or pivoted sashes.

Casement Stays. Metal bars to fix open casements in any required position.

Casern. A small building for military purposes.

Casings. 1. Linings used as cover boards for pipes, rough timbers, etc. See *Cradling.* 2. An alternative name for jamb linings. 3. The built-up boxes that carry the weights for vertically sliding sashes. Also see *Firrings.*

Castellated. Built in the style of a castle. Crenellated.

Casting. 1. The bending or buckling of timber lengthways of the grain. 2. Metal objects made by pouring molten metal into a mould. 3. Plaster or cement objects formed in a mould.

Casting Boxes. Forms for casting concrete for repetition work.

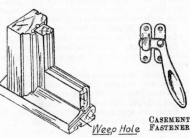

Weep Hole

CASEMENT WINDOW

CASEMENT FASTENER

Cast in situ. Concrete poured in a plastic state in its required position in a building.

Cast Iron. (C.I.) A compound of iron and carbon. It is hard and brittle and cannot be forged. Produced by smelting ores in a blast furnace, with limestone as a flux and coke for fuel. The quality varies considerably with the ore and variation in manufacture, and depends upon the proportion of carbon in chemical combination.

Cast-iron Drains. Drain pipes and fittings made of cast iron and coated with Angus Smith's solution, or by other processes, to prevent corrosion. They are used under buildings or where the ground is subject to movement or heavy traffic.

Castlewellan. See *Granites.*

Castor. A small swivelled wheel or revolving ball for the feet of furniture.

Cast Stone. Moulded concrete specially mixed or faced to have the appearance of natural stone. Also called pre-cast, reconstructed, patent, artificial, and imitation stone.

Catacombs. A subterranean burial place.

Catalin. A urea-formaldehyde thermo-setting plastic.

Catalpa. A greyish brown hardwood from the U.S.A., resembling Butternut. It is light, soft, stable, and durable. Used for construction and joinery. Weight about 24 lb. per c. ft.

Catch. The part of a fastening that secures the latch. See *Norfolk Latch.*

Catchment. An area of ground draining to a reservoir.

Catch Pit. A reservoir to receive surface drainage. A pit at the entrance to a sewer or drain to collect the solid matter that would probably cause a stoppage in the drain.

Catchwater Drain. Specially constructed drains to collect surface water to protect embankment walls, excavations, etc.

Catenary. The curve assumed by a flexible uniform chain supported horizontally from two points and acted upon by gravitation only.

Caterpillar Lights. A continuous series of lights, shaped like small span roofs, and lying on the inclined roof of a large factory. They give a larger lighting area than when lying flat in the plane of the roof, and require short glazing bars

Cat Head. A projecting cantilever used for hoisting purposes.

Cathedral. The see or seat of a bishop, possessing the cathedra or epis-copal chair. The principal church of a diocese.

Cathedral Glass. Patterned rolled glass for partitions, leaded lights, etc. There are many varieties: Double rolled, hammered, mottled, rough, stippled, etc. Also it may be tinted. *Cathedral sheet glass* is blown sheet glass with imperfections to give an antique effect.

Catherine Wheel. A rose, or circular window with radiating divisions.

Cathetus. The centre of the Ionic volute.

Cat Ladder. A vertical ladder for access to a loft, etc. Any arrange-ment on a sloping roof to give foothold and to avoid damage to the roof covering.

Catmon. A reddish brown hardwood from the East Indies. It is hard, heavy and durable; rather difficult to work. Used for constructional work and joinery. Weight about 45 lb. per c. ft.

Cat's Head. Same as beak head, *q.v.*

Caul. An appliance used in veneering. It is applied warm and keeps the veneer in position until the glue has set.

Caulicoli. Small volutes under the flowers on the abacus to the Corinthian capital.

Caulk. 1. To fill, or stop, an open joint tightly. 2. The cogging of a joist or binder on to the wall plate. 3. To split the ends of a chimney bar (*q.v.*) and turn them to anchor in the wall.

Caulking Gun. A mechanical appliance for sealing joints with caulking compounds.

Causeway. A raised footpath or road.

Causewaying. Forming roads of paving stones, or granite sets.

Cavetto. A hollow moulding in the form of a quarter circle, often called a scotia. See *Mouldings.*

Cavity Bricks. 1. Tubular hollow blocks. 2. Bricks with deep recesses instead of the usual frog, the walls and webs being about ¾ in. or 1 in. thick.

Cavity Lath. A long piece of wood, or batten, suspended in the cavity of a cavity wall during erection. It is used to catch the mortar droppings and is pulled up the cavity as the work proceeds.

Cavity Sub Frame. A patent frame for cavity walls to which metal windows are fixed.

Cavity Ties. See *Wall Ties*.

Cavity Walls. Walls constructed of two separate thicknesses with a 2 in. cavity between and tied together by tie-irons or tie-bricks. They are used to ensure dry interiors, more equable temperature, and for economy in facing bricks.

Cawk. Barytes. See *Barium sulphate.*

Caxton Floor. Patent reinforced hollow tiles used for fire-resisting floors.

C.C.N. Abbreviation for *close copper nailing.*

Cedar. A coniferous timber, very durable, fragrant, and with great resistance against insects and variation of temperature. Many varieties, both hardwoods and softwoods. The best known is red cedar (Juniperus). Used

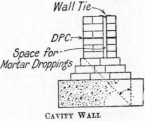

CAVITY WALL

for many purposes including pencils, fittings to furniture, shingles, etc. Nicaraguan and Honduras cedars are used for bank and office fittings and are excellent timbers. Some of the hardwoods are used as substitutes for mahogany in building. See *Western Red Cedar*.

Ceiling Boards. Plaster boards, *q.v.*

Ceiling Joists. Light joists, below the floor joists, to carry a lath and plaster or boarded ceiling. Also see *Eaves*.

Ceiling Laths. See *Laths*.

Ceiling Light. A lay-light. A borrowed light in a ceiling.

Ceiling, Matched. A ceiling boarded with tongued and grooved matchboarding.

Ceiling Straps. Narrow strips, nailed to floor joists or rafters, from which ceiling joists are suspended.

Ceiling Trimming. Framing ceiling joists round chimney breasts, and openings for trap doors or lay-lights.

Celcurising. A proprietary method of preservation for wood. It is effective against dry rot and insect attack.

Celery Top Pine. An Australian pine similar to Hoop pine.

Celestial. See *Blue Pigments*.

Cell. 1. A small room in a monastery or prison. 2. A small enclosed cavity. 3. A unit mass of living matter. 4. See *Rib and Panel Vault.* 5. See *Naos.*

Cella. The body of the temple, in early classic architecture. See *Temple.*

Cellactite. A proprietary asbestos-protected metal roofing, etc. It may be plain or corrugated.

Cella Die. Used for moulding hollow blocks of clay with closed ends.

Cellar. A basement or underground room, used for storage purposes.

Cellar Fungus. See *Dry Rot.*

Cell Line. The soffit line of a vault next to the ribs.

Cellular Bond. A bond for 9 in. walls leaving a cavity between inner and outer bricks as Rat-trap bond, *q.v.* Another form is to place the stretcher courses on edge and the heading course normally as in English bond.

Cellular Bricks. Cavity bricks.

Cellular Concrete. Gas-producing materials are mixed with the cement, which forms bubbles and makes the concrete light and insulating. Large cells, or pores, are formed to produce a light-weight concrete. The cells may be formed by chemicals or by making a foam in the mix by adding soap and albumen in the ratio of 1 to 2000. Also see *Aerocrete, Ice Concrete,* and *Bubblestone.*

Cellulose. A carbohydrate from the cell walls of plants. It forms the basis of numerous industrial processes: paper, artificial silk, lacquers, dope, celluloid, etc.

Cellulose Lacquer. Nitro-cellulose paint enamel. Quick drying, due to the evaporation of the *thinners.* Suitable for spraying, sponging, and stippling. Very adaptable for decorative work.

Celotex. A registered wood-fibre wallboard.

Cement. Materials that undergo chemical change, when moisture is applied, and set hard like stone. The most important is Portland cement, which is artificial. Natural cements are Roman, Medina, Rosendale, puzzolana. Other cements are slag, bituminous, sulphur, and ganister. The following differ very little from Roman: Atkinson's, Calderwood, Harwich, Mulgrave's, Whitby. Also see *Bauxite* and *Ciment fondu.*

Cementation Pile. One in which the hole is bored, filled with gravel or ballast, and grouted under pressure.

Cementation Process. A process of producing steel by first refining C.I. into almost pure W.I. and then adding the required amount of carbon for the steel.

Cement-bitumen. A jointless flooring compound of cement and aggregate with bitumen, coloured as required with pigments. An excellent flooring material but will not stand grease.

Cement Fillet. A triangular strip of cement to fill an angle or to make it water-tight. It is often used instead of lead flashings.

Cement Grout. Cement and water in a liquid state, sufficiently thin to be poured.

Cement Gun. An appliance for applying cement to walls, etc. The materials are forced through a nozzle and are hydrated as they emerge.

Cementite. Carbide of iron, Fe_3C.

Cementitious Bricks. Pumice bricks with Portland cement or some other hydraulic cement used as binding agents.

Cementone Products. The trade name for several waterproofing agents for brickwork, etc.

Cement Pressing. Moulding cement ornaments in plaster moulds. The model is piece-moulded. Since the introduction of artificial stone, cement pressing is seldom done.

Cement-wood Floor. A jointless flooring compound consisting of 1 part cement, $1\frac{1}{2}$ parts aggregate, and $1\frac{1}{2}$ parts specially treated sawdust, and required pigments.

Cemestos. A prefabricated asbestos-cement house.

Cemixene. A cement mixture prepared for use, with the addition of water only.

Cemprover. Proprietary liquids for improving the quality of cement work for particular requirements.

Cemseal. A proprietary colloidal waterproofer for cement and concrete.

Cenotaph. An empty tomb. A monument not over a grave.

Cental. 100 lb. of cement.

Centec. A portable electric router.

Centering, or Centring. Temporary timbering used for supporting the component parts of an arch or vault during erection.

Centering Slips. Wedge-shaped supporting strips of wood that allow for adjusting a square of glass or the position of a frame, etc. When the glass or frame is in the correct position the centering slips are fixed.

Central Heating. The heating of buildings by distributing hot water or steam through pipes and radiators or hot air through gratings. The water is heated at a convenient position, by means of coal, coke, gas, or oil, and the pipes may be buried in the walls or ceilings. There are five methods used: low and high pressure hot water, and low and high pressure steam or hot air.

Centre. A temporary wooden frame, or mould, used as a support while turning an arch.

Centre Hung Sashes. Pivoted sashes.

Centre Nailing. The nailing of slates near the middle of their length. The nails are covered by one slate only; but the method resists lifting by the wind better than *head-nailing*. See *Gauge*.

Centre of Compression. The centroid of the stress triangle, in beam design.

Centre of Curvature. The centre from which a circular arc is struck.

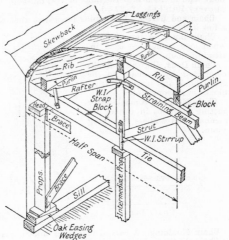

CENTERING FOR BRIDGE

Centre of Gravity. (C.G.). A point in, or near, a body about which all its parts balance each other. The weight of a body may be considered as acting at its C.G. The centre of an area is called the *centroid*.

Centre of Pressure. The point in a masonry bed joint through which the resultant pressure acts.

Centres. Pivots, especially applied to those for swing doors.

Centre to Centre. The method of spacing constructional members such as floor joists, studs, spars, etc., by measuring from the centres, or from *in to out*.

Centric. A registered highly-decorative blazoned glass.

Centric Load. An axial load. A load passing through the C.G. of a section, or along the axis of a structural member.

Centrix. A registered tile cutter.

Centroid. See *Centre of Gravity*.

Ceramic. Pertaining to pottery. Mosaic work with china clay surface.

Certificates. A statement by the architect that certain work on a building has been completed and that an agreed instalment is due to the builder.

Certosec. A proprietary anti-condensation paint.

Certus Glue. A proprietary casein glue. See *Glues*.

Cessing. See *Sissing*.

Cesspit. 1. A privy. 2. A cesspool, (2) *q.v.*

Cesspool. 1. A lead-lined box in a roof gutter to collect the water where it enters the down pipe. 2. A large chamber for collecting sewage from drains where no sewer is available.

C.G. Abbreviation for *Centre of Gravity*.

c.g.s. The centimetre, gram, second system of units.

Chadrac. A fire-resisting glazing by the electro-copper process.

Chain. 1. Used for measuring distances when surveying. The 100 ft. chain is divided into 100 links of 1 ft. each. Gunter's chain is 66 ft. and divided into 100 links of 7·92 in. each. A brass tag is attached at every 10th link. 2. 22 yd.

Chain and Barrel. A fastening for a door that allows the door to open a few inches only.

Chain Bond. A term applied when long stones are used instead of chain timbers, *q.v.*

Chain Dogs. Two steel hooks connected by a chain to the crane hook and used for lifting blocks of masonry, etc.

Chain Mortiser. A machine for cutting mortises in timber by means of an endless revolving chain.

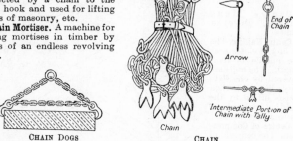

End of Chain

Arrow

Intermediate Portion of Chain with Tally

CHAIN DOGS   CHAIN

Chain Moulding. A carved moulding imitating the links of a chain.

Chain Riveting. Rivets placed in parallel rows and opposite to each other in adjacent rows.

Chain Timbers. Timbers used to bond and strengthen brick walls, usually circular in plan.

Chair. An iron socket forming a seating.

Chair Rail. A dado rail. A horizontal moulding fixed to a wall to prevent damage by chair backs, and for ornamentation.

Chalet. A cottage, or hut, for herdsmen and peculiar to Switzerland. A small house built in similar style.

Chalk. Carbonate of lime, calcium carbonate, or pure limestone, $CaCO_3$. It is added to clays for artificial marls, and calcined for lime.

Chalking. A film, on a painted surface, of unattached pigment due to the disintegration of the binding medium.

Chalk Line. A long string coated with chalk and used for setting out long straight lines on plane surfaces.

Chalky. Applied to a painted surface that has disintegrated, with the appearance of powder.

Challenge. A patent roof glazing.

Chamber. An assembly room. An apartment.

Chamber, Intercepting. A manhole separating the drainage system from the sewer.

Chamber Interceptor. A trap to a manhole to prevent the gases from the sewer entering the drains.

Chamf. Abbreviation for *chamfered*.

Chamfer. A corner bevelled so that the arris is removed equally on each face of the material. When it is unequal it is usually termed a bevel. See *Mouldings.*

Chamois. Soft pliable leather used for bedding glass as a protection against vibration.

Champ. A flat surface.

Champac. A yellowish-brown to olive-brown hardwood from India and E. Indies. It is fairly light, soft, strong, durable, lustrous, and polishes well, but difficult to season, unless girdled before felling. It is used for joinery, bentwood, etc. Weight about 36 lb. per c. ft.

Chancel. The area round a church altar enclosed by the communion rail or by a screen.

Chancel Screen. Ornamental parclose screens used to separate the chancel from the remainder of the church.

Chandelier. A decorative branched support for a number of lights.

Changing Station. An intermediate position in a survey where a change of levels takes place.

Channel. A duct, holllow, groove, tube, etc., especially if used to convey liquids.

Channel Iron. A British standard steel section shaped like a rectangular channel. See *Reinforced Concrete Cornice*, and *Rolled Steel Sections*.

Channelled Quoin. Rebates formed on the top edges of ashlar quoins.

Channeller. A machine used in quarrying. The machine cutters form channels to free the blocks of stone.

Channel Tiles. The under tile of Italian, or similar, tiling.

Chantlate. See *Sprocket Piece*.

Chantry. An endowed chapel for the singing of masses.

Chapel. A subordinate or private place of worship. A nonconformist church. A relatively small place of worship attached to a cathedral. A part dedicated to a saint.

Chapel-of-ease. A smaller church to relieve a larger one and which is easier of access for remote parishioners.

Chapiter. The upper part of the capital of a column.

Chaplash. A yellow to golden-brown hardwood from India, with variegated streaks. Fairly hard and durable, rather difficult to work smooth. Difficult to season unless girdled before felling. Weight about 34 lb. per c. ft.

Chaplet. A small cylindrical moulding, carved usually in the form of beads.

Chaps. Cracks or fissures in the bark of trees due to frost or very hot sun. The exposed cambium layer is liable to disease through exposure.

Chapter-house. A monastic building attached to a cathedral or abbey, and used for the assembly of deans and canons.

Chaptrell. An impost supporting an arch.

Charcoal. The carbonaceous residue, after partial combustion, of vegetable, animal, or mineral substances.

Charging Hopper. See *Concrete Mixer*.

Charles's Law. The volume of a given mass of gas, at constant pressure, increases by $\dfrac{1}{273}$ of its amount at 0° for each degree rise in temperature,

$$\therefore V_t = V_0 \left(1 + \frac{t}{273} \right).$$ The value $\dfrac{1}{273}$ is the coefficient of expansion.

Charley Forest. A Leicestershire stone, very good for oilstones.

Charring. The burning of the bottom of timber posts to coat them with charcoal as a protection against decay.

Chase or **Chasing.** A channel or groove formed or cut in the material.

Chase Form. A wooden mould for forming a channel, or chase, in concrete.

Chase Mortise. A stub mortise (*b*), with a chase (*a*) so that the tenon can be slid into position sideways. See page 58.

Chases. Cavities formed by omitting bricks in walls to allow *in situ* concrete to key into the wall, as in stairs, landings, etc. The required bricks are usually bedded in sand, so that the wall can be built conveniently, and removed when running the concrete.

Chatter. Limestone or granite chippings.

Chattering. The vibration of badly fitting plane irons, which causes ridges on the surface of the material.

Checkerboard. Ornamental brickwork consisting of a circular panel, or rowlock, and continued on each side in heading bond with a course, above and below, formed of headers on edge.

Checker Plate. A plate under the oven of a kitchen range to circulate the hot air.

Checkfire. Patent fire-resisting composite doors.

Checking. To test for accuracy : to compare with results previously obtained.

Check Nut. A lock-nut.

Checks. 1. A local name for rebates. 2. Oil or pneumatic *floor springs* (*q.v.*) for controlling swing doors. 3. Surface shakes in wood. 4. Slots in the edge of marble facings to receive lead or copper

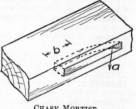

CHASE MORTISE

fastenings. 5. Snecks in snecked rubble masonry. 6. Same as checking, *q.v*

Chee. A reddish-grey hardwood from Indo-China, similar to ash but with silver grain, and marketed as Chinese Oak. It has many of the qualities of oak and used for superior joinery, etc. Weight about 35 lb. per c. ft.

Cheecol. A registered name for a lightweight concrete, a patent grouting process and several waterproofing agents registered as Permacem, Permacol, etc.

Cheek Boards. Formwork for the vertical sides of beams.

Cheeks. 1. The sides of a mortise, or the removed sides of a tenon. 2. The vertical sides of dormer windows. 3. The splayed sides to a fireplace opening.

Cheesewring. A light-grey Cornish granite. It has coarse grain and used for bridges and docks. Weight 168 lb. per c. ft. Crushing strength 1440 tons per sq. ft.

Chelura terebrans. A very destructive wood-boring insect. See *Gribble*.

Chemical Closets. Similar to earth closets, but a liquid reagent is used as a deodorant. They are used where there is no drainage system.

Chengal. A brownish-yellow hardwood from Malaya that darkens to dark brown. It has fine rays and ripple, is hard, very strong and durable. Used for construction, furniture, etc. Weight varies from 35 to 48 lb. per c. ft.

Chequerboard. A registered wovenboard fencing. See *Woven-oak* and *Rowlock*.

Chequered. A design consisting of differently coloured squares, like a chessboard.

Cherry Mahogany. See *Makore*.

Chest. A large strong box.

Chestnut (*Castanea*). A hardwood similar to oak except for the medullary rays, and often used as a substitute for oak. Very durable under ground. Weight about 38 lb. per c. ft.

Chevaux de Frise. A protection, with long spikes, on top of a fence wall.

Chevet. A feature in early Gothic churches in which the aisle is continued round the apsidal end of the choir. Small chapels were formed in the bay

Chevron. A carved zig-zag moulding used in Norman architecture.

CHEVRON

Chezy Formula. $V = c\sqrt{mi}$. A formula for calculating the flow in channels or pipes, where V = vel. in ft. per sec., m = hyd. mean depth, i = gradient, c = constant, for which values are given.

Chfd. Abbreviation for *chamfered.*

Chickrasey. A light-brown lustrous hardwood from India and Burma, belonging to the Mahogany family. It is fairly hard, heavy, and strong, but not durable. The decorative grain makes it suitable for furniture, veneers, panelling, etc. Weight about 37 lb. per c. ft.

Chil or Chir. The long-leaved pine from N. India. Used for construction and joinery. Weight about 35 lb. per c. ft.

Chile Pine. The Monkey Puzzle tree. The wood is similar to Hoop Pine, *q.v.*

Chilmark. A light brown limestone from Wiltshire. There are three varieties: Trough, Green, and Pinney. Trough is the best for general building. Weight 135 to 154 lb. per c. ft. Crushing strength 136 tons per sq. ft.

Chimney. A hollow structure of brick, concrete, or stone, forming a flue to convey smoke, etc.

Chimney Back. The wall at the back of a fireplace. It must be at least 9 in. thick in a party wall for a height of 12 in. above the lintol to the fireplace opening.

Chimney Bar. A cambered W.I. bar to carry the arch over a fireplace when the jambs are less than 18 in. The ends are split and turned up and down to tie in the brickwork.

Chimney Breast. See *Breast.*

Chimney Cap. An ornamental finish to the top of a chimney.

CHIMNEY BAR

Chimney Jambs. The vertical sides to a fireplace opening.

Chimney Piece. An alternative name for mantel-piece. A shelf over a fireplace.

Chimney Pot. A fireclay or terra-cotta terminal to a chimney to increase the height to prevent down-draught.

Chimney Shaft. The part of a chimney above the roof ; or a tall chimney.

Chimney Stack. The brickwork for grouped flues projecting above the roof.

Chimney Terminal. A chimney pot or cap.

China Clay. Kaolin, or primary clay. $Al_2O_3 2SiO_2 2H_2O$. Used for porcelain and pottery, and as a base or precipitant in paint.

China Wood Oil. Tung oil, *q.v.* A vehicle used in varnish. Pale amber in colour, and obtained from the seeds of the Tung tree.

Chinese Blue. A pigment for paints produced from a solution of iron salts and ferro-cyanide solution.

Chinese Red. See *Chromes.*

Chinese White. See *Zinc White.*

CHIMNEY STACK

Chipboard. The name applied to the numerous "man-made" timber boards, etc., such as Jicwood, Weyroc, Plimberite, etc. They are made from wood chips and plastic resin.

Chipping the Section. Removing the layer of blue plaster from a plaster cast.

Chisels. Steel cutting tools. There is a big variety used in woodwork, brickwork, masonry, and metalwork. See page 102.

Chloramine. A process for sterilizing water by treating it with chlorine and ammonia.

Chlorinated Phenolic Compounds. Used as antiseptics in paints to destroy bacteria and to counteract the action of moisture under painted surfaces.

Chocks. Small wooden blocks to resist thrust or pressure.

Choir. That part of a church reserved for the trained singers.

Choir Stalls. The pews in the choir.

Chopper. A slater's axe or zax.

Chord. 1. A straight line joining the ends of a circular arc. See *Circle*. 2. A long horizontal main member of a built-up girder. 3. *A Scale of chords*, *AD*, shows the lengths of the "chords of the angles" in a quadrant, and is used for plotting angles. Describe the quadrant *ABC*. Divide the arc into 9 parts of 10° each. Swing them on to *AB* for the required lengths of the chords.

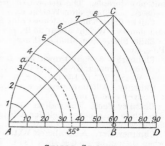

CHORDS, SCALE OF

Chord Method. A method of designing built-up beams in which the flanges only are considered to resist the bending moment. It is more approximate than the moment of inertia method.

Chromador. A structural steel containing chromium and copper. It is stronger than M.S. and corrosion resisting.

Chrome Bricks. Neutral fire bricks made from chromite.

Chromes, or Chromate of Lead. Yellow, to orange and red, pigment used in paint, etc. Poisonous, and changes colour when exposed to sulphurous atmospheres.

Chromite. Ore containing chromium and iron oxides. It is mixed with fireclay or bauxite as a binding agent in bricks.

Chromium Plating. An electric-plating process that produces a mirror-like finish for metal fittings.

Chuffs. Bricks full of cracks through exposure during burning, and unsuitable for constructional work.

Chuglum. An Indian hardwood used extensively for superior joinery and fittings, flooring, etc. It is distinguished as *black* and *white* chuglum. The latter is selected for false heart and marketed as Indian Silver Greywood. The colour varies from brownish grey to olive brown with purplish streaks.

Church. A building used for Christian worship.

Church Fittings. The internal wood, stone, and metal furniture in a church, such as pews, stalls, pulpit, altar table, screens, lectern, rostrum, kneeling board, litany stool, etc.

Churchill. A prefabricated factory-made sheet-steel house.

Church's Baryta Process. A stone preservative of doubtful value. The use of barium hydroxide converts soluble calcium sulphate into insoluble barium sulphate.

Chute. See *Shoot*.

Chuting Plant. See *Placing Plant*.

Chy. Abbreviation for *chimney*.

C.I. Abbreviation for *cast iron*.

Ciborium. The cover or canopy to an alter. A place for the host in the chancel of a church.

c.i.f. Abbreviation for *cost, insurance, freight*. The contract includes cost of materials, insurance, and conveyance.

Cilery. Foliage or drapery carved on the heads of columns.

Cill. See *Sill*.

Cimatium. See *Cymatium*.

Cimbia. See *Cincture*.

Ciment Fondu. Aluminous cement, from limestone and bauxite. Rapid hardening, and offers great resistance to sea water and sulphate solutions. It should not be mixed with Portland cement.

Cimex. A fumigant against domestic insects.

Cinch Anchor. A patent bolt anchor for walls, ceilings, and floors.

Cincture. A ring or listel, at the top and bottom of the shaft of a column.

Cinema. A place of entertainment where a bioscope projects moving pictures, or films, on to a screen.

Cinnabar. Native vermilion.

Cinquefoil, or Cincfoil. An ornament with five cuspidated divisions. See *Tracery Panel*.

Cippus. A small pillar used as a boundary, or mile, stone.

Circassian Walnut. *Juglans regia*. Walnut artificially burred to produce beautiful figure, and obtained from South Europe.

Circle-on-Circle. Work of double curvature, i.e. when it is curved in both plan and elevation.

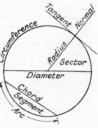

PARTS OF CIRCLE

Circle, Parts of. The various terms applied to the circle in geometry. Area of circle = πr^2, length of circumference = $2\pi r$.

Circ-perdu Method. An expensive method sometimes used in the casting of bronze. The model is made of wax which is melted out of the mould before the molten metal is poured in.

Circuit. 1. Completing a round. 2. The arrangement of the wiring for electrical installations.

Circuital Forces. In a system of forces in equilibrium the added vectors form a circuit. See *Bow's Notation*.

Circular-circular Sunk. A term used for sinkings in double curvature work.

Circular Saw. A machine-driven circular disc of steel with teeth round the periphery, used for cutting timber, stone, etc.

Circulating Pipes. The arrangement in hot water systems so that the hot water circulates through the secondary flow and return pipes, which gives an immediate supply of hot water to the draw-off taps.

Circulators. Gas-heated boilers to assist the circulation in hot water systems.

Circumscribing. To enclose a figure so that the line touches the figure without cutting it.

Circus. An open air structure for games, combats, etc., in Roman architecture.

Cir. F. Abbreviation for *circular face*.

Cirouaballi. A species of Greenheart from Guiana. It is hard and very durable and used chiefly for exterior work. Weight about 48 lb. per c. ft.

Cirrus Glass. See *Blazoned Glass*.

Cissing. Small holes, or pitting, in painted or varnished surfaces, caused by grease or smoke. An irregular coat of varnish. Ropey. The paint or varnish gathers into globules instead of spreading evenly caused by a dirty or too hard undercoat, or one not cut down sufficiently with an abrasive.

Cistern. A receptacle for holding water ; specially applied to the storage tank in hot water systems. The Metropolitan Water Board now require an 80 gall. tank for a domestic building with hot-water supply. See *Flushing Cistern*.

Citywood. Applied to the best quality mahogany from San Domingo.

Clack. A flap valve to a pump.

Cladding. One of the many names given to the edging strips for flush doors. Interior and exterior non-structural coverings to prefabricated houses.

Clairecolle. A mixture of whiting, size, and a little alum. It is used as a base for distemper to prevent suction by the plaster.

Claircolling. Preparing plaster surfaces for distemper.

Clamp. 1. A temporary wooden or iron cramp for securing two pieces together until the glue sets. 2. A batten fixed across the grain of timber to prevent warping. 3. A stack of moulded bricks and fuel, so that the whole can be fired, to produce *stock* bricks.

Clamp Burnt. Bricks burnt in clamps instead of kilns.

Clamping. Fixing or binding together by means of a clamp.

Clap Boards. Weather boards on the sides of a house. Roofing boards larger than shingles. Boards tapering in thickness for overlapping.

Clapeyron's Theorem. The "theorem of three moments" used in the design of continuous beams, when finding the bending moment.

Claratex. A registered type of tracing cloth.

Clark's Bit. An expansion bit used with the brace.

Clark's Process. A method of softening temporarily hard waters by adding lime water.

Clasp Nails. Wrought or cut nails with large flat heads to *clasp* the wood.

Classic Orders. See *Architectural Orders*.

Class Room. A room in a school or college where instruction is given.

Clavis. A keystone.

Claw Hammer. A hammer with a curved "pene" formed into claws, for withdrawing nails.

Claw Plate. See *Timber Connectors*.

Clay. Hydrous silicate of alumina ; the result of decomposition of felspar. The purest form is kaolin. Clay forms the bulk of brick earth, and is an essential constituent of cement. *Alluvial* clays are used for Portland cement. *Ball* clays (Devon) are very plastic and vitrify without changing shape and are used for pottery. *Boulder* clays are used for bricks and tiles. Brick clays are impure mixtures of clay and sand (loamy), clay and chalk (marly), etc. Other varieties are fireclays, marls, plastic, Kaolinite, pipe-clay, potter's clay, stoneware clay, terra-cotta, refractory, shales, sagger marls, siliceous, vitrifiable, carboniferous. The colour is influenced by chemical and mineral constituents : iron oxide (red), chalk (white), iron sulphide (buff), manganese (yellow). Diffused limes and alkalis act as flux, effect the fusion, and give strength. See *China Clay*.

Claying. Applying puddle, *q.v.*

Claying Model. Preparing the model for pouring the gelatine, in fibrous plaster moulding.

Cleadings. The timbering for an excavated shaft. Also see *Jacket*.

Clean Back. Applied to a through stone dressed on both faces.

Cleaning Eye. An opening in an intercepting trap to allow for easy access of the cleaning rods beyond the trap.

Cleaning Hinges. Special cranked hinges for casements opening outwards. They leave a large space between the hanging stile of the casement and the frame.

Cleaning Rods. Lengths of cane rods screwed together to form a long flexible rod for cleaning drains. A brush, scraper, corkscrew, or other tool, is fixed to the end to clear away any obstructions.

Cleaning Up. The finishing of joinery work. Smoothing, scraping and sandpapering.

Cleansol. A proprietary water softener.

Clean Stuff. Good quality wood free from knots and other defects.

Clearance. A small space between two surfaces to prevent contact, as between the edge of a door and the rebate.

Clearcole. See *Claircolle.*

Clear Space. The actual space between two members. See *Centre-to Centre* for alternative method of spacing.

Clear Span. The distance between the walls supporting a beam or truss.

Clear-story. Same as clerestory, *q.v.*

Cleats. 1. A bearing block nailed under a beam. Also see *Raking Shore.* 2. Wooden cramps used when jointing long timbers. 3. A small batten. 4. A small piece plugged to a wall to carry a bracket or shelf.

Cleavage Planes. The characteristic of slate and laminated stone which allows for splitting into thin plates. The planes of cleavage in slates are usually at an angle to the original bedding planes.

Clement's Shuttering. Patent steel uprights of 4 ft. by 1 in. by 1 in. tee-iron with bracing plates, used as shuttering for concrete.

Clench. See *Clinch.*

Clerestory. The top tier of windows lighting the choir and nave of a church. A line of windows near ceiling level in a high hall.

Clerk of Works. The person representing the owner and architect on a job. His duties are to see that the work is carried out efficiently and according to contract.

Clevis. The forked end of a rod used in steel connections.

Clg. Abbreviation for *ceiling.*

Climbing Shuttering. Metal forms for concrete. See *Metaforms.*

Clinch. The turning over of a nail point projecting through the material. Jumping up the head of a rivet when in position.

Clinker. Furnace slag.

Clinker Bricks. Vitrified bricks. Special types of specially hard-burned bricks used for paving.

Clinograph. An instrument used for drawing parallel lines. It has an adjustable arm held in position by friction.

Clinometer. A surveying instrument for obtaining direction, or angle of inclination.

Clip. 1. A lead tack, or tingle. 2. A curved piece of iron for connecting two spigot ends of C.I. gutters. 3. A *nail* with a curved head for driving into a wall to secure a pipe.

Clippy. See *Cripple.*

Clipsham. A cream-coloured limestone from Rutlandshire. It is an oolitic stone and very good for general building. Weight 150 lb. per c. ft. Crushing strength 290 tons per sq. ft.

Cloak. A piece of 3 lb. sheet lead or other impervious material over a lintel to a cavity wall. It is used to prevent moisture being transmitted to the inner wall by means of the lintel. See *Overcloak.*

Cloakscreen. A screen in a cloakroom having hooks for hats and coats.

Clockwise. A term applied to forces having a turning effect in the same direction as the hands of a clock.

Cloister. A covered arcade to a monastery or college. It is a form of corridor round a quadrangle.

Close. 1. An enclosed space

CLOISTERS

or court, as the precinct of a cathedral. 2. An entry from a street to a court.

Close Boarded. Applied to a roof that is covered with boards before laying the slates.

Closeburn. A red sandstone from Dumfries. Used for general building and dressings. Weight 124 lb. per c. ft. Crushing strength 478 tons per sq. ft.

Close Couple, or *Couple Close.* Roof timbers consisting of pairs of rafters tied together at the feet, to prevent spreading.

Close Cut Hip. A mitred hip, *q.v.*

Closer. 1. A *queen* closer is half of a brick, or 9 in. × 3 in. × 2¼ in., used for forming the bond. It is placed next to the quoin header. A *king* closer is used for the same purpose but is tapered from 4½ in. to 2¼ in. and is used at reveals. 2. The last block to be placed in a masonry course.

Close String. A string straight on the top edge. The treads and risers are housed into it. See *Stair Details.*

Closet. 1. Closets are classified as *water* or *conservancy.* The latter includes cesspools, privies, earth, and chemical closets. 2. A small private room.

Closing in. Glazing and hanging doors in a new building.

Cloudy. Variegated with darker shades or vague patches of colour.

Clout Nail. A short nail with a large flat head for fixing cords and felt. See *Nails.*

Clowring or **Clouring.** See *Wasting.*

Club Skew. A skew corbel. See *Gable* and *Kneeler.*

Clunch. A soft white limestone used for interior carvings.

Clustered Column. A column consisting of a number of small columns symmetrically grouped.

Clutch. Mechanism for throwing working parts in or out of action.

Co. Abbreviation for *course.*

Coach Screw. A large screw with a square head and driven by turning with a spanner. See *Bolts.*

Coak. A joggle or dowel in a scarfed joint to prevent one surface from sliding on another.

Coal Tar. See *Creosote.*

Coaming. A raised trimming to an opening, as to a hatch.

Coarse Stuff. The material forming the first coat for the plastering of walls and ceilings. It consists of sand, lime, and hair.

Coat. Each separate covering of paint and plastic materials.

Cobalt. A silvery-white metal produced from silver and copper ores. Oxides of cobalt and aluminium produces a blue pigment, *q.v.*

Cobbles. Rounded stones used for paving.

Coburg Varnish. A very pale oil varnish.

Coburn Fittings. Used for heavy doors sliding horizontally. The door hangs from runners sliding in a steel track. A floor channel with roller guide is provided at the bottom.

Cob Wall. A wall formed of blocks of unburned clay mixed with chopped straw. Layers of long straws are used occasionally to provide bond.

Cobweb Pattern. Applied to anything with ribs radiating and others forming concentric circles, as in grids and gratings.

Cock. To set anything out of the horizontal.

Cock Bead. A bead standing above the surface of the material. See *Mouldings.*

Cocking. 1. Cogging or caulking. 2. If hinges are fixed so that the nose of a door rises as it opens they are said to be *cocked.*

Cocking Piece. A sprocket piece, *q.v.*

Cockle. Wrinkle-like defects in glass, etc.

Cock Loft. A small space or room in a roof just under the slates and used for storage.

Cocks. Stone slates used at the ridge of a roof.

Cock's Comb. A thin steel plate with curved serrated edges for cleaning up mouldings in soft stones.

Cock-spur Fastener. A casement fastener having a small bell-crank lever rotating on a plate and engaging with a slot on the frame. See *Casement Fastener*.

COCK'S COMBS

Cock Stop. A tap or a tapped spout.

Cocuswood. A hard decorative timber from Burma and W. Indies. Yellowish brown colour with darker markings.

Coddings. The footings, or base, of ground floor chimney jambs.

Code of Practice. An agreed method of practice in building, structural engineering, etc.

Coefficient of Elasticity. See *Young's Modulus*.

Coefficient of Expansion. The expansion per unit length of a material when the temperature rises through one degree. Also see *Charles's Law*.

Coefficient of Friction. See *Angle of Friction*.

Coffer. 1. A deeply recessed panel in a soffit or ceiling. 2. A small chest.

Coffer Dam. See *Caisson*.

Cog. 1. A key to prevent lateral movement in a woodwork joint. See *Fish Plates*. 2. A tooth in geared wheels.

Coga. A patent door catch.

Cogging. A joint to prevent lateral movement between two timbers at right angles to each other, and where one supports the other, as a joist on a binder or wall-plate.

Coignet System. A system of reinforced concrete construction. Round mild steel bars are used to resist tension and, in the main beams, compression also. Stirrups resist shear and in the main beams act as distance pieces.

Coil Heating. Coils of pipes embedded in ceilings or walls, in the form of panels, for heating purposes.

Coin or Coillon. Same as quoin, *q.v.*

Colas. A proprietary bitumen for binding road surfaces.

Colcerrow. A light greenish grey granite from Cornwall. It is coarse grained and used for bridges and docks. Weight 168 lb per c. ft. Crushing strength 1340 tons per sq. ft.

Cold Chisel. A steel chisel. See page 102.

Cold Glue. See *Glues*.

Cold Press Resins. Thermo-setting, Urea-formaldehyde, resins, used for assembly work. They are fairly quick setting, within 15 minutes, and do not require heat for setting. See *Plastic Glues* and *Hot Press*.

Cold Short. Applied to wrought iron that is brittle when cold. See *Hot Short*.

Cold Shut. A defective joint in C.I. pipes cast in two sections. This method of casting is now out of date.

Cold Site. A term applied to water-logged ground because it is liable to keep the temperature of the surrounding atmosphere low.

Cold Store. A refrigerator. A building or room having a very low temperature, for the preservation of food. The temperature is lowered by various methods and agencies such as brine, ammonia, etc., and there are a number of patented methods of insulating the chamber.

Cold-water Supply. A term usually applied to the tap water supplied direct from the main, as distinct from the water stored for a domestic hot-water system, etc.

Colemanoid. A liquid chemical compound with a calcium base for water-proofing concrete, and hardening floor surfaces.

Collar. The annulet on a column.

Collar Beam, or *Collar Tie Roof.* Similar to *Close Couple,* but the tie beam is raised above the level of the wall plates.

Collaring. Pointing under the verges with cement mortar. A cement fillet under the overhanging slates.

Collarino. The cylindrical part of a column forming the lower part of the capital.

Collar Joint. The joint between the tie and rafter in a collar roof.

Collar Roof. See *Close Couple* and *Roofs.*

Collar Tie. See *Collar Beam.*

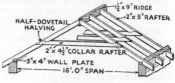

COLLAR TIE ROOF

Collimation Line. A line through the axis of the telescope of surveying instruments, such as the theodolite.

Collimation Method. A method of entering the field levels in surveying with a collimation level, or height of instrument, column.

Collodion Varnish. Used with aluminium paint, for interior work only.

Colloid. See *Gel.*

Collopakes. A colloidal compound for staining and waterproofing wall surfaces. It is a surface dye and not a paint.

Colly Weston Slates. Laminated stone split into thin slabs and used as roof coverings.

Colmorite. A printing paper for duplicating drawings.

Colonnade. A series of columns forming the side of a narrow covered path outside a building. The columns are bridged by lintels. See *Arcade.*

Colonette. An architectural arrangement of small columns and arches with orders.

Colorbitu. A proprietary bituminous water-paint.

Colorcrete. Buff or red coloured Portland cement, as sold by the Cement Marketing Co., Ltd.

Colorimeter. An instrument for measuring the intensity of colours.

Colorpavets. A proprietary coloured asphalt tile. It may be in various colours, and either on the surface or throughout.

Colorphalt. A proprietary coloured asphalt, green, grey, red, or brown.

Colorthrin. A proprietary pigmented solution for colouring concrete, etc.

Coloured Glass. Produced by introducing metallic oxides into the ingredients during manufacture, or by flashing. See *Flashed Glass.*

Coloured Lead Lights. Small panes of glass of varied colours and in different shapes secured in cames; also called stained glass.

Colour-filtering Glass. See *Daylight Glass.*

Colouring Concrete. Using pigments or coloured sand, with the cement, to give any required shade of colour. The colouring of concrete is becoming popular and a large number of pigments may be obtained, but they should be used under expert advice.

Colouring Drawings. The application of washes to architectural drawings to denote the various materials. Concrete, *Hooker's Green ;* Masonry, *Neutral tint or Sepia ;* Brickwork, *Scarlet or Crimson Lake ;* Carpentry, *Yellow Ochre ;* Joinery, *Burnt Sienna ;* Cast iron, *Payne's Grey ;* Wrought iron, *Prussian Blue ;* Steel, *Purple ;* Lead, *Indigo ;* Slates, *Neut. tint with Crimson Lake ;* Glass, *French Blue.*

Colour Wash. Distemper and water paint.

Colset. A proprietary bituminous waterproof compound in various colours.

Colt. A registered ventilator that automatically controls air change. A registered cowl for smoky chimneys.

Colterro. A registered copper wire and porous clay keys used instead of laths for plastered surfaces. It is similar to Bricanion lathing, *q.v.*

Columbarium. A building with tiers or niches for holding cinerary urns.

Columbian Pine. *Douglas Fir, Oregon Pine.* Similar to yellow deal, but with more open grain. Yellow with reddish markings. Strong, stiff, durable. Used for joinery and carpentry. Weight about 31 lb. per c. ft.

Column Guards. Lengths of $\frac{3}{10}$ in. M.S. "Rigifix" plates, bent at right angles, for protecting the corners of reinforced concrete piers against traffic. Narrow strips, or prongs, are stamped inwards to engage with the concrete. The guards are fixed in the angles of the shuttering before pouring the concrete.

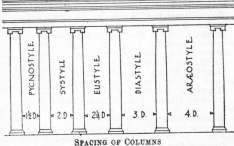

SPACING OF COLUMNS

Columns. 1. In reinforced concrete or steelwork the name includes any part of the construction that is resisting compressive stresses in the direction of its length, and the bending therefrom, whilst supporting or transmitting loads. 2. Any cylindrical post. 3. Usually applied to the columns in the different architectural orders, consisting of base, shaft, and capital. They are distinguished according to the particular style or order to which they belong. The following terms are used in classic architecture. Number of columns : Distyle, tetrastyle, hexastyle, octastyle, decastyle. Spacing of columns : Pycnostyle, systyle, eustyle, diastyle, araeostyle. Arrangement : In antis, prostyle, amphi-prostyle, peripteral, pseudo-peripteral, dipteral and pseudo-dipteral. See *Mouldings, Pillar,* and page 15.

Colura. See *Tester.*

Comb. The ridge of a roof. 2. See *Combing.*

Comb Hammer. A bricklayer's hammer with pene fitted with serrated edge.

Combination Beams. See *Flitched Beams.*

Combination Grate. A range built in an inside wall to serve for two rooms, kitchen and sitting room. A range easily convertible from kitchen range to sitting room grate.

Combined System. A drainage system serving for both rain-water and sewage. See *Separate System.*

Combing. 1. Devilling plaster surfaces. The surface is scratched with wire comb, scratcher, or a float with projecting nails, to provide a key for the succeeding coat. 2. Using a comb for graining painted surfaces.

Combinite. A proprietary bituminous roofing.

Comfort Index. A term used in heating. It is based on the cooling power of the air, the radiant energy of the sun, and the relative humidity.

Commode Step. A step with a curved, or bowed, riser at the foot of a flight of stairs. See *Curtail.*

Common Joists. The joists in a single floor, usually not more than 16 ft. span. Bridging joists.

Common Normal. A straight line acting as normal to more than one curve. See *Arches.*

Common Pitch. 1. The pitch of the fliers above and below winders.

2. The pitch of a roof when the rafters are about three-quarters of the span in length.

Common Rafters. The small rafters carrying the roof covering. See *Principal Rafter* and *Roofs.*

Communication Pipe. That portion of the service pipe, *q.v.*, that extends from the service main of the Water Board to the stop valve fitted on such pipe in the street; or to the point where such pipe passes the boundary of the street or enters any premises in or under the street, whichever is nearer to the service main.

Compartment Box. A junction for jointing pipes in different directions. It may be 3-way, 4-way, etc.

Compass. 1. A drawing instrument for describing circular arcs. 2. A surveying instrument for obtaining directions. 3. The limiting range of anything.

COMPASS PLANE

Compass Bricks. Specially made bricks tapered in width, for circular work.

Compasses. See *Compass* (1).

Compass Plane. A smoothing plane with adjustable sole for circular work.

Compass Rafter. A curved rafter.

Compass Roof. 1. A semicircular roof. 2. Rafters with inclined ties which also serve as struts. See *Scissors Truss* and *Roofs.* 3. A span roof.

Compass Saw. A saw for cutting circular arcs of small radii. See *Pad Saw,*

Compass Window. A bow window semicircular in plan.

Compass Work. Circular work.

Complementary Angles. Two angles that added together equal 90°, or a right angle.

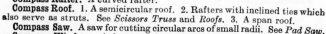

COMPASS SAW

Compluvium. A central opening in the roof of an ancient Roman building to which the rain-water flowed. It also provided light and air.

Compo. Cement-lime mortar.

Compo Boards. See *Building Boards.*

Compocrete. A proprietary composite floor-surfacing which may be applied to either old or new work.

Component. If a single force is resolved into two directions, then each of the lesser forces is a component of the original force. The components form two adjacent sides of a parallelogram of which the original force is the diagonal.

Composite. One of the "Orders" of architecture. The last of the five Classical Orders and composed of parts of the other orders, hence its name.

Composite Arch. A pointed, or lancet, arch.

Composite Beams. Beams constructed of two materials, as concrete and steel.

Composite Doors. Fire-resisting doors constructed of wood and metal.

Composite Truss. A roof truss with the members of wood and steel. Usually the tensile members are of M.S. or W.I., and the compression members of wood.

Composite Walling. Walls built of more than one kind of material as bricks and stone; or more than one style of arrangement, as coursed and uncoursed rubble. Brick walls faced with some other material.

Composition. 1. The design of a building to suit the requirements so that the whole is correctly and appropriately proportioned. 2. Plastic materials that set hard, used for jointless floors. 3. Plastic bituminous materials for roof coverings.

Composition Nails. Composed of 63 per cent copper. 33 per cent zinc and 4 per cent tin, and used for fixing slates and tiles. They are from 1¼ in. to 2 in. long. The 1½ in. nails weigh about 144 to the pound.

Compound. 1. A chemical combination of two or more elements. 2. An enclosure containing house and grounds.

Compound Arch. An arch consisting of a series of concentric arches placed successively behind and within each other. See *Concentric* and *Arch with Orders*.

Compound Girder or Beam. 1. One that is built up. 2. A flitched beam without the usual iron plate.

Compound Pier. A clustered column.

Compound Stanchion. See *Reinforced Concrete Cornice*.

Compressed Air. Used in caissons in under-water work to resist the

"VICTORIA" CONCRETE MIXER

pressure of the water on the walls of the caisson. Also used for drain-testing, and pneumatic appliances.

Compressed Cork. Manufactured from granulated cork by baking under high pressure. It is used for flooring.

Compression. A force tending to shorten a structural member by crushing the fibres lengthways.

Concamerate. To arch over, or vault.

Concave. A hollow curve; the reverse of convex.

Concealed Arch. A circular arc passing through the *crown* of a flat, or camber, arch. It is struck from the centre to which the voussoirs radiate, and is sometimes used for setting-out the thickness of the voussoirs.

Concealed Washer. A patent secret fastener for wall-boards. A star-shaped *expansion* washer spreads outwards into the material when hammered on to the *anvil* washer.

Concentrated Load. A load on a structure, or structural member, concentrated at one point.

Concentric. Circles of different radii with a common centre. *Concentric Arches* are parallel rings as in an *Arch with Orders*. See *Compound Arch*.

Conches. Shell-shaped roofs, or half domes.

Conchoid. The locus of a point on a line moving so that the line passes through a fixed point, and a fixed point on the line moves along an axis. The entasis of a column may be drawn by this method, which gives what is known as the conchoid of Nicomedes.

Concrete. A mixture of lime or cement, sand, and some form of aggregate, such as broken stone or brick, slag, pumice, clinkers, coke breeze, shingle, or ballast. The selection of the aggregate depends upon the requirements, such as strength, lightness, fire-resistance, etc. The whole is well mixed together, both dry and wet, so that the finer particles fill the voids and make a compact mass. The addition of water causes chemical action to take place in the cement, which sets hard and binds together the particles of sand and aggregate. The dry materials lose about one-third of the bulk when mixed. See *Standard Mix*.

Concrete Mixer. A hopper-like vessel in which concrete is mixed mechanically. A = charging hopper, B = rotating mixing drum, C = water tank, D = petrol-driven motor. See illustration page 69.

Condensation. The deposit of moisture due to humid atmosphere coming in contact with a cold impervious surface.

Condensation Groove. A groove to collect the condensation on the inside of a window. The moisture passes through a *weep-hole* to the outside. Also see *Glazing Bar*.

Conduction. The transmission of heat or sound from particle to particle of a body.

Conductor. A material that easily transmits heat, etc., by conduction. Metals are good conductors, but vary considerably. Copper is one of the best and is used for electrical conduction.

CONE

Conduit. 1. A channel or pipe to convey water. 2. Steel tubing used as a casing for electric cables, or wires.

Cone. 1. The fruit of pines and firs. 2. A solid generated by a line BB (generator) revolving about a fixed line AA (axis), so that the angle θ between the lines is constant, and the point of intersection O is fixed. Vol. = $\frac{1}{3}$ area of base × height = $\frac{1}{3}\pi$ $(AB)^2 × (OA)$. Surface area = $\pi r l$ = $\pi (AB)(OB)$.

Cone Hip Tiles. Roof tiles to cover the intersection at the hips. They are segmental in section and tapered in length, like a cone.

Conform. A patent steel shuttering.

Congé. A cavetto joining the base or capital to the shaft of a column.

Conglomerate. Applied to rocks formed of pieces of pebble-like older rocks cemented together.

Conical Roof. A roof shaped like a cone, as a steeple with a circular base. See *Roofs*.

Conic Groins. The groins produced by the intersection of conical vaults.

Conic Sections. Any plane section of a cone: circle, ellipse, parabola, hyperbola. These terms are also applied to the bounding curves, or perimeters, of the sections.

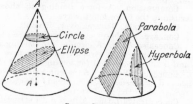

CONIC SECTIONS

Conifers. The cone-bearing, needle-leaved trees, or soft woods.

Coning. Constructing in conical form.

Conjugate Diameters. Any two lines bisecting each other with their extremities lying on the perimeter of an ellipse.

Connections. The fixing of the intersections of the various members in timber framing and steel frame construction. The various parts are fixed together by bolts, rivets, or welding.

Conoid, or Cuneoid. A surface or solid swept out by a horizontal line which moves so that it always touches a vertical line at one end and a semi-circle, or ellipse, at the other. Examples are common in double curvature work.

Conoidal Vaulting. Fan vaulting.

Conservancy Systems. Sanitary systems in which the foul faecal matter is on the premises for some time. See *Closet*.

Conservatory. A glass house for the display of plants. Usually an annex to a house.

Considère System. A system of reinforcing concrete introduced by M. Considère. The tension rods are inclined upwards at the supports to resist shear; and spiral reinforcement is used to resist compression.

Console. A large ornamental bracket, or corbel, usually having curves of contra-flexure. See *Balcony*.

Consolidating Piles. Short piles in soft ground to compress the ground and increase its bearing value.

Constant. That part of a formula that does not change so long as the materials or conditions are unchanged.

Construction Joints. The joints in concrete where new work is joined to concrete already set. Temperature stresses often cause cracks along these joints so that expansion joints are often used.

Consuta. A superior type of plywood. The veneers are sewn together in addition to cementing.

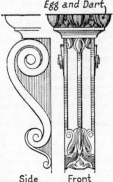

Egg and Dart

Side Elevation   Front Elevation

CONSOLE BRACKET

Contact Beds. Aërobic beds for the purification of sewage.

Continuity Stirrups. Steel rods, used in hollow tile floor construction. The under surface of the floor is strengthened by tension rods, and these are continued from the points of contraflexure over the top of the main beams as *continuity stirrups*.

Continuous Beam. A beam with intermediate supports. A long beam continuing over several spans.

Continuous Handrail. A handrail to a geometrical stairs. The returns are formed by wreaths instead of newels.

Continuous Impost. The term applied to the mouldings of an arch when they are continued to the ground without interruption.

Continuous Lights. A number of windows joined together and opened or closed by one piece of mechanism, which is geared to control all the lights.

Continuous String. The string to a geometrical stair. It is continued round the well without a break.

Contour Lines. A series of level lines on a map or plan, of a uniform difference of level. The difference of level is called the vertical interval. They are used to show the irregularity of the surface of the ground.

Contract. An agreement between builder and client that the builder will complete certain work satisfactorily for an agreed sum of money. The usual form of contract is the R.I.B.A. "Agreement and Schedule of Conditions of Building Contract."

Contraction. A loss of volume. Practically all building materials shrink either when losing moisture or when losing heat. See *Shrinkage.*

Contraflexure. *Contrary flexure.* 1. Points of contraflexure are the positions in a fixed or continuous beam or column where the tensile and compressive stresses reverse. 2. A curve of contraflexure is a continuous curve that reverses in direction, as a cyma recta.

Convection. The transmission of heat or electricity through gases or liquids by means of currents. As air and water rise in temperature they increase in volume, hence the density is less. The lighter air or liquid rises and causes circulation, as in hot water and ventilating systems.

Convection Currents. The movement of liquids and gases due to the expansion of the heated molecules. An *indirect method* of heating; as the flow of water transfers the heat from the source.

Convectors. Radiators.

Converting. Sawing timbers to smaller sections.

Conveyance. The sale, or transfer, of buildings or land.

Convolution. A fold, or revolution, of an expanding curve, such as a spiral, scroll, or volute.

Conwall Bracket. A patent bracket for supporting temporary timbering in centering, scaffolding, shuttering, etc.

Coolibah. A chocolate brown species of Eucalyptus from Australia. It is too hard and heavy for building purposes, and is one of the hardest and heaviest known woods. Weight about 66 lb. per c. ft.

Coom. 1. Centering for bridges. 2. See *Camp Ceiling.*

Combe Down. A Bath stone, *q.v.* from Somerset. It is light cream with fine grain and is very good for interior work. Weight 128 lb. per c. ft. Crushing strength 151 tons per sq. ft.

Co-ordinates. The quantities that determine the relative positions of points, lines or planes with regard to some source of reference. There are several systems used in geometry and science, such as polar, Cartesian, and Lagrange co-ordinates. See *Graphs.*

Copal. Fossil resin, used in the best varnish for external work.

Copalwood. The suggested standard name for Rhodesian mahogany.

Cope. 1. A covering. See *Monolith.* 2. The plane along which a block of stone is split by *coping.*

Coping. 1. Splitting a block of stone by drilling and driving in wedges. 2. Scribing the intersections of mouldings on framing instead of mitreing. 3. Brick, stone, terra-cotta, etc., used as a finishing to the top of a wall, as a protection and ornamentation See *Balcony.*

COPING

Copper. 1. A domestic washing boiler. 2. A malleable reddish-brown metal (Cu) used for exterior work because of its resistance to the atmosphere, and for ornamental purposes. See *Corrosion.*

Copperas. Sulphate of iron, or green vitriol. Used to set the colour when colouring brickwork.

Copper Bit. A piece of copper clamped in an iron rod with a wooden handle, and used by the plumber for soldering.

Copper Bit Joint. A lead pipe joint made by soldering with a copper bit. It is often used for joining brass unions to lead gas pipes.

Copperlite. A patent fire-resisting glazing. The glass is in small panes with I-section bars, welded together by the electro-copper process.

Cops. See *Merlons.*

Coptic Glass. A registered decorative blazoned glass.

Copyhold. A tenancy held for so long a period that the property has virtually become freehold. They are becoming extinct under the Law of Property Act, 1926, and being made freehold with new successions. Copy-holds are a relic of the feudal system, so that the right to tenancy is only shown on the original roll made by the lord of the manor's steward.

Coralin. See *Blue Prints*.

Coral-lac. A hard-wearing acid-proof wax for surfacing floors.

Corbel. A projecting support on the face of a wall. See *Balcony*.

Corbel Closet. A W.C. built into the wall on the cantilever principle, with no support from the floor.

Corbel Courses. The courses of bricks or stones forming a corbel, as in the support for an oriel window.

Corbel Pins. Strong W.I. brackets built into a course of the wall to carry a wall plate, instead of corbelling for the plate.

Corbel Table. A projecting course supported by carved dentils or brackets. Corbels supporting an oversailing moulding.

Corbie Step Gable. See *Crow Stepping*.

Corbulin. A proprietary waterproofer for concrete floors.

Cord. A timber measurement of 128 c. ft.

Cordelova Relief. A pressed wall-paper of the same type as *Cameoid*.

Cordon. A projecting course of stone in a wall, especially if the wall is curved.

Core. 1. The loose waste wood from a mortise. 2. That part of a cross-section, in masonry structures, in which if the line of pressure falls there is no tensile stress in the material. This gives rise to the "middle third rule." 3. Old plaster models used as a base for new models in fibrous plaster work. 4. The base, or foundation, for veneers and facings. 5. The segment of brickwork between a relieving arch and the lintel. See *Rough Arch*. 6. An iron bar along the underside of a handrail. 7. The sand, etc., for the interior of a hollow casting. 8. The interior portion of a column faced with other material.

Coresil. Proprietary cork insulating pads, etc.

Coring Out. 1. Clearing a flue of any projections and loose material after pargeting. A *sweep* is used for this purpose. 2. The running of a plaster moulding with a muffle on the mould; the final coat is called *gauging*.

Corinthian. The most ornate of the three main Architectural Orders. The chief distinguishing feature is the capital which is richly carved in two tiers of eight acanthus leaves.

Corite. A patent self-centering *hollow beam* for fire-resisting floors.

Cork. The bark of the cork oak. It is used for insulating purposes, and, in granulated form, in plastic mixes for tiles, floors and walls, for special purposes.

Corking. See *Caulking*.

Corkoustic. A proprietary acoustic material.

Corktex. A proprietary anti-condensation paint.

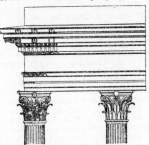

CORINTHIAN STYLE

Corkwood. The name applied to several very light tropical hardwoods. Usually they are known by some other name, except Uganda corkwood ; e.g. balsa, blolly, silk cotton tree, etc. Their use in building is chiefly for insulating purposes.

Corner Cramp. A cramp for mitred joints. A universal cramp.

Corner Post. The angle post in framed timber structures.

Cornice. The top part of an entablature. Horizontal projecting mouldings at the top of framing, or at the junction of wall and ceiling, or on façades of buildings. See *Mouldings.*

Cornice Brackets. Rough timber brackets or framing used as a foundation for large cornices.

Corning-Steuben. A patent moulded glass used for structural and decorative purposes.

Cornish Slates. Delabole slates. Thick and heavy and of rough texture. They are vari-coloured and mottled and often used for random, or irregular coursed, slating.

Cornstone. A limestone mottled with various colours.

Coromandel. See *Calamander.*

Corona. A wide overhanging flat member to an outside cornice, with throating or drip to throw the water from the face of the wall. See *Mouldings.* 2. A circle of lights, or a chandelier. 3. The apse to a church. 4. A proprietary sanitary waterpaint.

Top of Cornice, covered with Lead or Asphalt

Modillions.

Joint

Swan Neck Pipe

CORNICE

Corpse Gate. See *Lych Gate.*

Corpsing. A shallow mortise sunk in the face of the material.

Correnie. A bright-salmon coloured granite from Aberdeen. It is a Muscovite biotite and medium grained. Used for construction and decorative work. Weight 160 lb. per c. ft. Crushing strength 1300 tons per sq. ft.

Corridor. A passage leading to several rooms.

Corrosion. The rusting of metals due to the combined chemical action of oxygen, water, and carbon dioxide. Iron is the most easily affected, as the ferric oxide scales off and exposes fresh surface for the action to continue. In some metals (copper, zinc, lead) the oxidation forms a film that protects the metal, hence they are suitable for outside work without any other protective coat.

Corrugated. Wave-like in shape. A common form for galvanized iron and asbestos sheets for roofing.

Corsehill. A red sandstone from Dumfries. It is used for general building and dressings. Weight 150 lb. per c. ft. Crushing strength 640 tons per sq. ft.

Corsham. A light-cream Bath stone, *q.v.*, from Wiltshire. It has fine grain and good for inside work, only fair for exterior work. Weight 129 lb. per c. ft. Crushing strength 128 tons per sq. ft.

Cortile. The courtyard of a palace, round which the apartments are ranged. It is often surrounded by an arcade.

Corundum. A mineral of extreme hardness, consisting of nearly pure alumina. It is an important abrasive, *q.v.*

Costing. 1. Forming a schedule of labour costs for any particular piece of work. 2. Totalling the costs for completed work : materials, wages, plant, general charges, insurance, sub-contracts, etc.

Cot. 1. A protection for the hand when handling bricks. A piece of tyre inner-tube or boot-upper is used. 2. A small bed or crib. 3. A small cottage or hut.

Cot Bar. A curved bar in a semicircular sash.

Cote. A nesting place for birds.

Cottage. A small dwelling-house ; generally applied to one having four rooms only, living room, kitchen-scullery, and bedrooms.

Cottage Roof. A roof without principal rafters, consisting of wall plate, purlin, ridge, and common rafters.

Cottered Joint. A joint consisting of strap, gibs, and cotters. It is used for fixing and cramping up joints, as between a king or queen post and the tie beam. The cotters are wedged-shaped pieces of metal.

Cottonwood. American Poplar. Often marketed in this country as American whitewood. The name is applied to several soft and light hardwoods.

Coulisse. A grooved timber in which framing slides, such as the grooved posts for a sluice or for the side screens in a theatre. A gutter or channel.

Couloirs. External corridors to a theatre.

Coulomb's Theory. An early theory used for determining the stability of earth retaining walls, and from which most later theories have been evolved.

Coumarone Resins. From coal tar, naphtha and sulphuric acid. Used for varnishes.

Counter. A table in shops and business premises on which money is *counted* and goods displayed.

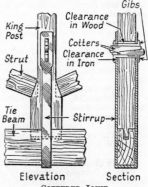

ELEVATION   SECTION
COTTERED JOINT

Counter Battens. 1. Battens fixed on the back of a number of built-up boards to prevent warping. They are fixed by slot screws, buttons, or they may be dovetailed into the boards. They are intended to allow free expansion and contraction. 2. Vertical battens on boarded and felted roofs.

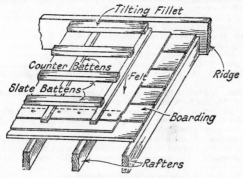

COUNTER BATTENS

Counter Brace. A brace used to counteract wind pressure, or change of stresses, as a diagonal across the rectangular space between the queen posts in a queen post roof truss.

Counter Ceiling. A second ceiling or one below the normal position to leave a space for pipes, ducts, tubes, etc., and for insulation.

Counter Cramps. Three wood battens fixed lengthways of the materials for cramping up and fixing end joints. Folding wedges, working in slots in the cramps, pull together the ends of the material.

Counter Drain. A drain running alongside a cutting or canal to intercept surface water.

Counter Floor. The bottom layer when two thicknesses of floor boards are used.

Counterfort. 1. An additional row of piles placed outside a coffer dam to increase the strength and to make more watertight. 2. A pier at the back of a retaining wall.

Counter Lathing. See *Brandering*.

Counter Screen. A low screen placed on a counter.

Countersink. To prepare a screw hole so that the head of the screw will be flush with, or below, the face of the material.

Counter Skylight. A ceiling light.

Counterweight. A prefabricated framed house consisting of tubular steel and tile-faced concrete.

Countess Slates. Slates 20 in. × 10 in. Requires 170 per sq. at 3 in. lap.

Couple. 1. Two equal and parallel forces acting in opposite directions. The magnitude of a couple is found by multiplying the amount of one force by the perpendicular distance between the forces. 2. A pair of rafters.

Couple Close. See *Close Couple* and *Roofs*.

Coupled Column. Columns arranged in pairs. The two columns forming a pair are usually half a diameter apart.

Couple Roof. A roof of small span consisting of pairs of rafters only. See *Roofs*.

Coursed Header Work. Masonry walling consisting of quoins, jamb stones, throughs, and binders of the same depth, with smaller stones built up to courses to the level of the larger stones.

Coursed Random. Rough stones used in walling and arranged to form courses about every 12 in. to 18 in. high.

Coursed Rubble. Roughly squared stones laid in regular courses, but with the courses of varying heights.

Courses. Horizontal layers of bricks or stones. A row of slates.

Coursing Joint. The joint between two string courses in an arch.

Court. A space enclosed, or partially enclosed, by the wings of a building, or by houses. It allows light and air to the surrounding buildings.

Courtrai Tiles. A type of interlocking tile from Northern France and Belgium.

Coussinet. 1. Part of the Ionic capital, between the abacus and echinus. 2. The top of a pier forming an arch springing stone.

Cove or **Coving.** 1. A hollow moulding in the form of a quarter circle or ellipse. An inverted scotia. 2. A metal or tiled lining round the opening over the hot plate of a kitchener.

Coved Arches. A term used when the intersections of vaulting form re-entrant angles, instead of salient angles, as in groined arches.

Coved Brackets. The brackets carrying the laths and plaster for a large coved moulding.

Coved Cornice. A cornice with a coved outline.

Cover Boards. Any horizontal boards acting as a cover.

Cover Flap. The panelled flap covering the boxings for window shutters.

Cover Flashing. See *Apron Flashing*.

Cover Plates. Steel or W.I. plates used at the joints in built-up, or plate, girders.

Cover Stone. A flat stone laid over the end of a girder, or over a pocket for any other purpose, to carry the walling above.

Covers. Slabs of Yorkshire laminated stone used for covering privies, coal-houses, etc.

Covey. A pantry.

Coving. The metal or tiled top to a range.

Cowdi Pine. See *Kauri Pine*.

Cowhouse. See *Shippon*.

Cowl. A revolving or louvred hood, fixed on a chimney pot to prevent down-draught.

Cownose. A brick with a semicircular end to avoid a sharp corner.

Crab. A portable windlass. A machine for hoisting weights. See *Winch*.

Crabwood. A British Guiana hardwood similar to mahogany and used for similar purposes.

Cracking. Applied to defective painted surfaces and due to the same causes as flaking, *q.v.*

Cradle. 1. Centering for culverts. 2. Heavy framework running on a slipway to receive a ship as the tide recedes. When the ship is settled in the cradle the whole is hauled above high water level. 3. A block, or frame, to steady small stuff when moulding timber on a spindle. 4. A movable adjustable scaffold for painting and repairing buildings. 5. A wood box with seive mounted on rockers and used for washing minerals.

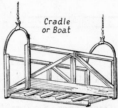

CRADLE SCAFFOLD

Cradle Roof. See *Wagon Roof*.

Cradle Vault. A term sometimes applied to a cylindrical vault.

Cradling. Rough brackets, or furrings, fixed to steelwork, etc., to carry plaster-work or linings. See *Cradle*.

Cradling Piece. A short joist between trimming joist and chimney breast to carry the ends of the floor boards.

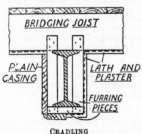

CRADLING

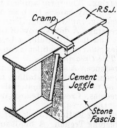

CRAMP FOR MASONRY

Craftplug. A patent fixing plug consisting of a hardened rubber hollow plug.

Craigleith. A light-grey sandstone from Edinburgh. It is a carboniferous stone and used for general building, steps, landings, etc. Weight 153 lb. per c. ft. Crushing strength 860 tons per sq. ft.

Cramp. An appliance for squeezing together, or cramping up, two or more pieces of material; or for permanently securing blocks of stone. See *Masonry Fixings* and page 78.

Cramped Flue. A flue reduced in size at any point, and often responsible for "smoky chimneys."

Crampet. A wall hook.

Crampon. 1. A rivet at the tails of asbestos slates to prevent lifting by the wind. 2. A grappling iron for lifting wood or stone. See *Stone Tongs.*

Crane. 1. A machine used for lifting weights. It usually consists of post or mast, jib, stays, winding gear, and tackle. There are many variations:

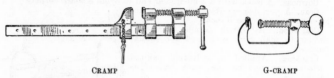

CRAMP   G-CRAMP

jib, wall, derrick, travelling, wharf, etc. They may be hand, steam, hydraulic, or electric driven. 2. A water cock. 3. A registered domestic boiler and heating equipment.

Crane Tower. See *Derrick Tower.*

Cranham. Patent porous terra-cotta partition blocks with good insulating properties.

Cranked Bolt. One in which the bolt is cranked, to shoot into a hole a little distance in front of the back plate; especially suitable for pivoted sashes.

Crapaudine Door. A door hung with centre pivots so that it revolves about its vertical axis.

Cratch. A fodder rack.

Crawl. Same as duckboard, *q.v.*

Crawling Paint. An uneven painted surface caused by applying too thick a coat, and the paint *running.*

Crazing. The term applied to fine cracks in cement surfaces. They are due to many causes such as a too wet mix, or too rich in cement, over trowelling, rapid curing, disturbance before setting, etc.

Creasing. A double row of tiles projecting about $1\frac{1}{2}$ in. over the face of a wall, and usually under a brick-on-edge coping, to throw off the water. A triangular cement fillet is run on the top to prevent the water from lodging. See *Tile Creasing.*

Credence. A small table at the side of an altar.

Creep. 1. Intercrystalline failure in metals due to high temperatures. 2. The movement of materials due to expansion and contraction.

Creepies. Creeping boards used by the slater as he proceeds up the roof fixing the slates.

Creeping Rafters. Jack rafters.

Creeping Trench. A small subway, containing a drain, and large enough to allow a man to crawl along to inspect the drain.

Creetex. A proprietary bituminous binding compound in plastic form for waterproofing surfaces.

Creetown. A light-grey Scottish granite from Kircudbright. It is a biotite stone with medium grain and used for heavy structural work. Weight 170 lb. per c. ft. Crushing strength 1630 tons per sq. ft.

Cremorne Bolt. Extension bolts for french windows; espagnolette bolts.

Crenellated. Mouldings that are notched, or indented, on the top edge. See *Battlemented.*

Crenelle. An embrasure, or loop-hole.

Creosote. A product of coal and wood-tar used as a preservative for timber, and as an antiseptic. It is obtained by distillation. See *Preservation*.

Crescent. Anything shaped like the moon in its first or last quarter.

Crescent Roof. See *Roofs*.

Cresset. An open frame for a beacon light.

Crest. The prominent part of a screw thread.

Crested Ridge. A ridge tile (*q.v.*) with a projection above the apex.

Crestings. Metal ornamental finishings to a terminal, as to a finial or parapet railings. Ornamental vertical projections to ridge tiles.

Crests. Ornamental work on the tops of buildings.

Crete-o-lux. A patent glass and reinforced concrete pavement light. The lights are "W" lenses of different types and sizes.

Crevasse. A registered decorative blazoned glass.

Cribs. 1. Circular ribs used as walings for deep shafts. 2. A rack in a stable. 3. A timber structure under a caisson.

Crib Work. Layers of logs, in the form of a skeleton frame, sunk in water as a dam.

Cricket. A projection formed in the roof to tie a tall chimney stack. It is somewhat like a gablet.

Crimped. Corrugated.

Cringle. The end of a rope formed into an eye, with a thimble.

Crinoidal. Limestones containing fragments of crinoids, or sea-lilies.

Cripple. A form of bracket, anchored by steel wire from the ridge, and supporting a slater's scaffold. Also for hanging on the rungs of a ladder. See *Jack Rib*.

Crippling Load. The load that will cause failure in a column or long strut.

Cripplings. Spars used as raking shores. See *Jack Rib*.

Cristol Glass. Registered moulded glass units assembled by the electro-copper process and used for lay-lights, domes, lamps, etc.

Critical Amplitude. The limit of horizontal displacement, due to vibrations, that a building will safely stand.

Crittal. See *Metal Windows*.

Crockets. Ornamental projections, placed at regular intervals, along salient corners of gables, canopies, windows, etc., and characteristic of Gothic architecture. See *Tracery Window*.

Crocodiling. Closely patterned cracks on plaster surfaces due to slow setting, so that evaporation takes place before hydration and setting. Similar markings on varnished surfaces in which the varnish has cracked due to exposure to the sun, and exposed the undercoat.

Cromlech. A prehistoric structure consisting of two vertical stones supporting a horizontal one.

Crook. See *Knee*.

Crop Beam. A horizontal member of the framing of a lock gate to a canal.

Crope. An obsolete word for *finial*.

Cropper. An appliance for cutting steel.

Crosettes. A right-angled projection at the top of a vertical architrave. Also called ears, elbows, or ancons. 2. The horizontal seating or shoulder on an archstone to rest on an adjacent stone. See *Voussoirs*.

Cross Bond. See *English Cross Bond*.

Cross Cut. Any type of saw with teeth specially shaped for cutting across the grain. The name is specially applied to the carpenter's hand saw and to machine saws for cross cutting, and sometimes to the double-handed saw.

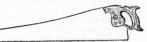

CROSS CUT OR HAND SAW

Cross Fall. The inclination, on a road, from the crown to the gutter where the water is collected.

Cross Garnet. See *Tee Hinge.*

Cross Grained. The grain of timber running in opposite directions; or not lengthways of the material. A term used to denote an inclination of the annual rings on a rift sawn piece of wood. Fibres shown endwise on the face of the wood. See *Grain* and *Slope of Grain.*

Cross Grained Float. A plasterer's float for scouring surfaces and rubbing up angles. The face is about 12 in. × 4½ in. × ¾ in.

Crossing. The space in a Gothic church formed by the intersection of nave and transept.

Cross Reeded. A rolled glass with wide ribs running in opposite directions on the two faces.

Cross Springer. The diagonal rib of a groined vault.

Cross Tongue. Used for jointing two pieces of timber together. They are cut diagonally from a thin wide board for strength. Plywood tongues are as effective.

Cross Vaulting. Vaults intersecting each other. Groin vaulting. See *Vaulting.*

Crotch. Curly grain or feather in mahogany due to the struggle for direction of the fibres between a branch and the trunk.

Crow Bar. A pinch bar. A steel rod bent and tapered at one end, and used as a lever for moving heavy weights.

Crowde. The crypt of a church.

Crown. 1. The highest point of an arch. 2. The top of any ornamental feature. 3. A spire formed by converging flying buttresses. See *Arches.*

Crown Bar. The top timber supporting the boards in the centering for a tunnel.

Crown Glass. Glass blown and spun in discs up to 50 in. diam. See *Glass.*

Crown Piece. A short piece of timber on a wall, to receive the foot of a strut.

Crown Plate. Horizontal piece of wood on the top of a post to give a greater bearing surface.

Crown Post. A king post ; but more usually applied to the post resting on the collar beam in hammer beam or bracketed roof trusses.

Crown Stone. See *Apex Stone.*

Crow's Nest. A small platform in a high position for observation purposes.

Crow Stepping. Forming a gable coping by a series of horizontal steps.

Crow Stone. The top stone, or apex, in crow-stepping.

Cruciform. Applied to churches having the plan in the form of a cross.

Cruck Construction. A primitive form of construction formed of bent branches with ridge tree.

Crusader. A proprietary petrifying liquid for waterproofing walls, etc.

Crush Room. A small hall to relieve the pressure of the crowd at the exit of a public building.

Crust. A term applied to the thickness of the material of a drain pipe.

Crutch. A forked contrivance used as a support.

Crutch Key. The T-shaped top to the plug of a bib cock, for turning the tap.

Crypt. An underground cell or chapel, generally used for burials.

Cryptic. A registered type of blazoned glass, *q.v.*

Crystal. A patent spraying equipment for paint, enamel, etc.

Crystalux. A type of glass lens used with reinforced concrete for pavement lights.

Crystopal. Thin sheets of vitreous, opaque, highly glazed material. It is fixed to rendered cement or wood surfaces by mastic.

Ct. Abbreviation for *cement.*

Cube Method. An approximate method of finding the cost of buildings, by multiplying the net area of the plan by the height from the bottom of the foundations to half way up the roof. The cubical contents so found are valued according to the type of building.

Cubicle. A small room, or compartment, for sleeping and dressing.

Cubiculum. A bedchamber in old Roman buildings.

Cubing. Working out the volumes of certain items when preparing a bill of quantities.

Cul-de-four. A spherical vault on a circular plan.

Cul-de-sac. A street closed at one end.

Cullamix. A proprietary coloured cement and aggregate.

Cullet. Scrap glass melted up and used in the manufacture of glass.

Cullis. See *Coulisse*.

Cullonen. See *Binburra*.

Cullum Floor. A fire-resisting floor formed of fire-clay blocks, or tiles, with tensile reinforcement. The blocks are supported by shuttering until the *in situ* concrete has set.

Culverhouse. A cote for birds.

Culvert. An arched channel for conveying water under a road or railway

Culvertail. A dove-tail.

Cumberland Slates. See *Westmorland Slates*.

Cuneate. Wedge-shaped.

Cuneiform. Wedge shaped.

Cuneoid. See *Conoid*.

Cup. A conical shaped receiver for the head of a screw. It is sunk and glued in the timber. See *Screws*.

Cupboard. An enclosed place or small room for storage purposes.

Cupboard Turn. See *Turnbutton*.

Cup Leather. A U-shaped ring of leather used as a packing for hydraulic machinery and pumps, to prevent leakage of water.

Cupola. 1. A spherical vault. Usually applied to the interior of a dome that is a separate structure from the outside framing. 2. A furnace for melting metal.

Cuprinol. A proprietary preservative for wood. It is an organo-metallic salts liquid, easy to apply and has no offensive smell.

Cupro-nickel. An alloy of copper and nickel with high corrosion resistance. See *Elio* and *Monel Metal*.

Cup Shake. A shake in timber caused by the lack of cohesion between successive annual rings, and due to lack of nutriment during a season of growth, or by exceptional winds or frost. See *Shakes*.

Curb, or **Curb Plate.** 1. A circular wall plate. 2. A frame round the mouth of an opening, or well. 3. The stones forming the edge of a footway. See *Kerb*. 4. The plate at the intersection of the two sloping surfaces of a mansard roof. Also see *Skylight*.

Curb Fenders. Curbs round the hearths of fireplaces, made of glazed tiles or bricks.

Curb Rafters. The upper rafters of a mansard roof.

Curb Roll. A roll used at the intersection of the two surfaces in a mansard roof.

Curb Roof. A roof resting on a curb. A mansard roof. See *Roofs*.

Curdling. Applied to paint that becomes jelly-like, or curls up into small pellets when applied. It is due to wrong or faulty driers.

Curf. See *Kerf*.

Curfew. A patent sliding-door gear.

Curing. Preventing the too-rapid drying of cement or concrete usually by covering with canvas and periodically spraying with water.

Curl. Same as crotch, *q.v.*

Curlew. A registered name for fireproof sliding doors and shutters.

Current. In electricity the unit of current is the *ampere* and the unit of

power the *watt*. The current is obtained from the formula: watts = volts ×
amperes.

Curstable. A moulded stone string course.

Curtail. A step with a spiral or semicircular end, at the foot of a flight of
stairs. It usually follows the outline of the
handrail scroll.

Curtain. 1. A fireproof screen between
auditorium and stage in a theatre, or between the sections in a mine. 2. Often
applied to a wide board that serves as a
screen, as at the front of a joiner's bench.
3. Suspended fabrics screening a window,
for furnishings.

Curtaining. The running, or sagging, or
tailing, of varnish, due to faulty brushwork.

Curtain Wall. An external wall not
carrying any loads and supported by a
bressummer.

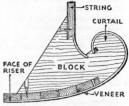

CURTAIL AND COMMODE STEP

Curtilage. The land adjacent and belonging to a domestic dwelling.

Curupay. A deep red hardwood with black stripes from C. America.
Very strong, durable, hard, heavy, and tough. Difficult to work, feels oily,
but polishes well. Used for structural work and superior joinery. Weight
about 60 lb. per c. ft.

Curved Batter. A battered wall with a vertical curve on the face.

Curved Rib Truss. A truss in which one of the main members is curved,
as in hammer beam, de Lorne trusses, etc.

Cushion. 1. See *Helmet.* 2. A stone on a pier serving as a springer for
an arch. 3. Anything forming a seating.

Cushion Caps. Capitals characteristic of the Anglo-Norman style consisting of plain square blocks with lower corners rounded. See *Bowl
Capital.*

Cushion Rafter. A second rafter under the principal rafter in wood trusses
of large span.

Cushioned Frieze. See *Pulvinated.*

Cusp. A point where a curve stops and returns on itself. The two
branches have a common normal. The meeting point of two foils in
tracery work. See *Tracery.*

Cuspidated. Pointed or terminated by a cusp.

Cussome. The under-eaves slate in stone slating, *q.v.*

Custodis. A system of building tall chimneys. Specially made radiating
bricks of large size are used. The blocks are highly refractory, very strong,
and perforated for uniform burning and insulation.

Cut. Implies *bevel* in carpentry. Also see *Deep Cut* and *Flat Cut.*

Cut and Mitred Beads. The beads forming rebates for a pivoted sash.

Cut and Mitred String. See *Bracketed Stairs.*

Cut Bracket. One moulded on the edge.

Cutch. A light to dark red hardwood from India. It is lustrous and
very heavy, hard, strong, and durable. Difficult to work. Used for structural work, furniture, etc. Weight about 60 lb. per c. ft.

Cut Graining. A method of graining with a special tool with notched
circular cutter, known as the Berries graining process. The wood is
cut as the required imitation and the pores filled with dark stain.

Cut Lines. The working-drawings used when preparing lead lights.

Cut Lock. A lock for which some part of the door has to be cut away to
receive it.

Cut Nail. A machine-made nail stamped from sheet metal, as a floor
brad.

Cut-out Valve. A valve to any apparatus that allows certain parts to be put out of action as required.

Cut Roof. A truncated roof. One terminated with a flat.

Cut Rubble. Squared rubble used for facing walls with random rubble backing.

Cut String. An open string to a flight of stairs. One shaped to the outline of the steps. See *Tread*.

Cut Stuff. Deals re-sawn into smaller sections.

Cutters. 1. Cutting blades, or knives, used in machines. 2. See *Rubbers*.

Cutting Box. A box with adjustable sides for cutting voussoirs to the required shape for brick arches.

Cutting Down. Cleaning of a polished or painted surface with an abrasive preparatory to re-polishing or painting.

Cutting Gauge. Like a marking gauge but with a thin blade instead of a pin. It is used for cutting thin material or for forming small rebates.

Cutting-in Lines. Forming the junction of different colours in painting.

Cutting List. A prepared list of the sizes of the materials required for joinery work. Three copies are prepared, for the setter out, machinist, and office respectively.

Cutting Plane. An inclined or oblique plane cutting a solid. A geometrical construction for finding the true shape of a section.

Cuttings. Trenches for drains, etc. A road between embankments.

Cutwater. Triangular framing protecting the end of a pier or jetty.

Cycloid. The path traced by a point on the circumference of a circle, as the circle rolls along a straight line.

Cyclone. A mechanical collector of the shavings and dust from woodworking machines.

Cyclopean. A term applied to the circular walls around towns, characteristic of the earliest examples of Grecian architecture.

Cylinder. 1. A prism of uniform circular section. Surface area $= 2\pi r$ and volume $= \pi r^2 l$. 2. A storage tank of cylindrical section used in domestic hot-water systems. See *Low-pressure System*. 3. A temporary centre, or drum, round which veneers are bent for blocking, for circular work. 4. A large vessel used for the preservation of wood under pressure.

Cylinder Glass. Sheet glass.

Cylinder System. The usual method of heating water in domestic buildings as distinguished from the tank system, *q.v.* The cylinder may be a copper cylinder or it may be rectangular and of galvanized iron. See *Low-pressure System*.

Cylindrical Vault. A plain barrel, or semicircular, vault.

Cylindro-Cylindric. A term applied to intersecting cylindrical vaults of unequal heights and spans.

Cylindroid. A "cylinder" with an elliptical section at right angles to the axis instead of a circular section.

Cylindro-spheric Groin. The intersection of a sphere with a cylinder of greater span and height.

Cyma Recta. A curve of contrary flexure. See *Mouldings*.

Cyma Reversa. Reverse of the cyma recta, having the upper part of the moulding convex and the lower part concave. See *Mouldings*.

Cymatium. The upper member of a capping or cornice. The cyma recta.

Cyp. *Princewood*. A yellowish brown decorative hardwood, with veins of darker colour, from Jamaica.

Cypress. A coniferous tree. The timber is yellow with reddish markings. Fine and uniform grain. Strong and durable. Many species of *Cupressus*, *Taxodium*, *Callitris*, and *Thuya*. Weight about 37 lb. per c. ft.

D

d. Abbreviation for *Dressed*.

Dabbing. Punched or pointed work in masonry. The dressed surface is covered with small holes by means of a specially faced hammer to give an appearance of strength.

Dabé. A light-brown hardwood from W. Africa. It has darker streaks and flecks and is fairly hard, heavy, and strong Suitable for superior joinery. Weight about 42 lb. per c. ft.

Dac or **Dawk.** A type of bungalow, or traveller's rest house in India.

Dado. 1. The lower part of a wall, from the skirting to the dado rail. 2. The plain part of a pedestal to a column. 3. An embossed wall-paper for dados.

Dado Base. The lowest members of a pedestal.

Dado Moulding. See *Chair Rail* and *Surbase*.

Dado Framing. Panelled framing fixed to the lower part of a wall. It is usually to about shoulder height.

Dagobas. See *Topes*.

Dahoma. Ekhimi. *q.v*

Dairy. A place where milk is converted into butter and cheese, or a shop where milk, butter, etc., are sold.

Dais. 1. A platform. Part of a floor raised above the level of the surrounding floor. 2. A seat with high back and canopy, for one in authority.

Dalbeithe. See *Granite*.

Dalkey. See *Granite*.

Dam. A retaining wall for water.

Dammer. An oleo-resin used in the manufacture of varnish.

Damper. An adjustable iron plate to check the draught to a flue.

Dampexe. A proprietary varnish for damp walls, or undercoating to prevent suction.

Damping. Wetting of cement surfaces to prevent too rapid drying, and so avoid crazing. Applying warm water to wood with woolly texture to raise the grain before finishing with sandpaper. Wetting the hollow side of a board to straighten it. Stopping vibration to lessen noise, in insulation problems.

Damp-proof Course. D.P.C. A layer of impervious material at least 6 in. above ground level and below any inside timbers, to prevent moisture from rising up the wall. See *Cavity Wall*.

Damp-proofing. Making waterproof by special materials or processes. The term is generally applied to the use

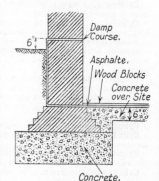

Damp Course.

Asphalte.

Wood Blocks

Concrete over Site

Concrete.

D.P. COURSE

of proprietary materials with cement, such as Pudlo, Colemanoid, Sika, Novoid, Super cement, Medusa, Ironite, Ferrolithic, Naylorite, Bareau, Ferrous hardener, Fluoric hardener, Cementone, Tricosal, Majauca, etc. The various processes are : surface coatings, mastic, integral, and membrane processes.

Damp-proof Sheeting. The term is usually applied to vertical layers of asphalte or hygeian rock used for waterproofing the base of walls. Sometimes a retention wall is used to enclose the material in a small cavity. Also the various proprietary reinforced bituminous sheetings used for d.p. courses, etc. : Ledkore, Callendrite, Basite, Dason, Ledbit, B.N.R. (Hessbit,

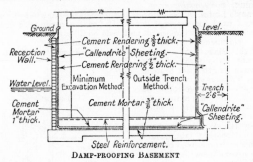

Ground

Level.

Reception Wall.

—Cement Rendering ⅜"thick.—

—"Callendrite" Sheeting.—

—Cement Rendering ½"thick.

Water Level.

Minimum Excavation Method.

Outside Trench Method.

Trench ←2'6"→

Cement Mortar 1"thick.

Cement Mortar ⅜"thick.

"Callendrite" Sheeting.

Steel Reinforcement.

DAMP-PROOFING BASEMENT

Ravenite, Blackwell), Anderite, Astos, Ruberoid, Permanite, Vulcanite Farotex, Hermatex, Masticon, Plastaleke, etc.

Dance Floor. A resilient floor specially surfaced for dancing upon. There are numerous registered types.

Dancette. The zig-zag or chevron moulding characteristic of Norman architecture.

Dancing Cairn. A bluish-grey granite with black specks from Aberdeen. Used for constructional work. Weight 171 lb. per c. ft. Crushing strength 850 tons per sq. ft.

Dancing Steps. Winders not radiating from one common centre. The treads of the steps are increased in width at the narrow end to give increased foothold. Also called balanced winders.

Dangerous Structure. A building certified by a local authority to be in a dangerous condition. The owner must repair or pull down the structure, according to instructions.

Danta. A Gold Coast hardwood somewhat like mahogany and suitable for superior joinery, etc.

Dantzig Oak. A term sometimes applied to selected oak cut radially. Wainscot oak. Oak shipped from the Baltic.

Dao. A light brown hardwood with darker markings from the Philippines. It is fairly hard, heavy, strong, and very durable. Resembles Queensland walnut in texture and grain. Used for superior joinery, flooring, etc. Weight about 42 lb. per c. ft.

Dapping. A term used for *notching* in timber bridge construction.

Dappled. Variegated ; patchy in colour.

Darby, or **Derby.** A long floating rule used in some districts by plasterers.

Darley Dale. A grey pinkish brown sandstone from Derbyshire. A micaceous stone used for general building and engineering. Weight 148 lb. per c. ft. Crushing strength 670 tons per sq. ft.

|←——— 6" ———→|

DAWNAY LINTEL

Dason. Bitumen on a tough hessian base, used for damp-proof courses.

Data. The preliminary information to be obtained before proceeding with any piece of work : dimensions, conditions, etc.

Datum Line. The line representing a level plane which serves as a base for obtaining levels. A line on a building site from which heights are measured. Ordnance datum, in Great Britain, is mean sea level at Newlyn, Cornwall.

Daubing. See *Dabbing*.

Dawnay's Lintol. A solid fireclay lintol, shaped somewhat like a cast iron beam and used for fireproof floors. It forms a permanent centre upon which the concrete is deposited, and rests on filler joists.

Day. A division of a window between two mullions.

Daylight Glass. A patent glass that "corrects" artificial light by modifying the rays with high wave length (red, orange, yellow). These rays overbalance short waves (blue, violet) of the spectrum. The glass reduces the illumination but gives a more natural light.

Day-work. Work valued on the actual cost of labour and materials with an agreed percentage to cover supervision, plant, and profit.

D.C.F. Abbreviation for *deal-cased frame*.

Ddt. Abbreviation for *deduct*.

D.D.T. An insecticide destructive to most insects and pests.

Dead Air. Air trapped in cells, which is the basis of most insulating materials.

Dead Bolt. The bolt in a lock actuated by a key, as distinct from the bevelled bolt, or latch, controlled by the knobs.

Deadening. 1. The act of making floors, partitions, etc., sound proof. 2. Patches on varnished surfaces due to unequal porosity.

Dead Knot. One not joined firmly to the surrounding wood. See *Knots*.

Dead Load. The forces acting on a structure due to gravitation. The weights of the different materials. A steady load gradually applied. See *Uniform* and *Distributed*.

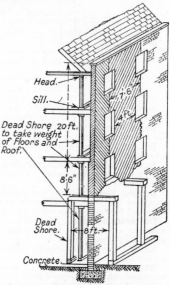

DEAD SHORES

Dead Lock. A lock operated by a key only. One without a latch.

Deadmen. 1. Temporary brick pillars, built with bricks on edge, to carry the pins for the line when building arches to keep them vertical. 2. Stone, concrete, timber, or plates, forming the anchor for a land tie. See *Camp Sheeting*. 3. A strutted plank, reared on end, to serve as a profile in bricklaying.

Dead Shore. Vertical shores carrying dead loads, as when supporting needles. See *Shores*.

Dead Sounding. See *Pugging*.

Deafening. Sound-proofing. See *Pugging* and *Deadening*.

Deal. A term applied to soft timbers, pines and firs, from 9 in. to 11 in. wide and from 2 in. to 4 in. thick. The sizes vary at different ports. Also applied to the more common firs and Scotch pine, as white deal and yellow deal.

Death Watch Beetle. *Xestobium sp*. Beetles very destructive to old constructional timbers in buildings. They make a ticking noise, hence the name. Several proprietary preservatives are claimed to be cures : Cuprinol, Hope's Destroyer, Solignum, etc.; but the beetles are difficult to eradicate, when firmly established, as the burrows of the larvae are so deep. Fumigation by gas, carbon disulphide, etc., is usually effective. Infested timber loses its structural value. The insect is about $\frac{1}{4}$ in. long and mottled dark brown in colour.

Decagon. A rectilinear plane figure with ten sides. Area = sq. of side × 7·694.

Decastyle. A portico with ten columns, in classic temples.

Decatone. A proprietary water-paint or distemper.

Decay. Disintegration of wood tissue due to the action of bacteria and fungi. See *Dry Rot, Wet Rot, Diseases,* and *Preservation.*

Decayed Knot. See *Knots.*

Decibel. A unit of sound measurement.

Deciduous. Trees that shed their leaves in the autumn. The timber is classified as hardwood.

Deck. A platform extending from side to side in a ship and serving as a floor and covering.

Deck Cants. Timbers fixed on decks to receive wood framing.

Decking. 1. The working platform on the top of a derrick tower that carries the derrick crane. 2. Shuttering for horizontal surfaces.

Decorated Style. Also called Curvilinear, Middle Pointed, or Geometrical. It is the second of the pointed, or Gothic, styles of architecture ; rich in ornamentation, with tracery of geometrical and flowing, or flamboyant, form. Crockets, grotesques and diaper wall decorations were extensively used. Fourteenth century. Examples : Westminster Abbey, York, Ely, Lincoln, Salisbury, and other cathedrals.

Decousto. Proprietary insulating materials : plaster, tiles, and synthetic stone.

Deeping or **Deep Cut.** Resawing timber parallel to its wider faces, as converting deals to thin boards.

Deep Seal. See *One-pipe System.*

Deeside. See *Granite.*

Defects. The principal defects in timber are heart, star, and cup shakes, loose or dead knots, upsetts, waney edge, rind galls, burrs, twisted grain, and wandering heart. See *Timber Tree.*

Deflection. Distortion of a structural member due to lack of stiffness. Bending downwards. See page 366.

Deformed Bar. Any form of reinforcing rod for concrete, not having a plain surface.

Deformeter. A method of analysing the effect of external forces on a proposed structure by means of celluloid scale models.

Defurring. Removing lime incrustation from hot-water pipes and boilers.

Degame Wood. A variegated yellowish-brown hardwood, from the West Indies, used for decorative purposes. A good substitute for lancewood. Weight about 50 lb. per c. ft.

Dehumidifier. Part of the installation to remove moisture in an air-washing or air-conditioning plant.

Delabole Slates. Excellent multicoloured slates from Cornwall. They may be obtained as randoms, rustics, or as sized slates.

De Lank. A light greenish-grey granite from Cornwall. It is used for bridges and dock work. Weight 165 lb. per c. ft. Crushing strength 1170 tons per sq. ft.

Dellanite. A proprietary rubber flooring.

De Lorme. A system of roofing. The trusses are built up of laminated semicircles keyed together.

Delph Stone. Yorkshire stone from the coal measures.

Delta Metal. An alloy of copper, zinc, and ferro-manganese.

Demi-relievo. Sculpture in relief in which half of the figure projects from the surface of the material.

Demolition. The taking down of old buildings. This is usually performed by specialists called *housebreakers*. See *Unbuttoning*.

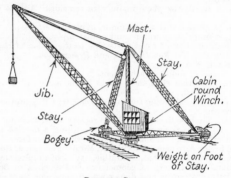

DERRICK CRANE

Density. The weight of a material per unit volume. See *Specific Gravity*.

Denticulated. A moulding with dentils.

Dentils. Rectangular projections on the face of a moulding. They are usually formed by removing alternate parts of a fillet of the moulding. See *Mouldings* and *Mantel*.

Denya. A Gold Coast hardwood suitable for heavy constructional work, flooring to resist hard wear, etc. It is greenish-brown and gold in colour, hard, tough, and with interlocked grain.

Depeter. Rough plastered or cement surfaces produced by pressing small stones, spar, flint, crockery, or marble, into the material before it has set. It is similar to rough cast. Also plasterwork to imitate stone.

Depot. A building for storing goods; a warehouse or depository. A railway station. A military headquarters.

Depretor. A plaster surface representing tooled stone.

Depth Gauge. An attachment for measuring or controlling the depth of a sinking.

Derby. See *Darby*.

Derbyshire Spar. See *Fluor Spar*.

Dermas. A proprietary asphalt emulsion for floor coverings.

Dern. A door-post or threshold.

Derrick Crane. A crane in which the jib has both horizontal and vertical movement.

Derrick Towers. The three towers supporting the staging, or decking, for a derrick crane. The towers are strong, braced, timber structures. The king, or crane, tower supports the crane, whilst the two anchor towers tie down the sleepers of the crane by means of chains. See page 89.

Described Circle. A circle that encloses, and touches without cutting, one or more plane figures.

Descriptive Geometry. Geometry dealing with points in space. Three dimensional, or solid, geometry.

Desiccation. The drying of timber in artificially and scientifically heated chambers. See page 90, and *Seasoning*.

Design. To plan the form and characteristics of a building. To calculate the correct dimensions of structural members.

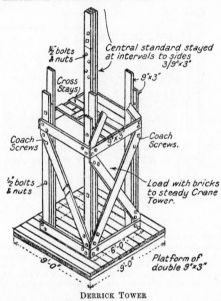

½" bolts & nuts

Central standard stayed at intervals to sides 3/9"x3"

Cross Stays)

9"x3"

Coach Screws

9"x3"

Coach Screws.

½" bolts & nuts

Load with bricks to steady Crane Tower.

6'-0"

9'-0"

9'-0"

Platform of double 9"x3"

DERRICK TOWER

Desolva. A proprietary material for removing the fur from hot-water pipes.

Destructors. Furnaces for the combustion of household refuse collected by local authorities.

Detailer. A designer of details for steelwork construction. The engineer who calculates the sizes of the members and connections.

Detail Paper. A thin semi-transparent paper generally used for detail drawings, and as a substitute for tracing paper.

Detent. A catch for locking mechanical movements.

Detrusion. The shearing of the fibres along the grain of timber. Thrusting out.

Develling. Scratching a floated surface to provide a key for the plaster setting coat. See *Devil Float*, page 90.

Development. The laying-out of a surface, to show the true shape ; or the rebatment of a line into the planes of projection to show the true length and inclinations.

Devil. A portable fire grate used by plumbers for heating irons, etc.

Devonian. Geological strata between Silurian and Carboniferous.

De-watering. Removing the water from a caisson, in under-water work.

Dewdrop. A registered obscured rolled glass for artificial lighting.

Dexion. A meccano-like system of unit construction with numerous applications in building.

Dextrine. A gummy substance obtained from starch. A powder mixed with water and used as an adhesive, especially for embossed wallpapers.

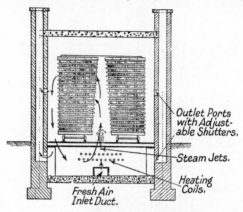

Outlet Ports with Adjustable Shutters.

Steam Jets.

Heating Coils.

Fresh Air Inlet Duct.

DESICCATION, OR SEASONING, CHAMBER

D.H. Abbreviation for *double hung*.

Diaglyphic. Sculptured ornamentation below the face of the material.

Diagonal. A line from corner to corner across a rectilinear plane figure, quadrilateral or polygon; or across a solid.

Diagonal Bond. Used to give lateral bond to thick walls. About every fifth course the interior headers are placed diagonally across the wall. See *Raking Bond*.

Diagonal Grain. Applied to wood when the fibres do not run parallel to the axis of the piece. It is caused by faulty conversion or a tree being crooked and is a serious defect for structural work. It is limited in stress-graded timber. See *Slope of Grain* and *Oblique Grain*.

Diagonal Ribs. The ribs in a *rib and panel* vault, running from the angle between the walls to the crown.

DEVIL FLOAT

Diagonal Slating. Slates laid diagonally on a roof. The method is mostly applied to asbestos cement tiles.

Diagraph. An instrument for enlarging drawings mechanically.

Dialines. A process of printing, producing brown lines on white paper. See *Blue Prints*.

Diallage. A foliated silicate of magnesia; a pale variety of hornblende and dolomite found in serpentine, or soapstone.

Diameter. A straight line passing through the centre of a circle, or polygon, cutting the figure into equal parts, and terminated at both ends by the perimeter. A line through the centre of a sphere.

Diametral Plane. A plane cutting a cylindrical or spherical solid along the diameter or axis.

Diamond. 1. Used by the plumber for cutting glass; and fixed round the periphery of a circular saw, for cutting stone. 2. A rhombus.

Diamond Fret. An ornamental moulding shaped like diamonds, or rhombuses, by means of intersecting fillets, and used in Norman architecture.

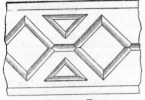

DIAMOND FRET

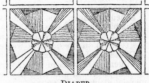

DIAPER

Diamond Jumper. A mason's chisel with the edge in the form of a cross.

Diamond Mesh. Expanded metal formed by shearing and punching flat steel plate into a network of strands. Used for reinforcing concrete, etc. Also see *Lattices*.

Diamond Saw. A circular saw used in stone cutting. The "teeth" consist of diamonds secured in sockets which in turn are fixed in the periphery of the saw at regular intervals.

Diaper. 1. Flat surfaces ornamented by lozenges or squares, and foliage. The pattern is repeated over an area, in regular formation. 2. Chequered patterns in brickwork and masonry.

Diaphragm Plates. Stiffeners placed between the webs, at intervals, in a box girder.

Diaphragms. 1. Spider webs, or lines, ruled on optically-worked glass discs for focusing surveying instruments. 2. The webs in hollow terra-cotta blocks. See *Terra-cotta*.

Diastyle. Columns spaced with three diameters between them. See *Columns*.

Diathermancy. The quality of a substance that allows radiant heat to pass through the substance.

Diatomaceous Earth. Infusorial earth, which is a siliceous marine deposit consisting of minute single-celled algae. It is baked into bricks or hollow blocks because of its insulating properties.

Diazomata. The passages round the seats of a theatre (in ancient Greece).

Dicalcium Silicate. A constituent of cement and the chief contributor to the strength of concrete after 28 days.

Dicotyledons. Deciduous or broad-leaved trees that supply nearly all commercial hardwoods.

Dictator. A registered name for a door spring and door check.

Die. 1. The dado, or cubical part, of a pedestal. 2. Baluster seatings, in masonry, worked on the plinth and handrail capping. Also see *Balcony*. 3. A tool for cutting a screw thread on a bolt or pipe.

Diespeker. A patent fire-resisting floor. Hollow burnt clay blocks are supported by reinforced concrete beams cast *in situ*.

Die Square. Timbers square in section, between quarterings and balks in size.

Differdange Beam. Broad flange beams. They have a greater lateral strength than the ordinary I-beam section, and are used for stanchions.

Differential Pulley. This consists of an upper block (which is one wheel having two grooves of different diameters with toothed grooves) and a lower block (which is a single movable pulley). An endless chain operates the pulley. It has a high velocity ratio and is capable of lifting very heavy weights. Also called a Weston pulley block.

Diffulume. Patent lenses for diffusing light in pavement lights.

Difusaire and **Difusic.** Registered louvre ventilators for walls, ceilings, and doors.

Digest. A small room or lounge attached to a restaurant and where liquors are served.

Diglyph. Similar to triglyph, but with two channels.

Dihedral Angle. The angle between two intersecting planes.

Dilapidations. Damages that accrue to premises during a term of tenancy. They are legally divided as voluntary waste (due to the action of the tenant) and permissive waste (due to neglect).

Dilatation Joints. Expansion joints.

Diluents. Liquids in paint, varnish, knotting, distemper, that allow for easy application, and that dry out later: turpentine, white spirit, commercial alcohol, and water.

Diminished Stile. A door stile diminished in width at the lock rail to give increased glass area above. Also called a gun-stock stile (*q.v.*).

Diminishing Course Work. Slating in which the gauge of each successive course is diminished. Graduated courses.

Diminishing Piece. A tapered pipe for decreasing the sectional area of drains, etc.

Dimmers. Adjustable shutters, or louvres, to regulate the passage of sound.

Dinas Bricks. Fire-bricks practically composed of sand only, with a small percentage of lime.

Dinette. A nook in a kitchen fitted with fixed seats and small table for informal meals.

Dinging. 1. Forcing the mortar into joints by means of a jointer. 2. One coat plastering; usually cement and sand applied to rough walls and marked with a jointer to imitate brickwork.

Dining Kitchen. A kitchen in a small house or flat that is also used for dining in.

Dining Nook or Recess. A small part of a sitting room or kitchen, often curtained off, used for dining in.

Diorite. Greenstone. A crystalline, granular, igneous, unstratified rock of felspar and hornblende or some other ferro-magnesian silicate. Excellent for road metal.

Dip. 1. A drop below the regular inclination of a pipe, to clear an obstruction, which increases the resistance of flow. 2. The inclination of a stratum of rock from the horizontal.

Dipstone Trap. A mason's trap. The trap is formed by a slab of stone placed on edge and dipping below the surface of the water. It is not used in good practice.

Dipteral. 1. A building having double wings. 2. A temple surrounded by a double row of columns.

Dipteros. A double-winged temple.

Dip Trap. One in which the discharging pipe dips below the water level of a drainage trap. It is an obsolete form of trap.

Diptychs. Folding doors or shutters; especially when above an altar.

Direct Heating. Heating by radiation, as when a room is warmed by a heated body within it, such as a coal, gas, or electric, fire.

Directing Point. A term used in perspective drawing for a point where a line meets the directing plane, which is a plane parallel to the plane of the picture and passing through the point of sight.

Direct Labour. Productive labour as distinct from administration and supervision. The term is also used when local authorities engage workmen instead of throwing the work open to specialist contractors.

Directors. A pair of compasses with an additional leg, attached with a universal joint, for setting out 3 points at once.

Directrix. A line at right angles to the axis of a conic section directing the construction of the curve. The ratio of the distances of any point ,on the curve, from the focus and the directrix is constant.

Disc or Disk. A thin cylindrical piece of material.

Discharging Arch. See *Relieving Arch.*

Disconnecting Chamber. A manhole separating a drainage system from the sewer. The separate drains for a building are usually collected in this chamber and the sewage conveyed by one pipe to the sewer.

Disconnecting Trap. See *Interceptor.*

Disc Piles. A pile with a flat plate bolted to it for extra bearing strength. The pile is sunk by means of a *water-jet.*

Diseases. The diseases in timber are decay, druxiness, doatiness, foxiness, plethora, wet rot, dry rot. These are distinct from defects, which are not always detrimental, whilst diseased timber should be condemned.

Dished. The term applied to a hole rounded or chamfered on the edge, or sunk to receive the flange of a pipe.

Display Cabinet. See *Show Case.*

Dist. Abbreviation for *distemper* and *distributed.*

Distance Piece, Block, or Strap. Wood, concrete or metal used to keep two members the required distance apart, as for the reinforcements or the shuttering in concrete work.

DISHED FOR SPITTER

Distemper. A mixture for colouring walls, consisting of whiting, size, and colouring matter. See *Washable Distemper.*

Distemper Brush. A large flat brush from 6 in. to 8 in. wide for applying distemper. See *Knot Brush.*

Distributed Load. One that is spread over an area or length and not concentrated. It may be uniform or otherwise. See *Uniform Load.*

Distribution of Pressure. The pressure, due to the load, across a horizontal section of a wall or base of any structure.

Distribution Pipe. Any pipe conveying water, supplied by the Water Board, from a feed or storage cistern or from a hot-water apparatus. The water is under pressure from the apparatus or cistern.

Distribution Rods. Transverse rods in reinforced concrete.

Distyle. Two columns to the portico of a classic temple. See *Columns.*

Dividers. Instruments like compasses but with pin points, used for stepping out distances or dividing lines by trial.

Division Walls. The interior walls of a building, or a wall serving two houses. A party wall.

dl. Abbreviation for *deal.*

D.M. Abbreviation for *disconnecting manhole.*

Doatiness. A disease in timber denoted by a speckled appearance and indicating incipient dry rot.

Dock Gates. Large watertight gates to docks. For dry docks they also act as sluices.

Docks. Artificial enclosures for ships whilst loading or unloading or undergoing repairs. There are *tidal docks*, in which the level of the water is the same as outside; *wet docks*, in which the level is kept constant by means of gates; dry, or *graving docks*, from which the water is pumped when the gates are closed.

Dodecagon. A rectilinear plane figure with 12 sides. Area = sq. of side × 11·196.

Dodecahedron. A solid with twelve equal pentagonal faces. See *Polyhedra*. Vol. = (linear side)³ × 7·663. Surface = (linear side)² × 20·646.

Dodecastyle. A portico with twelve columns.

Dog. A strong iron fastening used for heavy timbers in temporary work. It consists of a length of iron with the ends pointed and bent at right angles. A small type is used for temporarily holding glued joints. Also see *Fire Dog*.

Dog-eared Fold. Folding the end of the lead into a crease for the end of a gutter, instead of bossing.

Dog Grate. A grate consisting of a loose fire-box. It is usually on legs and is not fixed or built in, but placed in a tiled recess, with direct access to the flue.

Dog Leg. A bricklayer's templet for the skewback of an arch. Also called a *gun*. A specially shaped brick for a squint return, or splayed corner for 4½ in. work.

Dog Legged. Stairs in which the returns are made from the same newel so that the outer strings are in the same vertical plane and there is no well.

Dog Shore. A horizontal shore, distinguished from a flying shore by the absence of braces.

Dog Tooth. An ornamental moulding with carved projections in the form of small pyramids, etc.

Dog Tooth Bond. Masonry in which the headers extend about ⅔rds into the wall and alternately from each side.

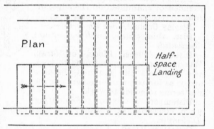

DOG-LEGGED STAIRS

Dog Tooth Course. A course of bricks laid diagonally and with one corner forming a triangular projection over the face of the wall. It is placed under an oversailing course.

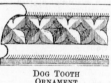

DOG TOOTH
ORNAMENT

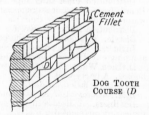

DOG TOOTH
COURSE (D

Dolerite. Basaltic rock used for road metal. Like greenstone.

Dolly. 1 See *Puncheon*. 2. A temporary hardwood block placed between the head of the pile and the pile hammer to protect the pile.

Dolmen. A cromlech.

Doloment. A composition for jointless floors.

Dolomites. Limestones with equal molecular proportions of carbonates of lime and magnesia (45·65 per cent and 54·35 per cent by weight).

Dolomitic Lime. Lime containing a percentage of magnesia; if it exceeds 40 per cent it is called dolomite.

Dolphins. See *Cutwaters*.

Dome. A hemispherical roof. The term is also used for roofs with elliptical or polygonal plans. A cupola. See *Gore*.

Domestic Architecture. The designing of dwelling-houses.

Domestic Heating. The arrangement for supplying hot water to houses. It usually consists of a boiler behind the fire, hot-water cylinder or tank, cold water cistern, and circulating pipes. Geysers, califonts, etc., with or without injectors, are used with gas or oil. Also electric immersion heaters.

Domical. Shaped like, or related to, a dome.

Donjon, sometimes **Dungeon** (*q.v.*). The keep of a castle.

Donkey. Applied to auxiliary apparatus, such as donkey-engine, donkey pump, etc.

Donkey's Ear. A mitre shoot, for woodworking, that can be fixed vertically in the vice.

Donnacona. A registered wood-fibre wallboard.

Donsella. A variegated reddish brown hardwood from C. America, and Mauritius. Very hard, heavy, and durable. Used for superior joinery cabinet work, etc. Weight about 58 lb. per c. ft.

Dook. A wood plug.

Door Bar. A fall bar, *q.v.*

Door Case. The frame enclosing a door. See *Casings*.

Door Chain. A chain with a stud engaging with a horizontal slot, and allowing the door to open a few inches only.

Door Check. A mechanical appliance for controlling the closing of doors. It is usually fixed at the head.

Door Cheek. The stile of a door frame. A door post or jamb.

Door Finishings. The ornamental features about a door such as linings, architraves, plinth blocks, moulded grounds, pediments, etc.

Door Frame. A solid frame for a door, consisting of two stiles and a head and sometimes a transom. They are made from about 5 in. × 3 in. stuff, and, usually, for external doors.

Door Furniture. Hinges, locks, latches, bolts; but more especially applied to handles, knobs, escutcheons, and finger plates, which may be of wood, glass, plastics, or metal. There are over 20 different alloys and finishings in metal. Other door fittings include springs, closers pulls, letter-boxes, etc.

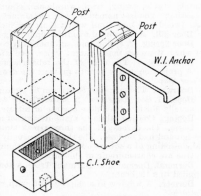

DOOR FRAME FIXINGS

Door Head. 1. A level projection over a door. A hood or canopy. 2. The top rail of a door frame.

Door Knobs. The attachments to the spindle of a latch, to provide leverage for the hand. There are several patent methods of securing the knobs to the spindle : Mace's, Pitt's, Kaye's, etc.

Door Latch. See *Latch*.

Door Nails. Studs or nails with large heads to imitate bolts.

Door Posts, or **Door Cheeks.** The jambs, or stiles, of a solid door-frame.

Doors. The usual types of doors are : ledged, ledged and braced, framed and ledged, solid, flush, and panelled. The last named may have any number of panels ; and they are named according to the number of panels, or to some special feature, such as double margin, sash, Gothic, etc. A panelled door consists of stiles, rails, muntins, panels, and mouldings. The standard sizes are 6 ft. 4 in. × 2 ft. 4 in. to 6 ft. 8 in. × 2 ft. 8 in. Special types are : dwarf, jib,

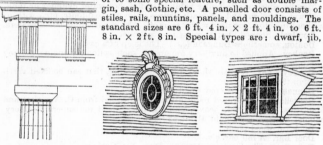

DORIC   DORMERS

hatch, stable, swing, trap, warehouse, sliding, folding, revolving, fire-resisting. The last named may be metal, metal-covered wood, or composite (wood and metal), of which there are many registered designs.

Door Sill. A horizontal member connecting the feet of door posts.

Door Spring. A spring that automatically controls the closing of a door, of which there are many patent types. See *Floor Spring*.

Door Stop. 1. The edge of the rebate against which the door closes. 2. A projection on the floor to prevent the door from opening too far. See *Floor Stop*.

Doorway. The entrance to a room or building.

Dope. 1. See *Retarders*. 2. Materials for waterproofing canvas. Dope is usually chiefly of cellulose acetate or nitrate, with pigments if required.

Doping. Coating fabric to make it airproof and waterproof.

Doric. The earliest of the three Greek Orders. It was the simplest in design and had no base. The shaft was usually fluted and had a plain capital consisting of a square abacus and an echinus. Triglyphs, metopes, and guttae are characteristic of this order.

Dormant, Dormand, or **Dormant Tree.** A sommerbeam, or sleeper. Also applied to a tie beam.

Dormer. A window in a sloping roof, with vertical casements. Also see *Lucarnes*.

Dormer Cheeks. See *Cheeks*.

Dormitory. A sleeping chamber containing many beds.

Dorry Machine. An abrasive machine to test the wearing properties of a material.

Dorse, or **Dorsal.** A canopy. Wall hangings. The back support for a canopy.

Dortoir. A sleeping apartment in a monastery.

Doseage Plant. Chemical water-softening plant.

Dossoret. A deep abacus to a capital in Byzantine architecture.

Dote. Incipient decay in wood. A general term for rot or decay in timber.

Dots. 1. Small patches of coarse stuff which are plumbed, or levelled, for the ends of the floating rule to run the screeds for plastering. 2. See *Lead Dot.*

Dotting On. A term applied to similar items added together, when "taking off" for a bill of quantities.

Double Bellied. Applied to a turned baluster with both ends alike.

Double Bolection Moulding. See *Fielded Panel.*

Double Boxed Mullion. A mullion prepared for two pairs of weights for sliding, or balanced, sashes.

Double Doors. A pair of doors rebated on the middle, or meeting, stiles. Sometimes called folding doors, *q.v.*

Double Curvature. See *Circle-on-Circle.*

Double, or Doubling, Eaves Course. See *Eaves Course.*

Double Faced Architrave. One moulded on both edges. Also built-up architraves and skirtings stepped to form two plane faces.

Double Flemish Bond. Walls showing Flemish bond on both sides.

Double Floor. A floor consisting of binders and common joists. The length of the binder should not exceed 24 ft.

Double Framed Floor. A floor consisting of girders, binders, and common joists.

DOUBLE FLEMISH BOND

Double Hung Sashes. A *sash and frame,* in which both sashes are hung by balanced weights.

Double Lath. A lath twice the ordinary thickness. See *Laths.*

Double Lean-to, or **Double Pent.** A V-shaped roof consisting of two lean-to roofs meeting with a V or parallel gutter. See *Roofs.*

Double Margin Door. A wide door with four stiles, having the appearance of two folding doors. The inner stiles are keyed together.

Double Measure. Moulded on both sides.

Double Partitions. Partitions constructed with a cavity to receive sliding doors, or for sound resistance. See *Hollow Partitions.*

Double Pitched Roof. A mansard roof, *q.v.*

Double Rebated Linings. Wide linings rebated on both edges.

Double Return Stairs. A staircase with a main flight and return flights to both right and left.

Double Rolled Cast. A rolled glass with hammered surface.

Double Roman Tiles. Large interlocking tiles, $16\frac{1}{2}$ in. × $13\frac{1}{2}$ in., requiring about 65 to a square. They are laid to about 15 in. gauge and weigh 6 to 7 cwt. per square.

Double Roofs. Roof trusses combining more than one type. Usually applied to hammer beam, and similar, trusses.

Doubles. Slates 13 in. × 10 in. down to 13 in. × 6 in. About 415 per sq. at 3 in. lap are required.

Double Shear. Applied to a structural member that is resisting shear along two section planes, as in riveted joints.

Double Roman.

Double Skirting. A wide skirting built up of two pieces. See *Soldiers.*

Double Step. A joint for heavy constructional timbers. It provides greater resistance against detrusion and does not weaken the member so much. See *Step Joints*.

Double Tenons. Two tenons side by side not in the same plane, as used on a lock rail to receive a mortise lock. See *Tenons*.

Double Tier Partition. A framed partition extending through two storeys.

Double Vault. One vault built over another, with a space between.

Double Window. See *Storm Window*.

Doubling Course. See *Eaves Course*.

Doubling Piece. See *Tilting Fillet*.

Douglas Fir. *Pseudo-tsuga taxifolia.* Columbian or Oregon Pine. Similar to redwood but coarser and more open grain. Large sizes, straight grained, strong and fairly durable. Yellow with reddish markings. It is used for structural and constructional timbers, piling,

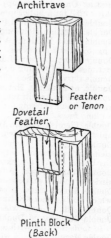

Architrave

Feather or Tenon

Dovetail Feather

Plinth Block (Back)

DOVETAILS (SECRET)   DOVETAIL FEATHER

oinery, plywood, flooring, wood blocks, etc. Weight about 30 lb. per c. ft.

Doulting. Two varieties, cream and buff, limestone from Somerset. The cream, fine-grained, stone is used for interior carving, and the Chelynch, or buff-coloured, is used for general building. Weight 125 and 150 lb. per c. ft. Crushing strengths 103 and 210 tons per sq. ft.

Doulton's Joint. A patent joint for drain pipes in wet ground. An outer canvas casing protects the cement whilst setting.

Dovetail Feather. A dovetailed thin slip to fix two pieces together.

Dovetail Groove. See *Grooves*.

Dovetail Key. A key with the section shaped like a dove's tail as used for counter battens, end to end joints, etc. See *Hammer-head Key*.

Dovetail Moulding. A Norman moulding having a continuous band in the form of dovetails.

Dovetails. Used to joint pieces end to end at right angles to each other. Fan shaped pins are formed on one piece to fit in similarly shaped sockets in the other. There are many variations of dovetailed joints : common, lapped, secret, etc.

Dovetail Saw. Like a tenon saw, *q.v.*, but smaller and with an open handle.

Dovetail Sheeting. A patent steel shuttering that can be adapted to any form of circular work.

Dowel. Cylindrical pieces of wood or metal used for fixing one piece of material to another. For masonry they are rectangular in section and often made of slate. Wood dowels are often used in woodwork instead of a mortise and tenon joint. See *Fastenings* and *Bar Tracery*.

Dowelling Jig. A tool used to guide the bit when boring for dowels.

Dowel Screw A double-ended screw. See *Screws*.

Down Comer. See *Down Pipes*.

Down Draught. A current of air down a chimney, due to a cold or faulty flue.

Down Pipes. Stacks of vertical wood, iron or lead pipes to convey the rain-water from the roof to the drains. Also called *down-spouts*.

Dowsing. Using a divining rod to find water below ground.

Dozer. A mechanical excavator. See *Mechanical Shovel*.

D.p. Abbreviation for *drain pipe*.

D.P.C. Abbreviation for *damp-proof course*.

Draft. Narrow margins or channels worked on the surface of a stone, and made out-of-wind by sighting straightedges; they are used when preparing a plane surface.

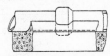

DOWEL JOINT

Stiles · Top Rail 2″ × 4½″ · ⅝″ Dowels not exceeding 3½″ apart · Bottom Rail

Drafted Margin. A narrow margin worked round the edges of the face

DRAG

DRAIN PIPES AND FOUNDATION

of a stone. The remainder of the face is worked in some special manner, and usually projects over the margin.

Draftex. A registered type of tracing cloth.

Drag. A thin steel plate toothed like a saw, with straight or curved edges, for facing up cast plaster work or stonework.

Dragged, or Combed. Stone worked to a finished surface by means of a comb, or drag.

Dragging Tie. See *Dragon Tie*.

Dragline. A mechanical excavator used as a skimmer, trencher, and grab. See *Mechanical Shovel*.

Dragon Beam, or Piece. The timber accompanying the dragon tie and carrying the foot of the hip rafter.

Dragon's Blood. The red resin from the dracaena tree, used as a colouring pigment.

Dragon Tie. See *Angle Tie*.

Drain Chute. A drain pipe with one end considerably increased in depth. The oval end is placed in the manhole to facilitate the entry of cleaning rods.

Drain Cock. A tap at the bottom of a tank or hot-water system, to empty the tank or system.

Drainer. See *Draining Board*.

Drain Ferret. Patented thin glass phials containing a substance that produces dense smoke and pungent odour when broken in a drain.

Drain Hole. Same as *weep hole, q.v.*

Draining Board. A hardwood (usually teak) board, grooved and inclined, to collect the water from crockery, and to convey it back to the sink. The D.B. may be of other materials than wood.

Drain Pipes. The cylindrical pipes for conveying the sewage away from a building. They may be salt-glazed earthenware or stoneware, coated

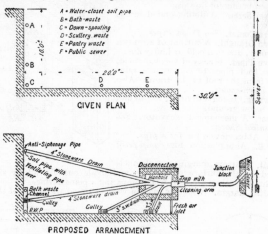

DRAINAGE SYSTEM

cast-iron, or concrete. They must be straight, smooth, and have good socketed joints. The joints are made watertight by gaskin and cement, bitumen and cement, or caulked with lead if they are iron pipes. There are a number of patented joints.

Drains. Three main types: (*a*) Foul, or soil, for sewage. (*b*) Storm, or water, for rain-water. (*c*) Subsoil drains. Types (*a*) and (*b*) are usually combined. The following are some of the requirements for a good drainage system. Smooth joints; straight from point to point; even gradient (multiply diam. of pipe by 10, i.e. 4 in. diameter $\times$ 10 = fall of 1 in 40); good foundations; self-cleansing (minimum velocity 3 ft. per sec.); inspection chambers or eyes at every change of direction, or to collect branches, or at every 200 ft. in long straight lengths; disconnecting chamber between main drain and sewer; branch drains min. diam. 4 in.; main drains min. diam. 6 in.; C.I. pipes for drains under buildings, manholes each end; F.A.I. and O.V.; gulleys securely trapped; soil pipes direct to drain.

Drain Sentinel. See *Intercepting Trap*.

Drain Testing. The testing of drains to see if they are air and water-tight. They may be tested by smoke, water, scent, or compressed air. See *Drain Ferrets* and *Hydrostatic Test*.

Drapery, or Linen-fold, Panel. One moulded and carved to imitate the folding, or draping, of cloth. See *Linen-fold Panel*.

Drapes. Heavy curtains used in film and music studios to regulate the resonance, and on theatre stages to hide the non-acting area.

Draughting. Setting out and preparing a drawing in orthographic projection.

Draw, or Draught. The clearance allowed to pull up the joint in gib and cotter fastenings, and when drawboring.

Draw Back Lock. One in which the bolt is operated by a cranked lever on the inside and a key on the outside.

Draw-Boring. Tightening a joint at the shoulder by means of a dowel. The hole in the cheeks of the mortise is a little in advance of that in the tenon.

Draw Bridge. A bridge hinged at one end so that it can be raised when required. The term is also applied to swing bridges that move horizontally.

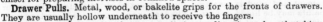

DRAW BORING

Drawer Lock Chisel. A small cranked chisel for preparing mortises in confined spaces.

Drawer Pulls. Metal, wood, or bakelite grips for the fronts of drawers. They are usually hollow underneath to receive the fingers.

Drawing. Straightening the straw to lie in one direction, when thatching roofs. 2. The mechanical method of manufacturing ordinary window glass.

Drawing Appliances and Instruments. Board, T-square, set squares, pencils, rules, protractors, French curves, pens, compasses (ink and pencil and spring bows), dividers, etc. These form the minimum requirements of a draughtsman. There are many variations of the above, and many other aids to mechanical drawing.

Drawing Board. A thin board squared round the edge, and clamped to prevent warping. It is used with T-square and set squares in mechanical drawing. The sizes vary according to the paper.

Drawing Paper. There are numerous kinds and qualities. *Cartridge* is used for ordinary work. It is a machine-made paper and made in over twenty different qualities. *Whatman* is used for good work and for coloured drawings. Other kinds are Arnold, Michallet, Creswick, David Cox, Van Gelder, Varley, Joynson, Canson, etc. Standard sizes are as follows: Emperor (68 × 48), Antiquarian (52 × 31), Double Elephant (40 × 27), Atlas, Columbian, Elephant, Imperial (30 × 22), Half Imperial (22 × 15), etc. The sizes are in inches, and are only given for those in common use.

Drawing-room. A sitting-room or parlour in domestic dwellings. A room to which the company withdraws after dining.

Draw Knife. A long cutting tool cranked at the ends for two handles.

Drawn Glass. Ordinary window glass. It is gathered from the molten glass by a bait, or net, and drawn between asbestos rollers, through lehrs. The temperature is controlled at each stage to suit the process. The sheets are then cut to size.

Draw-off Pipes. The pipes in a hot-water system conveying the hot water from the cylinder to the taps.

Draw-off Tap. Applied to any tap for drawing water, but especially to the one in a dwelling-house supplying drinking water direct from the main, which empties the system when the main stop cock is shut. A special tap to the boiler, to empty the hot-water system for repairs. See *Low-pressure System*.

Draw Pin. An iron, tapered, pin cranked at the top for withdrawal, and used for pulling up a shoulder in a mortise and tenon joint. The tenon is left projecting and a hole is bored through the tenon partly inside the mortise so that the pin can be driven in.

Dreadnought. 1. A registered fire-resisting composite door. 2. A patent roofing consisting of sheet asphalt and pre-cast pumice blocks, rendered or tiled.

Dream Hole. An opening in the wall of a tower to admit light.

Drencher System. Sprinklers turned on by hand, in the event of fire.

Dress Circle. The lowest gallery of a theatre, or a part of the gallery.

Dressed Brick. One beaten with a dresser before burning, to improve the shape and make it tough.

Dresser. 1. A kitchen fitment combining drawers, shelves, and small table. 2. A hornbeam or boxwood tool used by the plumber for working sheet lead. 3. See *Dressed Brick*.

Dressing. 1. Preparing and finishing stone. 2. Beating lead to the required shape with a dresser or bossing stick.

Dressings. Ashlar quoins, sills, heads, etc., in brick or rubble buildings. The term is applied to the prepared stones round openings, and to mouldings projecting over the face of the wall.

Dreston. A patent absorbent plaster for positions of excessive condensation.

Dri-crete. A registered hollow block concrete construction.

Driers. Any material for assisting plastic material to set hard. Dry cement and sand is scattered over cement work as a drier. Driers for paints may be liquid or paste. The former are usually compounds of lead, manganese, or cobalt, in volatile liquid. The latter are the same metals with barytes or whiting added and ground in oil. Turps, terebene, zinc sulphate, red lead, litharge, lead acetate, are the best-known driers.

Drift. An appliance for removing the core from a mortise made on a hand-mortiser. A large punch for removing fastenings of any description.

Drift Bolt. Round or square bars of iron or steel used as fastenings in heavy timber construction. A common size is 30 in. × $\frac{3}{4}$ in. diam. with one end pointed and the other formed into a head. They are driven like spikes and prevent lateral movement.

Drift Plate. A rectangular steel plate used by the plumber when working overcloaks on roof coverings.

Drilling. Boring circular holes in metal, stone, or other hard materials by means of a drill. The drill rotates whilst boring and may be actuated by hand, brace, bow, or machine.

Drip Box. A cesspool to a gutter.

Drips. 1. The steps in lead gutters to allow for joining the lengths of lead without interrupting the flow of water. 2. The projecting edge of a throated sill or moulding.

Drip Shield. A projecting guard, or hood, to casement windows, to throw the water away from the casements.

Dripstone. A moulding projecting over an opening, such as a door or window. A *label moulding* in Gothic architecture. It is intended to throw the water away from the opening.

Drop. An ornamental terminal to the bottom of a suspended post. Small truncated cones or cylinders used in the mutules, and under the triglyph, in the Doric order. Guttae.

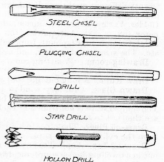

STEEL CHISEL

PLUGGING CHISEL

DRILL

STAR DRILL

HOLLOW DRILL

DRILLS AND CHISELS

Drop Arch. A pointed, or Gothic, arch, with radius less than span. See *Arches*.

Drop Black. See *Black Pigments*.

Drop Bottom Bucket. A receptacle for depositing concrete by means of the bottom opening outwards.

Drop Dry. A patent roof glazing. It consists of a zinc bar mounted on wood or metal core with zinc or copper capping.

Drop Handle. A hinged drawer pull or handle to drop down when not in use.

Drop Moulding. A planted panel moulding lying below the surface of the framing.

Drop Point. A method of laying asbestos slates diagonally on a roof. See *Honeycomb Slating*.

Drop Sidings. Weatherboarding on timber buildings. See *Sidings*.

Drop System. A hot-water system. The boiler is placed as low as possible in the building and the hot water is conveyed by a flow pipe to the highest required point. It is then distributed in branch pipes as required, which flow downwards and join the return pipe near the boiler.

Drop Test. A method of testing the comparative bearing values of earths. A steel ball is dropped from a given height and the diameter of the impression measured.

Droved, or **Droving.** See *Boasting* (2).

Drum Curb. The cylindrical lining for a well in course of construction.

Drum Head. A round-end step at the foot of a flight of stairs. See *Round-end Steps*.

Drums. 1. Temporary centres, or cylinders, upon which veneers are bent and secured until they are staved, or blocked. 2. Hollow masonry columns threaded over steel stanchion or cores. 3. The stylobate, or vertical part, under a dome, or cupola. 4. The *bell*, or solid part, of a Corinthian or Composite capital.

Drunken Saw. A circular saw packed so that it runs eccentrically. It is used for cutting plough grooves and removing the waste between double tenons.

DRUM-HEAD STEP

Druxy, or **Druxiness.** A disease in timber caused by a wound in the cambium layer being attacked by fungus. The timber has a speckled appearance.

Dry Area. See *Area*.

Drybilt. A prefabricated wooden house.

Dry Conservancy. Earth and chemical closets used where water supply and drainage are not available.

Dry Dock. See *Docks*.

Dry Feet. Brick, stone, or concrete foundations to a *pisé de terre* building.

Dryflex. A proprietary, anti-corrosive, graphite paint.

Drying Stresses. See *Case-hardening*.

Dry Kiln. A chamber for seasoning wood.

Dry Mix. Mixing the gauged materials for concrete in a dry state before adding water. Also applied to concrete with just the correct water-cement ratio, to distinguish from a *wet-mix*.

Dry-out. Applied to a plastered surface in which excessive evaporation of moisture has taken place before setting, leaving the plaster soft and chalky.

Dry Rot. Decay in timber caused by contagious fungoid growth. There are many different forms but the most destructive is *Merulius lachrymans,* or weeping fungus. The attacked timber is covered with a blanket-like covering having an unpleasant odour, and is distinguished by red and brown stripes. The disease spreads rapidly in moist stagnant atmosphere. The preventives are well-seasoned timber and ventilation. The cure is difficult, and requires drastic treatment. All infected timber must be removed and burnt; the walls should be treated with a blow-lamp; mortar joints raked out and re-pointed; the remaining timbers treated with strong preservative, such as sulphate of copper, creosote, hot lime or one of the patent preservatives; ventilation must be provided.

Other forms include *Coniophora cerebella,* cellar fungus, very dark thick strands; *Poria vaporaria,* creamy fine strands; *Paxillus panuoides,* similar to *Coniophora.* These require more moisture than *Merulius lachrymans* and are usually found in cellars and mines.

Dry Rubble. Rubble walling built without mortar, usually used for boundary walls.

Drys. Fissures in stone at various angles to the bed making the stone useless for structural work.

Dry Walling. Same as dry rubble, *q.v.*

Dry Wedging. Wedging-up temporarily, without glue or paint.

D.S. Abbreviation for *drop siding* and *double sunk.*

D.S.P. A proprietary adhesive for all purposes. Abbreviation for *Dunlop Special Products.* Applied to several compounds used in building.

D-trap. A lead trap like the letter D in shape. An obsolete form of intercepting trap for drains.

Dubbed off. 1. Removing arrises to allow for easy entrance, as for a tenon into a mortise. 2. A badly shaped or abrupt easing to a moulding.

Dubbing. 1. Levelling up uneven plastered surfaces. 2. Using pieces of brick or tile in a thick coat of cement or plaster. 3. Cutting with an adze.

Duchemin Formula. A formula for calculating the wind pressure on inclined surfaces.

Duchess Slates. Sizes 24 in. × 12 in. Requires 115 per sq. at 3 in. lap.

Duck Board. An inclined scaffold board with cleats to provide foothold.

Duck Foot. A drain pipe with a right-angled bend to receive a vertical pipe. It has a flat base to provide a seating.

Duct. A cavity for pipes. A tube or canal for the conveyance of anything.

Ductile. The reverse of brittle. The property of a material that allows a decrease in sectional area by a tensile force.

Ductube. A patent rubber tube that is inflated to the required diameter and placed in position in formwork for concrete before casting. When deflated it releases itself and is removed to leave a duct through the concrete for wires, etc.

Dullrae. An invisible system of continuous electric heating under thermostatic control.

Dulux. The trade name for Nobel chemical finishes, either gloss or flat. A special synthetic vehicle gives it hard setting and lasting properties.

Dumboard. A proprietary insulating material.

Dumbrick. A patent sand-cement brick with extra deep frog.

Dumb Waiter. A small lift for conveying food.

Dumb Well. A cesspool or cesspit.

Dummy. 1. A long iron or cane handle with a bulbous iron head for working out irregularities in bends in large lead pipes. 2. Applied to any useless imitation, as a dummy doorway, window, etc. See *Blind.*

Dummy Doorway. A brick panel shaped like a doorway for reasons of symmetry.

Dumper. A special type of tip-up lorry for removing excavated material, etc.

Dumpling. 1. A wooden block, like an inverted basin, used on a spindle moulding machine, for double-curvature work. 2. The solid part of earth, near to an excavation, able to withstand a load, such as the pressure from a shore.

Dumpy Level. A surveying instrument for obtaining levels.

Dums. A local term for casings and frames.

Dungeon. A prison in or under a tower. The keep of a castle.

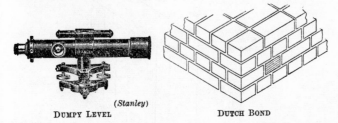

(*Stanley*)

DUMPY LEVEL   DUTCH BOND

Dungun. A dark chocolate brown hardwood from Malay. It is hard, heavy, tough, strong and durable, and used for piling, structural work, etc. Weight about 48 lb. per c. ft.

Duodecagon. See *Dodecagon*.

Duodecimals. Calculations in which subdivisions of 12 are used, instead of 10 as in ordinary decimals, to conform to feet and inches and shillings and pence.

Durability. The resistance of a material against the action of the atmosphere, decay, and wear.

Duralumin. An alloy of aluminium, copper, manganese, magnezium, silicon, iron. It is strong, light, and resists corrosion.

Duramen. The inner part of a tree. Heart wood. See *Timber Tree*.

Duresco. A registered name for a large range of paints, etc.

Durex. A bituminous roofing felt.

Durobestos. A proprietary asbestos-cement product.

Durogrip. A patent non-slip flooring consisting of ceramic tiles with raised dots. It is glass-hard and of various colours.

Durok. A bituminous flat roofing system. It consists of two layers of *Rok* with adhesive mastic between. The first layer may be nailed to boarding.

Duromit. A very hard crystalline floor and road surfacer for cement surfaces. It is claimed to be waterproof, dust proof, non-slippery, and able to resist the heaviest traffic.

Dust Board. A cover board. Also the divisions between drawers.

Dusting. The term applied to the wearing of a cement surface.

Dusting Brush. A painter's brush with soft bristles, from 4 in. to 6 in. long, used for dusting down before applying paint; may be circular or flat.

Dutch Arch. The voussoirs are parallel in thickness and radiate in opposite directions from the centre. It is a flat arch and constructionally weak. See *Arches*.

Dutch Bond. Similar to English bond except that it is formed by a ¾ bat in the stretching courses and no closers are used. The header, shown hatched, is placed in alternate stretcher courses to spread the line of fracture over a longer area.

Dutch Clinkers. Small vitrified paviors. Usually 6 in. × 3 in. × 1 in.

Dutch Metal. An alloy of copper and aluminium or zinc. Also called leaf gold, or foil. A substitute for gold leaf.

Dutch White. White lead and sulphate of barium.

Dwangs. Nogging pieces. Also the strutting between joists.

Dwarf Door. One less than 5 ft. 6 in. high. A screen door.

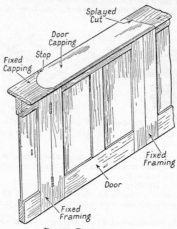

Splayed
Cut

Door
Capping

Stop

Fixed
Capping

Fixed
Framing

Door

Fixed
Framing

DWARF PARTITION

Dwarf Partition. A low partition with capping, often used as an enclosure in offices.

Dwarf Wall. Usually applied to the sleeper walls supporting ground floor joists, or to one less than a story in height.

Dyeline. A method of printing by semi-dry development. It produces brown or black lines on white paper or cloth. See *Blue Prints*.

Dyke. 1. A trench. 2. A dry-rubble wall. 3. Wall-like masses of igneous rock. 4. A mound to prevent inundation.

Dynamics. The section of mechanics dealing with moving forces.

Dynamometer. An instrument for measuring force or power.

Dyne. The unit of force in the centimetre—gram—second, or c.g.s. system, as used in dynamics, physics, etc.

E

E. A symbol for *modulus of elasticity* and *efficiency*. **e.** A symbol to denote the eccentricity of pressure of a load on a structural member.

Eagle. A registered precast floor unit.

Early English. Thirteenth century. The first of the Gothic, or pointed, styles following Norman architecture. The chief characteristics were the tall narrow windows with lancet arches. Examples: Westminster Abbey, Southwark, Salisbury, Rochester, and Wells Cathedrals.

Ears. 1. The projecting lugs on down-pipes to receive the nails for fixing. 2. See *Crosettes*.

Earth Closet. A conservancy system in which earth is applied, to the excreta in the pail, as a deodorant.

Earthenware. Sanitary ware made from ordinary brick earths. They are more porous and not so satisfactory as stoneware.

Earth Pressure. The consideration of the pressure and resistance of earth when designing retaining walls, foundation, etc. Various assumptions and theories are adopted in the design of retaining walls, but they are based on the assumption that earth is a granular mass, lacking in cohesion, and that each kind has its natural slope after weathering. The principal theories are (1) wedge, (2) Rankine's, (3) Scheffler's. Research is still being conducted for a more satisfactory solution.

Earth Table. A plinth, or the lowest course above ground level. Also called ground-table and grass-table.

Easements. A right whereby the owner of one property (dominant tenement) has certain privileges or conveniences over property (servient tenement) belonging to another owner.

Easing. 1. The curving of an angle to avoid an abrupt change of direction. 2. Slackening the folding wedges supporting a centre or shore, when the work is completed, to allow the work to settle, and to see if it is satisfactory. See *Centering*.

Easing Wedges. Wedges arranged for easing a centre or shore from the superincumbent load.

Eaves. The bottom edge of a sloping roof overhanging the face of the wall, where the water is collected in the gutters.

Eaves Board. A tilting fillet in the form of a feather-edged board.

Eaves Catch. A local name for tilting fillet.

Eaves Course. 1. The bottom course of tiles or slates. 2. A projecting course of tiles, brickwork or masonry immediately under the eaves.

Eaves Fascia. A board nailed to the feet of the spars, and usually carrying the gutters.

Eaves Gutter. An arrangement at the eaves to collect the rain-water and convey it to the down-pipes.

Eaves Plate. A beam for carrying the feet of rafters. It is used where there is no supporting wall below, and rests on piers or posts.

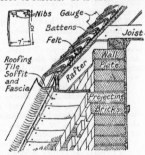

Eaves Pole. A tilting fillet.

Eaves Soffit. The horizontal surface under projecting eaves.

Ebanoid. A proprietary bitumen for setting wood-block flooring.

Ebnerite. A proprietary composition for jointless floors.

Ebonite. A hard black, non-plastic, elastic material produced by treating rubber with sulphur, zinc oxide and carbon black. It is used for insulating.

Ebony. A hard, tough and heavy decorative timber, sold by weight. Dark green (with brown stripes) to black, in colour. Heavier than water.

EAVES

Eccentric Load. A force that does not pass through the centre of gravity of a section, or along the axis, of a structural member. A non-axial load that produces a non-uniform stress.

Echinus. An ornament carved in the form of an egg. See *Egg and Dart*.

Eclipse. A patent roof glazing. A patent telescopic scaffold-board. A registered metal-faced plywood. A patent smoke machine for drain testing.

Ecofix. A patent zinc-plated steel strip for all types of clips, fittings, etc.

Economic Ratio. The best proportion of steel to concrete in reinforced concrete.

Economizer. A nest of small tubes containing cold feed-water for a boiler and which is heated by waste gases from the furnace.

Eddy's Theorem. "The B.M. at any point in an arch = horizontal thrust × vertical intercept between centre line of arch and line of pressure."

Edge Nailing. Secret nailing for boarded surfaces.

Edging. 1. A border or fringe. 2. Edging strips, *q.v.*

Edging Strips. The strips on the edges of flush doors, to hide and secure the edges of the plywood faces. There are numerous types according to the quality of the doors. They are also called bandings, claddings, clashings, railings, and slamming strips.

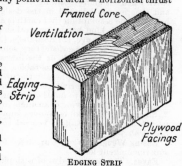

EDGING STRIP

Edward's Tiles. A registered design of terra-cotta sill tile, with special seating blocks for jambs and mullions.

Eel Grass. A plant with grass-like leaves used for sound deadening. The grass is cured and laid at all angles to form a thick cushion full of irregular cells of dead air.

Effective Bond. The bond in brickwork formed by a closer, i.e. $2\frac{1}{4}$ in.

Effective Depth. The depth to the C.G. of the tensile reinforcement in concrete beams.

Effective Pillar Length. The length upon which the ratio of pillar length to least radius of gyration is calculated.

Effective Span. The distance between centres of bearing; or from centre to centre of supports for continuous beams. The term is used to distinguish from *clear span*.

Efficiency (E). The ratio *useful work* to *total work*, in a machine. E = mechanical advantage ÷ velocity ratio.

Efflorescence. A chalk-like appearance on buildings, due to the evaporation and crystallization of the alkaline salts contained in the bricks and mortar and lime plaster. Dilute hydrochloric acid removes it; but if left to the weather the salts will gradually disappear.

Effluent. Purified liquid sewage.

E.g. Abbreviation for *Eaves gutter*.

Egg and Dart, or **Egg and Anchor.** An ornament consisting of a moulding carved in the form of eggs, or echinus, separated by vertical arrows or anchors.

EGG AND DART

Egg-shaped Sewer. An oval-shaped brick sewer with terra-cotta or concrete invert. It is now seldom used owing to the efficiency and cheapness of concrete.

Eghoin. A yellow lustrous hardwood from Nigeria. It is light but fairly hard and strong. Used for match-boarding, shelving, etc. Weight about 30 lb. per c. ft.

Egyptian. The style of architecture prevailing in ancient Egypt. Ornamentation characteristic of Egypt, such as the lotus flower, date palm, and the papyrus.

Eidograph. A pantograph. An appliance for mechanically copying a drawing to any scale.

Ejectors. 1. A special chamber for collecting sewage from buildings in flooded areas. The sewage is lifted by compressed air and conveyed to a filter bed above flood level or to a sewer. 2. Any mechanical appliance for ejecting or raising liquids, as used in some of the patent water heaters.

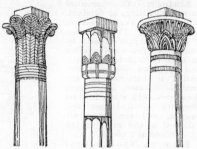

EGYPTIAN COLUMNS

E.J.M.A. The trade mark of the English Joinery Manufacturers Association.

Ekeing. Adjusting or "making do" something not quite right for its purpose. Lengthening a piece of wood that is too short.

Ekhimi. A West African hardwood resembling plain oak except for the interlocking grain. It is suitable for structural work or interior joinery.

Ekki. A very hard, heavy, strong, and durable hardwood from Nigeria. It is purplish-brown and used for flooring, superior joinery, or structural work. Weight about 64 lb. per c. ft.

Ekpaghoi. A reddish brown hardwood, with darker markings, from Nigeria. Rather hard, heavy, and difficult to work smooth. Used for superior joinery and veneers. Weight about 47 lb. per c. ft.

Elasticity. That quality of a material that allows it to regain its original condition after distortion. See *Modulus of Elasticity*.

Elastic Limit. The greatest stress a material can resist and retain its elasticity. Beyond that limit *permanent set* takes place.

Elbow Linings. Interior linings for splayed window jambs. The linings, below the boxings for the shutters, extending to the floor.

Elbows. 1. The vertical returns for panelling in recesses. 2. Right-angle bends for pipes. 3. See *Crosette*.

Elbow Socket. A small L-shaped piece to receive a gas burner.

Eldorado. A registered cork tile flooring or panelling, or for insulating.

Ele, or **Eling.** Same as aisle, *q.v.*

Electrical Resistance. The resistance of a conductor is measured in *ohms*. It depends upon the material (usually copper), and temperature, and is proportional to the length of the wire, and inversely proportional to the sectional area.

Electric Bell. A signalling device consisting of a gong and an electrically vibrated hammer. See page 110.

Electricity, Terms Used. Accumulator, alternating current, ammeter, armature, battery, booster, cable, cells, charge, circuit, conductor, Coulomb's law, direct current, dynamos, electrodes, E.M.F., Faraday's law, fuse, generator, induction, insulation, Joule's law, meter, motor, Ohm's law, rectifier, resistance, short circuit, switch, voltmeter.

Electric Units. Unit of pressure = *volt*. Unit of current (practical) = *ampere*. Unit of power = *kilowatt-hour*.

Electric Welding. Joining metals by heating them to the required temperature by an electric current. It may be electric-arc or electric-resistance welding.

Electro-copper Glazing. Fire-resisting glazing. The minimum thickness of glass is ¼ in. and the maximum size of squares 16 sq. in. The glass is bedded in electro-copper bars. The squares are arranged with copper strips between. The whole window is then placed in a copper bath where copper flanges are electrically deposited to fix the squares.

Electrolier. An ornamental suspended fitting, or pendant, for carrying two or more lamps.

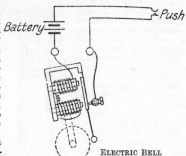

Electrolysis. The resolution of compounds into elements by electricity. Tubes are inverted in slightly acidulated water, and over two electrodes connected to batteries for the analysis of water. The current decomposes the water into its elements of hydrogen and oxygen, and the gases are collected in the tubes in the proportions of 2 and 1.

Element. The simplest known constituent of any substance. It allows for no further

ELECTRIC BELL

chemical analysis, and each one is represented by a different symbol. Only 92 are known, but these comprise 99 per cent of the earth's crust.

Elephant. A moulding machine with an overhead spindle. See *Drawing Paper.*

Elevation. The front view of an object. The façade of a building.

Elevator. A lift or hoist. Also machinery for raising material from a lower to a higher level. Buckets are used on an endless belt for plastic or granular materials.

Elicoust. Registered partition units that allow for speedy erection.

Elio. An alloy, similar to monel metal, used for sheathing under-water work, where the action of the sea-water is exceptionally severe.

Elitherm. A proprietary jointless flooring material specially suitable for re-surfacing old floors.

Elizabethan. The style of architecture prevailing during the reigns of Queen Elizabeth and James I. A transitional period between Gothic and Renaissance. It is best represented in domestic architecture, and in a number of colleges at Oxford and Cambridge.

Elk. A local term for ballast.

Ellicem. A proprietary cement paint.

Ellipse. A section across a cone not at right angles to the axis. The sum of the distances of any point on the circumference from the foci is equal to the major axis. Area $= \pi \times \frac{1}{2}$ major $\times \frac{1}{2}$ minor axes. See *Conic Sections.*

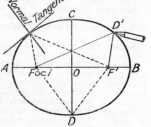

Ellipse of Stress. An ellipse from which the magnitude and direction of the resultant stress of two principal stresses may be found. It is used in structural design.

ELLIPSE

Ellipsoid. A solid of which every plane section is an ellipse. See *Spheroid.*

Elliptical Arch. One in which the intrados is a semi-ellipse. The extrados is drawn parallel to the intrados and is not a semi-ellipse. See *Arches.*

Elliptic Grate. A firegrate with hobs but no back.

Elm. *Ulmus campestris.* A tough flexible hardwood with coarse texture. Shrinks and warps freely. Very durable under water. Reddish brown colour with large knots. Many varieties and wide spread. Used for structural and constructional work, piling, weather boards, etc. Weight about 40 lb. per c. ft.

Elsan Chemical Closet. A patent conservancy closet in which the deodorant is a liquid reagent.

Embanking. Forming an embankment, which is usually intended to support a road at the top or to protect a cutting at the bottom.

Embankment Walls. Retaining walls with a surcharge. A wall supporting an embankment. See *Angle of Repose* and *Wing Wall.*

Embattlemented. An indented parapet. The indents, embrasures, or crenelles, were originally intended for the discharge of missiles, but later became a decorative feature only.

Embellishments. Decorative ornamentation.

Emboss. To carve or mould in relief. Carvings prominent over the face of the material. Raised ornamentation.

Embossed Paper. Wall paper with raised ornamentation, used for dadoes, etc.

Embrasure. 1. A narrow opening with interior splayed jambs. 2. The intervals, or crenelles, between merlons.

Emdecca. Patent imitation tiling on zinc sheets.

Emerald Green. A bluish-green pigment. It is poisonous and also used as an insecticide.

Emerald Pearl. See *Granites.*

E.M.F. Abbreviation for *electro-motive force.*

Emissivity. The radiating qualities of a source of heat.

Empirical Formulae. Formulae derived from experimental work and not from mathematical reasoning.

Emporium. A large store or mart.

Empress Slates. Sizes 26 in. × 16 in. Requires 79 per sq. at 3 in. lap.

Empty-cell Process. A term used in the preservation of wood in which the surplus antiseptic is withdrawn from the cells. The wood is impregnated, the cylinder, or tank, emptied, and a vacuum created in the cylinder. This removes the free preservative, but leaves the cell walls impregnated. Also called Open-cell Process.

Emulsion. See *Retarders,* and *Bitumen Emulsion.*

Emy's Roof. A system of roofing in which the main member is a laminated semicircle.

Enamel. An opaque glass-like surface applied to bricks, tiles, terra-cotta, etc. See *Enamels.*

Enamelled Bricks. Bricks faced with enamel, or opaque glaze.

Enamels. Good quality paint and varnish combined in the manufacture, and used to give a hard glossy surface. It is usual to first give the surface a coat of undercoating to suit the particular enamel. Also see *Cellulose Enamel.*

Encallowing. Removing the callow, overburden, or feu, from brick fields, before digging the clay.

Encarpa. A festoon of flowers or fruit on a capital or frieze.

Encase. To enclose with a case or linings.

Encastre. Beams fixed at the ends. Fixed beams. See *Fixed Ends.*

Encaustic Tiles. Ornamental tiles with the pattern formed by mixing different coloured clays and then burning. They may be obtained in various shapes and sizes.

Enciente Wall. An outer wall enclosing a temple.

Enclosures. The framing enclosing shop windows. It usually goes from window board to ceiling.

Endecagon. A polygon with eleven sides. Area = sq. of side × 9·366, for a regular figure.

End Fixing. A term used in the design of columns. The fixing of the ends influences the *buckling factor*. See *Pin Joints*.

End Matched. Applied to matched boarding that is tongued and grooved on the ends.

Endogens. Plants, such as palm trees, in which the new growth takes place on the inside of the stem.

Ends. Short lengths of battens, deals, etc.

Eng. A hard, heavy, strong, dark-brown hardwood from India and Burma. It is used for superior joinery, flooring, heavy constructional work, etc. It is difficult to season and work and stains with corroding iron. Weight about 50 lb. per c. ft.

Engaged Column. A column attached to, and partly concealed by, a wall.

Engineering Bricks. Strong, durable, vitrified bricks. The term is usually applied to Staff. blues and pressed reds, such as the Accrington bricks.

English Bond. The strongest bond for brick walls, consisting of courses of headers and courses of stretchers alternately. The bond is formed by a closer next to the quoin header. *English Cross Bond* is the same as English bond except for a header next to the quoin stretcher in alternate stretcher courses.

ENGLISH BOND

English White. See *Pigments*.

Enlarging Mouldings. Increasing the dimensions of a moulding but retaining the characteristics. The method is to select a number of points on the outline, project the points to a base line, and, by means of similar triangles, divide the required new base line in the same proportions, and then reverse the procedure.

Enneagon. A *nonagon, q.v.*

Enrichments. Carvings, or other embellishments.

Ensonit. A registered wall-board prepared for staining and polishing.

Ensowal. A registered wall-board.

Enstyle. The spacing of columns 2¼ diameters apart. See *Columns*.

Entablature. The horizontal members carried by the columns in classic architecture, consisting of architrave, frieze, and cornice. See *Mouldings*.

Entablement. A horizontal platform, above base and dado, to support a statue.

Entail. The elaborate and delicate parts of carved work.

Entasis. The gradual swelling towards the middle of the shaft of a column. The increased diameter increases the strength and corrects the illusion of being concave.

Entasis Reverse, or Rule. A templet cut to the outline of the entasis, and applied by the mason when shaping the stone.

Enterclose. A passage or corridor between two rooms.

Entresol. A low story between two main stories of a building. See *Mezzanine*.

Entry. A narrow passage between buildings.

E.O. Abbreviation for *extra only*.

Eonit. A proprietary concrete block with pumice aggregate. It has ow thermal conductivity.

Epergne. A central ornamentation with branches, especially for table decoration.

Epheta. A proprietary lime-incrustation remover.

Epinaos. The vestibule at the rear of the naos in a classic temple.

Epistyle or Epistylium. The architrave of an entablature.

Epitithydes The upper member of the cornice to an entablature

Equilateral. A triangle with three equal sides.

Equilateral Arch. A Gothic, or pointed, arch with radius equal to span. See *Arches*.

Equilateral Roof. A roof in which the span and the rafters form an equilateral triangle.

Equilibrium. A state of rest, produced by action and reaction, of a system of forces. A condition in which no change tends to take place. *Equilibrium moisture content* of wood is when the moisture content is in equilibrium with the humidity of the surrounding atmosphere, and the wood is stable. See *Moisture Content*.

Ercaline. A proprietary lacquer for metals.

Erector. One engaged in erecting steelwork.

Eremacausis. Slow combustion or oxidation. The gradual decay of timber exposed to the atmosphere.

Erg. The unit of work in the *centimetre-gramme-second* system.

Erinoid. A casein product used for the same purposes as Bakelite. Rennet is added to pure milk, raised in temperature, and the curd is collected and dried. It is ground, pigmented, filled, and forced through extruding machines to the required shape.

Erun. A chestnut-brown hardwood, with lighter streaks, from Nigeria. It is coarse and liable to pick up, and is very hard and heavy. Used for structural and constructional work, flooring, etc. Weight about 50 lb. per c. ft.

Escalade. A ladder fixed vertically to a wall.

Escalator. A continuous moving stairs. Used instead of lifts, or elevators.

Escalloped. See *Scalloped*.

Escape. 1. The apophyge between column and base. 2. An iron staircase on the outside of a building, only used in the event of fire.

Escarp. A steep slope or embankment.

Escoinson. See *Sconcheon*.

Escritoire. A writing desk with chest of drawers.

Escutcheon. 1. An ornamental plate to protect a keyhole. 2. An armorial shield. 3. A boss at the centre of a vaulted ceiling.

Espagnolette Bolt. A bolt that fastens in three places with one operation. It is fixed on the meeting stile of french windows, and secures the top, bottom, and middle.

Espalier. Trellis work on which plants are supported and trained.

Espavé. A variegated light-brown hardwood from C. America. It is somewhat like mahogany and is called Pesége Mahogany, but it is inferior. Used for carpentry and joinery, etc. Weight about 33 lb. per c. ft.

Esplanade. A terrace along the sea-side. Any level space used as a public walk or drive.

Essex Board. A proprietary wood-fibre wall-board.

Establishment Charges. Overhead expenses. Administrative and office expenses that are apportioned to a job and added to the prime cost to obtain the total cost.

Estaminet. A tap-room. A restaurant where smoking is allowed.

Estate. A piece of landed property under one ownership. A portion of land for the erection of a number of dwelling-houses. A building estate.

Estimate. An offer to complete certain work at an agreed cost. The usual methods adopted to prepare estimates are: 1. Bills of quantities. 2. Schedules. 3. Rough quantities. 4. Cubing.

Estrade. A low platform. An elevated part of the floor of a room.

Etching. Patterning or texturing the surface of glass with acid.

Eternit. Proprietary asbestos cement tiles.

Ethyl Silicate. Used in solution with alcohol as a preservative for stone. The process, after evaporation, deposits silica in the pores. Access of water into the joints, however, may cause exfoliation.

Etruded Metal. See *Extruded.*

Eucalyptus. The most important Australian timber trees. There are about 300 species including the largest known trees, except Sequoia. The exported woods are hard, heavy, strong, durable, and fire-resisting. The best known are Argento, Blackbutt, Bloodwood, Coolibah, Gimlet, Grey Box, Gum, Ironbark, Ironbox, Jarrah, Karri, Mallee, Mallet, Maiden, Merrit, Messmate, Mountain Ash, Peppermint, Red Mahogany, Santavere, Stringybark, Tallow Wood, Tasmanian Oak, Tuart, Wandoo, White Mahogany, Woolybutt, Yate, York Gum, and Yorrell.

Euler's Theory. A formula used in the design of long columns and struts, $P = \dfrac{\pi^2 E I}{l^2}$. P = breaking load, and ends pin-jointed. Multiply by 4 for fixed ends.

Eupatheoscope or **Dummy Man.** An instrument for registering changes of temperature.

Euphon. A proprietary glass quilt for insulation. The glass quilt is between specially treated Kraft paper.

Eureka. A patent valve to prevent back flow in sanitary appliances. Also the registered name for numerous proprietary articles.

Eustyle. The spacing of columns $2\frac{1}{4}$ diameters apart. See *Columns.*

Evaporation. A method sometimes adopted for disposing of sewage.

Everard. Patent building sheets to imitate brickwork or pebble-dash.

Everite. A registered name for various asbestos cement products.

Everseal. A proprietary bituminous waterproofer.

Evolute. A plane figure forming the eye of an *involute, q.v.*

Evos Door. A registered door and casings, in which everything is prepared before delivery. The door is hinged, lock fitted, and architraves mitred.

Ewery. Same as pantry or scullery, but reserved for bowls or ewers.

E.W.T. Abbreviation for *elsewhere taken.*

Ex. & ct. Abbreviation for *excavate and cart away.*

Excavating Plant. Mechanical equipment for excavating sites, trenches, etc.

Excelate. A registered name for natural slate fascias, pilasters, stall risers and other exterior decorative work.

Excelsior. 1. Frazzi terra-cotta wall-blocks with cavities, from $1\frac{1}{2}$ to $4\frac{1}{2}$ in. thick. They are light in weight, fire-resisting, and have good sound-proof properties. 2. Woodwool, *q.v.*

Exeau. A proprietary liquid cement waterproofer.

Exedra. Applied in ancient architecture to places appropriate, or set apart, for conversation. The apse of a church. A porch or vestibule.

Exfoliate. To scale off. To separate into scales or laminae.

Exhaust Shafts. Used in ventilation systems to carry away vitiated air. The smoke flues are often used for this purpose by means of mica flap outlets.

Exmet. 1. Expanded sheet metal made from about 22 B.W.G. and suitable for reinforcing brickwork. See *Diamond Mesh.* 2. A patent wall plug.

Exogens. Plants in which the new growth takes place in the cambium layer just under the bark. All *timber* trees belong to this class.

Expamet. Expanded metal for roads, and positions requiring great strength.

Expanded Metal. Steel reinforcement formed by cutting and expanding plain or ribbed sheets of steel. The term, however, is often applied to any form of mesh reinforcement, such as Expanded metal, Hyrib, Trussit, Jhilmil, Self-centring, B.R.C. fabric, Barmesh, Walker-Weston, etc.

Expanding Bit. An extension bit. A brace bit in which the cutters can be adjusted to bore any size of hole, from ½ in. to 1½ in., and from ⅞ in. to 5 in. diam.

Expanding Plug. A plug for stopping drain pipes when drain testing.

Expandite. 1. A proprietary tar-base paint for iron, steel, etc. 2. A patent caulking gun.

Expansion. An increase of volume due to an increase of temperature or moisture content. Nearly all building materials require consideration, when designing, against the effects of expansion and contraction. See *Coefficient of Expansion*.

Expansion Joints. Special joints for hot-water pipes to allow for expansion and contraction with change of temperature.

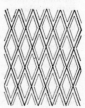
EXPANDED METAL

Expansion Pipe. An open pipe bent over the cold storage tank, for a hot-water system, to provide an outlet for air and steam, and to allow for the free expansion of the water. Also called vent pipe. See *Low-pressure System*.

Expansion Washer. See *Concealed Washer*.

Ex. sur. tr. and ct. Abbreviation for *excavate surface trenches, and cart away*.

Ext. Abbreviation for *external*.

Extenders. Same as fillers. See *Inert Fillers*.

Extensometer. An instrument for measuring minute increases of length.

External Wall. An outer wall of a building not being a party wall.

Extract Duct. A tube for extracting the vitiated air from rooms.

Extrados. The outside curve, or back, of an arch. See *Arch*.

Extruded Metal. A method of producing long hollow tubes, solid bars, hollow mouldings, etc., by forcing the metal, from a solid billet, through dies, or an orifice, of the required section. The pressure may be from 18 to 5000 tons, and the speed of extrusion from 10 to 1000 ft. per minute, according to the metal and the section. Copper, tin-base, and magnesium-base alloys are extruded hot, but zinc, tin, lead, and lead-tin alloys only require normal temperatures. The hollow bronze mouldings used for shop fronts and external fittings are about 12 B.W.G. in thickness.

Eye. 1. A general term for the centre of anything, as the eye of a volute. 2. A small staple to receive a hook for securing doors. 3. An inspection opening in a drain pipe. 4. A horizontal opening at the top of a dome.

Eyebrow Window. Applied to a window over which the eaves are raised in the form of a flat segment.

Eyelet. 1. The forged end of a rod punched or drilled to receive a bolt or rivet. 2. A small eye.

Eytelwein's Formula. A formula for calculating the velocity of sewage in drains. $V = 50 \sqrt{\dfrac{dH}{L + 50d}}$. Head, length and diameter in feet, and velocity in ft. per sec.

F

F. Abbreviation for *face* or *flat* and for *Fahrenheit*, also for *factor of safety*.

f. A symbol denoting the maximum intensity of stress allowed on a material. The kind of stress is shown by the addition of a small letter, i.e. f_b = bearing, f_c = compression, f_s = shear, and f_t = tension.

Fabric. The structural part of a building : walls, floors, and roof.

f.a.c. Abbreviation for *feet average cube*.

Façade. The front or face of a building.

Face Bedded. Applied to stone in which the natural bed is vertical and parallel to the face of the wall. The face of the stone is liable to flake away with weathering. See *Joint Bedded*.

Face Mix. A special mix of cement and stone dust for facing concrete blocks for imitation stone.

Face Mould. A developed templet or pattern for applying to the face of wood, stone, etc., when shaping the material. It is required for voussoirs, wreathed handrails, etc., and gives the outline to which the material must be cut. See *Moulds* and *Wreath*.

Face Side. The exposed face of the material. It is prepared first and used as a base of operations for further processes.

Facets, or Facettes. The projections, or fillets, between the flutings on a column or pilaster.

Facia. Same as fascia, *q.v.*

Facing. 1. An architrave. 2. Preparing the face of the material.

Facing Bricks. Selected bricks for the external face of a wall. They are selected for their weather-resisting properties and pleasing appearance, and are usually backed by common bricks. There is a large variety, both in colour and texture, and in quality.

Facings. 1. Special materials used for the face only of a wall. See *Walls*. 2. Finishings. 3. Thin wrought boards used to cover inferior or rough surfaces.

Facing Slips. Narrow sand-faced bricks, 2 in. instead of $4\frac{1}{2}$ in., for refacing old brickwork.

Factable. A coping.

Factor of Safety. Ultimate strength ÷ working stress. A number, by which the breaking load is divided, to give the safe load. For ordinary structural work the value varies between 3·5 and 10 according to the material and the conditions.

Factory. A building specially designed for manufacturing purposes.

Factory Acts. The various Acts controlling the erection and conducting of factories and workshops. Their chief function is to safeguard the workers. See *Acts*.

Factory Made. A term applied to mass-produced goods, prefabricated houses, etc.

Faggot. A bundle of pieces of iron, 120 lb. weight. A bundle of sticks.

Fagus sylvatica. See *Beech*.

Fahrenheit. A method of graduating thermometers, in which the freezing point is 32° and the boiling point 212°.

F.A.I. Abbreviation for *fresh air inlet*.

Faience. Architectural glazed ware. The term is usually applied to all glazed terra-cotta, but more especially to majolica and similar soft opaque glazes.

Fair Ends. The ends of projecting masonry that require to be worked to a finished surface.

Fair Cutting. Cutting brickwork to the finished face of the work. In estimating it is only allowed for $4\frac{1}{2}$ in. from the face, and is measured by the foot-run.

Faldstool. 1. A Litany stool in churches. 2. A folding chair.

Fall. 1. The inclination of a drain, gutter or any flat surface, to throw off the water. See *Drains*. 2. The free rope in pulley tackle. See *Mast*.

Fall Bar. 1. A bar for securing batten doors. It is pivoted at one end, controlled by a keeper, and rests in a stop. A finger hole replaces the handle and latch of a Norfolk, or thumb, latch, *q.v.* 2. A pivoted top bar of a kitchen range.

Falling Mould. A thin templet for marking the depth of a handrail wreath, etc., after it has been cut to the face mould. See *Wreath*.

Falling Stile. The stile of a gate opposite to the hanging stile. Especially applied to gates that rise at the nose as they open, due to the cocking of the hinges.

Fall Pipe. Rain-water down-pipes.

False Attic. A style of architecture, used to crown a building, resembling the "Attic Order" but without balustrade or pilasters.

False Ceiling. A second ceiling formed to leave space for pipes and wires, etc.

False Ellipse. An approximate ellipse formed of circular arcs. It is generally used for brick arches because of the convenience when preparing the voussoirs. See *Arches*.

False Header. A half brick, as used in some cases in Flemish bond.

False Tenon. An inserted tenon, *q.v.*

Falsework. Shuttering or centering for concrete.

Fan. 1. A sloping protection or screen to a scaffold. 2. A revolving wheel with vanes, used in artificial ventilation. 3. A metal plate used to spread the water for flushing a W.C. pan.

Fane. 1. A temple or church. 2. See *Vane*.

Fanlight. A sash over the transome in a door frame. Originally it was only applied to a semicircular sash with radiating bars.

Fanlight Opener. See *Quadrant*.

Fan Tracery. The elaborate ribs and veins used in fan vaulting.

Fan, or Conoidal, Vaulting. Consists of main ribs at equal angles to each other and of the same curvature, lying on the surface of an inverted concave cone. It was used in the Perpendicular Style and is peculiar to this country.

F.A.Q. Abbreviation for *fair average quality*.

Farad. Electro-magnetic unit of capacity.

Farleigh. A light cream Bath stone from Wiltshire. It is fine-grained and used only for carving, and interior work. Weight 120 lb. per c. ft. Crushing strength 62 tons per sq. ft.

FAN TRACERY VAULTING

Farotex. A proprietary, plastic, bituminous waterproofer.

Farrowmastic. A proprietary glazing compound for metal windows.

Fascia. A wide, flat, level board standing on one edge, as between the sash and cornice in a shop front. The term is applied to other materials used for similar purposes, for covering the structural members across a wide opening. Also see *Mouldings, Eaves fascia* and *Masonry Fixings*.

Fascine Buildings. Buildings formed of logs and boarded surfaces.

Fascines. Bundles of twigs about 20 ft. long by 10 in. diam. used as a foundation for roads in marshy ground, or as a protection to river banks.

Fasteners and Fastenings. Nails, screws, bolts, coach screws, dowels, holdfasts, plate dowels, spikes, staples, dogs, anchors, etc. Also the various arrangements for securing doors and windows. See *Sash Fastener*.

Fastigium. The ridge of a house or the apex of a pediment. A pointed summit.

Fast Sheet. See *Fixed Sash*.

Fat. That portion of a cement and sand mix having more than the ordinary proportion of cement, making it rich.

Fat Board. A piece of board used by the bricklayer for accumulating the fat when pointing.

Fathom. A timber measurement of stacked wood, of 216 c. ft.

Fatigue. The deterioration of the strength, or resistance, of a material due to continuous overstrain.

Fat Lime. Pure lime. See *Lime*.

Fatty. Plasterer's skimming stuff, having insufficient sand.

Fauces. A passage, in ancient Roman architecture.

Faucet. A socket to a pipe.

Faucet Ears. The projections to the socket of a pipe, for fixing to the wall by nails.

Fauld's Tools. A mason's chisel that allows for the interchanging of steel blades having double cutting edges.

Faulting. Break in continuity of strata due to earthquakes.

Fauteuil. The stalls of a theatre.

Fauton. A metal rod embedded in concrete.

FASTENINGS

Fawcett's Lintels. Hollow fireclay lintels for floors. They are placed diagonally between steel joists. The end lintels are split to complete the rectangular bays. The section is flat on the soffit and arched at the top, and it has dovetailed keys for a plastered ceiling.

Fayalite. Iron silicate causing black fused spots in bricks.

Faying. See *Snape*.

f.b.m. Abbreviation for *foot board measure*.

F.C. Abbreviation for *fair cutting*.

Fd. Abbreviation for *framed*.

Feather. 1. A thin strip of wood used as a tongue. 2. Ornamental figure in wood, especially mahogany, due to the confusion of the fibres at the junction between branch and trunk. 3. The pendulum slip between the weights of sliding sashes. Also see *Dovetail Feather*.

Feather Edged Boarding. Boards tapering, or bevelled, in thickness and used for weather boarding. The thick edge of one board overlaps the thin edge of the preceding board. See *Weather Boarding*.

Feather Edged Coping. A coping bevelled in one direction only, as used on parapet walls to throw the water into the gutter.

Feathered Washboard. A figured rolled glass, sandblasted and silvered for special lighting effects.

Featherings. The cusps in foliated tracery work. See *Tracery Panel*.

Feathers and Plug. Two thin pieces of iron between which the plug is driven for coping, or splitting, hard stone.

Fee. See *Callow*.

Feed Cistern. A storage cistern that supplies cold water to a hot-water apparatus. See *Low-pressure System*.

Feet-run. Estimating from the length only of certain items, in a building.

Felloe. The outer rim of the framework of a centre. It carries the lags.

Felly. See *Felloe*.

Felspar. A widely distributed mineral consisting of silica and alumina with other constituents such as soda, lime, potash, etc. Anhydrous silicates. It forms an important constituent of granite and other stones.

Felstone. A compact felspar.

Felt. 1. Silver grain; obtained by cutting timber along the medullary rays. 2. Fibrous material impregnated with some form of waterproofer usually bituminous. There are numerous kinds and they are used for many purposes such as roof sarking, damp-proof courses, etc. See *Inodorous Felt* and *Counter Battens.*

Felting Paper. Flour paper, *q.v.*

Femerell. A ventilator in a roof. See *Louvre.*

Femora. The spaces separating the glyphs in the triglyph.

Femur. The surface between two channels in a triglyph in the Doric order.

Fen. Low-lying marshy ground.

Fence. 1. A guide for a tool. See *Plough.* 2. A guard or protection.

Fencing. 1. The material for forming a fence. 2. Erecting a fence. 3. A fence or railing.

Fender. 1. A horizontal balk protecting the foot of a scaffold from road traffic. 2. A curb to a fireplace. 3. See *Camp Sheeting.*

Fender Piles. An outer layer of piles to protect work from moving bodies.

Fender Posts. Protecting posts to a refuge, or "safety island."

Fender Wall. A dwarf wall enclosing and supporting the hearth to a ground floor fireplace.

Fenestella. A niche on the south side of the altar in a church, containing the piscina.

Fenestral. A small window. Pertaining to windows, window blinds and shutters.

Fenestra Joint. A patent intersecting joint for bars in metal windows.

Fenestration. The arrangement of the windows to a building.

Feretory. A tomb or shrine.

Ferraris. A registered ceramic mosaic.

Ferreko. A proprietary oil-resisting paint.

Ferric-. Applied to compounds containing iron.

Ferro-concrete. Concrete reinforced with iron or steel.

Ferrocrete. A rapid hardening Portland cement.

Ferrodor. A proprietary anti-corrosive paint.

Ferrodur. A proprietary waterproofer for concrete.

Ferroform. A patent adjustable shuttering for concrete.

Ferro-glass. Translucent glass blocks assembled and reinforced by steel rods and concrete. It is a patented system and used for pavement lights, stallboards, roofs, etc.

Ferrolithic. A metallic preparation in powder form for making concrete floors waterproof, dustproof, and oilproof.

Ferrophalt. A proprietary reinforced asphalt.

Ferroput. A proprietary putty for glazing metal casements.

Ferrous-. Applied to compounds containing iron.

Ferrous Floor Hardener. A metallic powder sprinkled and trowelled into the cement, for hardening floors.

Ferruginous. Limestones containing iron. Usually brownish or reddish in colour.

Ferrule. 1. A brass band round a chisel to help to fix the tang of the blade and prevent the wood from splitting. 2. A thimble of brass or copper used for a joint between a lead pipe and one of some other material.

F.E.S. Abbreviation for *feather edged springer.*

Festoon. A carved ornament in the form of a garland suspended from the two ends. See page 120.

Fetcham Injector. A patent appliance for injecting preservatives in worm-infested timber in buildings.

Feu. See *Callow*.

Fiberlic. A proprietary wood-fibre wall-board.

Fibre Board. Wall-boards made from fibrous material such as wood pulp, cane, straw, bamboo, etc. There is a large variety and many different finishes for the surfaces. They are used for their insulating properties and as a substitute for plaster.

Fibreglass. A proprietary insulating material of spun glass. It is water-fire-, and vermin-proof.

Fibre Saturation Point. A term used in the seasoning of wood. It is the point at which the cells contain no free moisture but the cell walls are saturated. Further seasoning causes shrinkage. The moisture content, at this point, varies with different woods but it is usually about 30 per cent. The physical properties of the wood do not change with change of moisture content when it is above fibre saturation point.

Fibrotile. Corrugated asbestos-cement tiles, 4 ft. × 3 ft. 10 in., some what resembling Roman tiles.

Fibrous Concrete. Fibrous aggregates are used instead of, or in addition to, sand and gravel. Asbestos, wood-wool, straw, sawdust, etc., are used

FESTOON

to give the required characteristics to the concrete such as lightness, insulation, elasticity, or as a fixing for nails.

Fibrous Plaster. Plaster castings made of plaster reinforced with canvas and wood. They are cast in moulds prepared from *models*, which allows for repetition work. The castings are ready for fixing in position on the building.

Fid. A conical plug, or small thick wedge.

Fiddleback. Fine regular waves in maple and sycamore. It is so named because of its use in the manufacture of violins. Also applied to a similar grain in mahogany. See *Ripple*.

Fidler's Gear. Hoisting "tackle" having a tilting mechanism so that it lowers large blocks of stone at any required angle. It is especially useful for under-water work.

Field Book. A land-surveyor's book in which he records his measurements and observations.

Field Drains. Half round or round pipes made from brick clay or terracotta, without sockets, and used for draining fields.

Fielded Panel. A raised panel with a wide flat surface.

Field Rivets. Rivets inserted on site.

Field Work. Land surveying.

Figure. Ornamental grain in wood.

Figured Plate. Plate glass not polished, but having a specially prepared surface. See *Glass*.

Figured Rolled. A patterned translucent glass produced by an engraved roller.

Files. Described according to their shape, the nature of the abrading surface, or their use. The various kinds are : cotter, feather edge, float, half round, knife, parallel, pillar, rat-tail, riffler, round, safe edge, square, triangular, warding. The *cut* may be bastard, dead smooth, middle, rough, second cut, or smooth, according to the number of teeth per inch, which may be from 14 to over a 100. Rough files are single cut, smooth files are double cut, and the intermediate grades may be either. See *Rasp*.

Filler. A paste for filling the grain of timber before polishing. Plaster of Paris may be used; but specially prepared pastes may be obtained, stained to any required colour. Materials for levelling, or evening up, a surface before painting or polishing. See *Inert Fillers* and *Stopping*.

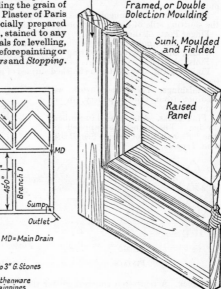

FIELD DRAINS

FIELDED PANELS

Filler Joists. Small rolled steel joists fixed between the main or secondary beams and carrying hollow blocks, etc., in fire-resisting floor construction.

Filler Slabs. Purpose made slabs, as for the underside of a terra-cotta coping to a balustrade between the balusters, when the blocks fit over a steel core.

Fillet. 1. A small flat moulding, rectangular in section. See *Mouldings*. 2. A triangular strip of cement to make an angle waterproof. 3. A strip of wood, small in section. 4. Narrow strips of glass used as margins.

Filling Station. A place for supplying motorists with petrol, etc.

Fillister. An adjustable rebate plane for woodworking.

Fillistered Joint. A rebated joint.

Filter. Any arrangement to purify water to make it fit for consumption. The two types for domestic purposes are high- and low-pressure types. The former requires a head of water but the latter is simply a container. The medium of filtration may be sand, charcoal, spongy iron, silicated carbon, etc. Two popular types are the Berkefeld and the Pasteur Chamberland. A device to cleanse air or liquids. Glass that intercepts the unwanted rays of light.

Fin. 1. The cement that runs between the joints of the formwork. 2. A thin projecting surface to give greater radiation, etc.

Final Set. See *Setting*.

Fine Set. The reverse of rank set, *q.v.*

Fine Stuff. The final, or setting, coat in plastering, consisting of plasterer's putty and fine sand.

Finger Plates. Ornamental glass, wood, plastic, or metal plates fixed on the shutting stile of a door to protect the paint.

Finger Post. A directional sign post.

Finial. An ornamental projection at the apex of a gable, spire, etc.

Fining Off. Applying the setting coat in plastering.

Finished Ground. One that is partly exposed and wrought.

Finishing Off. Preparing the finished surface of joinery work. See *Cleaning Up*.

Finishings. The joinery work, as distinct from the carpentry, in a building.

Fink Truss. A truss used in timber bridges. The term is also applied to the "French" roof truss. See *Roofs* and *Trusses*.

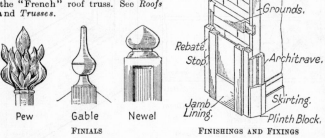

Pew	Gable	Newel

FINIALS   FINISHINGS AND FIXINGS

Finlock. A registered concrete gutter that incorporates with the lintol for the eaves of a building.

Fir. Applied to several species of coniferous trees, but more correctly to the genus *Abies*, or single-leaved conifer. The timber is called white deal or spruce, but the latter name implies the better qualities, as silver spruce (Sitka). It is nearly white in colour with yellowish markings and very strong for its weight. Used for inside constructional work and cheap joinery. Spruce is used for kitchen fittings because of its clean appearance. See *Whitewood, Douglas Fir, Spruce.*

Fireback Boiler. The boiler to a kitchen range, supplying the hot-water system. It may be of cast-iron, cast copper, or welded wrought iron. A boot boiler is angle-shaped with vertical and horizontal branches to expose a bigger surface to the fire. See *Saddle Boiler*.

Fire Bars. The bars carrying the fire in a grate or furnace.

Fire Barriers. The arrangements in the planning of a building to prevent the spread of fire. These are controlled by Factory Acts and Insurance regulations. Limited undivided floor space, enclosed stairways and lift shafts, fireproof communicating doors, are among the provisions.

Firebasket. A receptacle for the fuel for an open-hearth fireplace.

Fire Bricks. Bricks to resist high temperatures. They are made from (*a*) fireclays, (*b*) rocks of silica and clay (gannister), (*c*) rocks of nearly pure silica, (*d*) basic materials, as magnesia, alumina, chromite.

Fire Cement. Refractory cement to resist high temperatures. The Kestner types of refractory hydraulic cement may be used for temps. up to 1300° C.; but above 1300° high refractory cement (gannisters, chromite, silica) is used.

Fireclay. Refractory clay used for firebricks and for setting grates, boilers, etc.

Fire Cracks. Fine cracks in plastering due to unequal contraction of the different coats.

Fire Dogs. Iron supports for burning logs on an open hearth. Andirons.

Fire Doors. Solid wood, composite, asbestos between sheet metal, or metal, doors to resist fire. See *Doors.*

Fire Drenchers. Fire extinguishers operated by hand. See *Sprinklers.*

Fire Escape. Emergency stairs for use in case of fire. There are numerous regulations governing the type, position, and construction.

Fire Knees. The projecting part of a party wall above the roofs of dwelling houses, when in terraces.

Fire Polish. Glass blown or spun in intense heat acquires a more vitreous surface and brilliant lustre which is called fire polish.

Fireproof Aggregates. Used in fire-resisting concrete: brick waste, crushed firebricks, fused clinkers, calcined iron ore, pumice, slag, etc.

Fire-resisting Floors. Any form of floor conforming to the regulations for fire-resisting buildings. The types include: reinforced concrete; steel joists and concrete; steel joists and lintels (reinforced hollow blocks or beams, or arched construction); concrete with mesh reinforcement. The steelwork is arranged as main beams, secondary beams, and filler joists. See *Hollow Blocks.*

Fire-resisting Glass. Glazing that conforms to the regulations for fire-resisting construction. There is a large variety of patent reinforced glass, laminated glass, etc. Also see *Electro-copper Glazing.*

Fire-resisting Timbers. The following are recognized as fire-resisting by the London County Council: Acacia, Beech, Crabwood, Douglas Fir, Ash, Greenheart, Gurarea, Gurjun, Idigbo, Iroko, Jarrah, Karri, Keruing, Laurel, Meranti, Mora, Nigerian Walnut, Oak, Odoko, Okan, Padauk, Pyinkado, Secondi Mahogany, Silver Greywood, Sweet Chesnut, Sycamore, Tasmanian Myrtle, Teak, White Olivier, Yew. The list is periodically extended. Chemical treatment will make any wood fire-resisting. The chemicals crystallize in the pores, from which the air has been extracted, and in the event of fire some of them generate non-inflammable gases and so prevent the access of air; others fuse and glaze the cell walls. In all cases the object is to prevent the access of oxygen to support combustion. The chemicals used include ammonium chloride or phosphate or sulphate, borax, alum, and calcium chloride. There are several proprietary fire-resisting paints for surface application.

Fire Retardent. Surface coatings that make combustible materials fire-resisting. The specification BS/ARP 39 gives the necessary ingredients. See *Fire-Resisting Timbers.*

Fire Stop. Any structural arrangement, doors, etc., that cuts off one part of a building from another part in case of fire.

Fire Survey. A survey of a premises on behalf of the insurance company before issuing a policy, or to see that the original conditions of the insurance policy are being conformed to, both with regard to the building and the surrounding buildings. If the hazards have changed, the premiums are revised.

Fire Wall. See *Fire Stop.*

Fir Fixed. A term applied to unwrought timbers fixed by nails only.

Fir Framed. Applied to unplaned timbers with prepared joints at the intersections, as in roof trusses.

Firkin. A measure of volume of 9 gallons, or 1.44 c. ft., but it often implies any wooden vessel, of no fixed capacity.

Firola. A patent fireproof revolving door or shutter. The doors are controlled by a lever handle.

Firrings, or Furrings. Fillets used to level up the edges of joists or studs or the surface of a wall, to provide a plane surface for boarding, etc. A non-structural framework of wood or metal to provide a ground for decorative treatment. See *Hearth Details*. Also see *Cradling*.

First Fixings. The preparation of a building for the joinery work: grounds, plugs, etc.

First Floor. The floor immediately over the ground floor.

Firstings. The first *gauging* of plaster for plaster casts.

Fish. See *Vesica*.

Fish Bellied. Built-up girders varying in depth to conform to the variation in bending moment.

Fishing. Lengthening timbers with the assistance of fish plates.

Fisholow. A registered stainless steel sink.

Fish Plates. Plates of metal or wood placed across a lengthening joint and bolted to the two pieces.

Fisk Window. A patent sound-resisting window. A series of sashes are

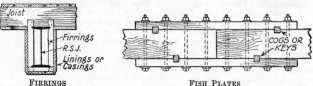

FIRRINGS   FISH PLATES

scientifically arranged so that whe n the window is open external noise are reflected outwards.

Fistuca. The monkey of a pile-driving machine.

Fitch. A small flat or round paint brush used on mouldings and small surfaces.

Fitments. The various kitchen labour-saving devices, cupboards, built-in furniture, etc., characteristic of modern domestic buildings.

Fittings. Wood or metal furniture additional to the ordinary finishings of a building, such as cupboards, shelves, etc. They are distinguished according to their particular uses, such as bank, church, office, and shop fittings.

Fit-ups. Large pieces of shuttering framed together for repetition work in concrete.

Five-centred Arch. An approximate elliptical arch consisting of circular arcs described from five centres. See *Arches*.

Fixatile. A Purimachos cement for fixing tiles to fireplaces, walls, etc.

Fixed Beam. A beam fixed at the ends. See *Encastré*.

Fixed Ends. The term applied to pillars and beams when the ends are adequately restrained in position and direction.

Fixed Sash. A fast or stand sheet. A window having no frame. Also applied to sashes permanently fixed in solid frames.

Fixing Block. A block of any material that will hold nails and screws built in a wall for securing joinery, etc.

Fixing Bricks. Bricks made of special materials to hold nails or screws, for fixing joinery work.

Fixing Fillet, or Pallet, or Pad. A thin slip of wood, 9 in. × 4½ in., built in the joint of the wall to serve as a fixing for joinery work.

Fixings. Grounds, plugs, etc., to which joinery is fixed. They are described as first, second, and third fixings, according to the progress of the plastering and other trades. See *Finishings*.

Fixton Pads. Hardwood pads for W.C.'s, used in place of wooden seats, or covers.

Fixture. Anything attached to a building. Landlord's fixtures are those fixed by the owner or a previous tenant. Tenant's fixtures become the property of the landlord if their removal will cause material damage to the property.

Flags or **Flagstones.** Stones used for pavements. They are usually of laminated limestone, or York stone, or reconstructed stone.

Flaired Head. A reinforced concrete column with the head shaped like an inverted cone or pyramid. Spread out like a fan or cone-shaped. See *Mushroom Construction*.

Flaking. The cracking and breaking away of paint, distemper, etc., due to unsuitable, badly prepared or dirty foundation. Common causes are water paint on glue size, flat oil paint on old oil paint without cutting down and washing thoroughly, finishing coat too hard or inelastic.

Flamboyant. A style of Gothic architecture so called because of the wavy, flame-like tracery work associated with it. Prevalent at the end of the fourteenth century.

Flange. 1. The upper and lower parts of an I-beam resisting compression and tension. 2. A projecting rim to a pipe, or a raised edge to the rim of a wheel.

Flank. 1. A local term for a roof valley. 2. The side of a building.

Flanking Windows. Windows placed at the side of an external door or as wings to the door. They are often used with an entrance door to light the hall

Flanning. The splaying of the jambs on the inside of an opening.

Flaps. 1. The leaves, or folds, of window shutters. 2. A part of a counter or desk top hinged to open. 3. See *Back Flap*.

Flap Trap. A hinged iron flap over the outlet of a drain.

Flap Valve. An air valve with mica flap as a fresh-air inlet, *q.v.*, to a drainage system.

Flare Kilns. Kilns for burning lime, in which the fuel is kept below the limestone so that only the flame affects the stone. This method is expensive in fuel and labour but produces a more uniform lime than the tunnel kiln.

Flashed Glass. Glass coloured on one side only by a thin layer of tinted glass. See *Glass*.

Flashed Opal. A blown two-ply diffusing glass. A thin face of white opal is fused to a coloured or clear glass base.

Flashing. 1. Burning bricks with and without air alternately, to give irregular colour. 2. Dull patches in glossy painted surfaces, due to faulty underwork, or turpentine in brush before the application of gloss paint.

Flashing Board. A board to carry flashings. See *Lear Board*.

Flashing Hooks. Wall hooks to carry

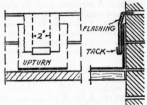

FLASHINGS

steel rods from which expanded metal is suspended for ceilings.

Flashings. Sheet lead, zinc, or copper used to prevent water from penetrating the joints where a vertical surface projects through a roof. See *Stepped Flashings*.

Flat. 1. A domestic suite of rooms on one floor of a building of more than one story.   2. A flatting coat of paint is often called a *flat coat*, and the surface a *flat surface*.

Flat Band. A square plain impost.

Flat Brush. The usual type of brush used for painting. They vary in width from 1 in. to 4½ in. and have replaced the knot brush, *q.v.*

Flat Cost. Wages of workmen and cost of materials for a job. Prime cost.

Flat Cut. Sawing wood through its thickness. See *Deep Cut*.

Flatels. Self-contained flats with service and restaurant. Flats with the amenities of an hotel.

Flat Joint. A mortar joint flush with the bricks. See *Pointing*.

Flat Joint Jointed. A flat mortar joint ironed with a bricklayers' jointer. See *Pointing*.

Flatlet. Small flats consisting of bed-sitting room, bath, W.C., and kitchen recess, or kitchenette.

Flat Roof. A level roof or one with an inclination of not more than 20° with the horizontal. The surface is usually of lead, zinc, or asphalt.

Flat Slab Construction. Reinforced concrete floors and columns without beams. Mushroom construction, *q.v.*

Flatting. Paint without oil to give a dull surface; turps or white spirit is used for both solvent and vehicle. A straight oil paint that dries with a dull finish.

Flatting Varnish. An oil varnish with a high proportion of resin. It cuts down easily with sandpaper and provides a smooth surface for the final coat.

Flaunching. Weathering the top of a chimney stack with cement mortar.

Fleaking. Thatching with reeds.

Fleche. A tall spire with a small base. It is often of wood in Gothic architecture. A small spire-like ventilator on the roof of a building.

Fleetweld. A registered name for an electrode used for welding.

Flemish. An obscured glass with high light-transmission value. It is described as either large or small Flemish.

Flemish Bond. Brickwork bond showing alternate headers and stretchers in each course. *Single Flemish* shows the bond on one face of the wall only. *Double Flemish* shows the bond on both sides (*q.v.*). *Flemish Garden Wall Bond* has a header to every three or four stretchers in each course.

Flettons. Bricks made by the semi-dry process from Peterborough shale. They are yellowish red colour. The clay contains nearly enough fuel to burn the bricks, hence they are cheap, and are the best known variety of pressed bricks for interior work.

Fleur. A separate piece to a ridge tile, for ornamentation.

Fleur-de-lis. A carved ornament in the form of a conventional lily, or iris.

Fleuron. Foliage as used in the abacus of the Corinthian capital.

Flewing. Same as flueing, *q.v.*

Flex. Flexible insulated wire for electric currents.

Flexine. A proprietary rubber mastic for attaching floor coverings to concrete.

Flexoply. A proprietary flexometal plywood used for shuttering, etc.

Flextol. A registered name for numerous types of electrically driven hand tools.

Flexure. A bend, as in an overloaded beam, or any curved outline.

Flexwood. Very thin veneer mounted on fabric. It may be obtained in many different woods in sheets up to 10 ft. by 2 ft., and is flexible and water-resisting.

Fliers. The steps in a straight flight of stairs. Parallel treads.

Flies. The wings to a stage in a theatre.

Flight. A series of fliers from one level to another. A complete unit of a stairs.

Flimsy. Anything that does not appear to be strong enough for its purpose. Frail or badly constructed.

Flinching. See *Snape*.

Flintkote. A proprietary bituminous waterproofer.

Flint Walling. Walls built of flints or split flints and bonded with tiles, bricks, or stones. The flints are very hard siliceous stones found in chalk beds.

Flitch. 1. The separate pieces of a flitched beam. 2. A case of veneers of about 500 sheets.

Flitched Beam. A rectangular beam built up of two or more pieces bolted together. When the beam is built up of three pieces the middle one is usually a wrought iron or steel plate.

Flitched Plate. An iron or steel plate placed vertically between two timbers to form a flitched beam. The plate is usually $\frac{1}{12}$th the thickness of the beam.

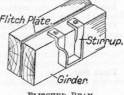

FLITCHED BEAM

Float. 1. To apply the second coat of plaster in "render, float and set" or "lath, plaster, float and set." 2. A form of wooden "trowel" used in plastering and cement work. 3. The boards or buckets, or vanes, on which the water impinges to turn a waterwheel. 4. The ball to a ball valve. 5. A single cut file. 6. A measure of timber of 18 loads.

Floating. The second coat in "three-coat" plastering. See *Float* (1).

Floating Bays Filling in a plastered surface between the screeds.

FLOAT

Floating Bricks. See *Rhenish Bricks*.

Floating Floor. A patent sound-resisting floor, in which rebated wood battens rest on rubber isolators, and carry loading slabs to prevent spring. The battens are boarded in the usual way.

Floating Foundation. A reinforced concrete raft designed to be monolithic in all directions and to carry a heavy load such as a building on poor ground.

Floating Rules. Long straightedges for levelling plaster and cement surfaces.

Float Stone. A stone for rubbing bricks, to clean or level them.

Flogging. Cleaning up floors by hand tools. Abrading machines are now generally used. Producing fleck in graining by a badger softener.

Flood Lights. Lights directed upwards to illuminate the façade of a building; or a system of lighting that gives an even intensity throughout the lighted area.

Floor Cramps. Any form of cramp used for squeezing up the joints of floor boards before nailing them to the joists.

Floor Finishings. May be wood, cement, asbestos, bitumen, rubber, glass, metal, brick, tiles, stone. There are also numerous patent and proprietary compounds. See *Flooring, Jointless Floors* and *Sorel Cement*.

Floor Guides. Any form of groove arranged on the floor to control sliding doors, partitions, etc.

Floor Hinge. See *Floor Spring*.

Flooring. Materials for forming the surface of floors. The factors for selection are cost, appearance, resistance to wear, and stability. Floor boards should be rift sawn. A large number of woods are suitable, including Redwood, Whitewood, Oak, Maple, Gurjun, Rhodesian Teak. See *Floor Finishings*.

Flooring Clips. Metal clips fixed in concrete floors to carry the battens for the floor boards.

Floor Joists. The timbers in a floor to which the floor boards are nailed.

Floor Line. A line marked on the feet of finishings, door posts, etc., to denote the floor level.

Floors. The horizontal divisions forming the storys of a building. They may be of timber, timber and steel, reinforced concrete, or patent fire-resisting floors, and they are named according to their position in the building, in an ascending order : basement, ground floor, first, second, etc. Timber floors may be single, double, or framed.

Floor Slab. That portion of a reinforced concrete slab and girder floor between the beams. See *Secondary Beam*.

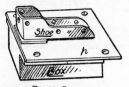

FLOOR SPRING

FLOOR STOP

Floor Springs. Springs housed into the floor and controlling the opening and shutting of swing doors. The plate *p* is level with the floor.

Floor Stop. See *Door Stop*.

Floor Strutting. Short timbers nailed between joists to stiffen and prevent them from canting or buckling. The strutting may be solid or herring-bone, *q.v.*

Floor Timbers. Beams, binders, and joists in timber floors.

Florentine Arch. An arch with a semicircular soffit and a pointed, or Gothic-shaped, extrados to increase the strength at the crown. See *Arches*.

Florentine Blind. An exterior blind forming a hood. It slides on iron side rods and reefs into a box and can be used for windows circular in plan.

Floriated. Having florid, or elaborate, ornamentation.

Florid. 1. A term applied to the Tudor style of architecture, fifteenth to sixteenth centuries, because of its profuse ornamentation. 2. Highly ornate.

Flour. The dust resulting from crushing stone, etc., for aggregate.

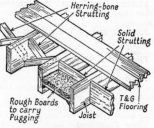

FLOOR STRUTTING

Fine sawdust, or wood dust, used with numerous compositions for floors, synthetic compounds, etc.

Flow Pipe. The pipe along which the hot water flows from the boiler to the cylinder in a hot-water system.

Fluate. A chemical preservative for stone, the basis of which is *fluo-silicate, q.v.* The silico-fluoride reacts with the calcium carbonate and forms a siliceous skin.

Flue. An enclosed passage for smoke, air, gases. Smoke flues may be lined with fireclay pipes or pargeted. The minimum size for brick smoke flues is 9 in. × 9 in., and they should be enclosed by not less than 4 in. walls.

Flue Gathering. See *Gathering* (2).

Flue Grouping. Collecting the flues to a stack before it penetrates the roof. Grouping increases the efficiency, because of the mutual warmth, decreases cost, and tends to better architectural treatment.

Flueing Soffit. A twisted flush soffit to geometrical stairs.

Flue Linings. Fireclay pipes 2 ft. long and square or circular in section. See *Pargeting* and *Flue.*

Fluid Pressure. The pressure of a liquid is directly proportional to the depth, is equal in all directions, and acts perpendicular to any surface. Hence the pressure at a depth of 10 ft. in water is $62\frac{1}{2}$ lb. × 10 ft. = 625 lb. per sq. ft.

Fluing. The splaying of window jambs. Same as flueing.

Flume. A conduit. A channel or wooden trough to convey water; especially applied to one conveying water to a mill wheel.

Flumes. The metal shutes to a placing plant.

Fluoric. A metallic fluo-silicate. It is mixed with water and applied to cement floors, when it combines with the free lime and forms fluor spar in the surface voids, making a hard, dustless floor.

Fluor Spar. Calcium fluoride, CaF_2. Derbyshire spar. A mineral found in crystals or in compact masses like marble. The crystals are very varied in colour and shade. It is used for decorative purposes, and as a flux in metallurgy.

Fluosilicate. Salts of hydrofluosilic acid, Na_2SiF_6, which are prepared by the action of gaseous silicon fluoride on solid fluorides.

Flush. 1. Surfaces in the same plane, or level. 2. To clear a drain by a full flow of water. 3. The crushing and breaking away of the edges of stone due to uneven pressure on the bed.

Flush Bolt. A bolt for securing a door, let into the material so that it is level with the surface.

Flush Doors. Solid or semi-solid doors, or framed doors with plywood faces. There are numerous types and they have superseded panel doors in nearly all types of buildings. The usual type consists of a deal framed core with plywood faces, and edging strips to hide and

FLUSH DOOR

secure the edges of the plywood. The success of these doors depends upon the bonding material, and synthetic resin is used instead of glue.

Flushed Joint. A spalled joint in masonry.

Flushing. 1. The fracturing, or spalling, of small pieces from the edges of stone. 2. Cleaning drains by a flush of water.

Flushing Cistern. A tank for supplying the water to flush drains, W.C. pans, etc. They may be automatic or non-automatic. The former are used for urinals and act at regular intervals, whilst the latter are operated by hand. See *Waste Preventer.*

Flushing Valve. A device for flushing a W.C. instead of a service box. See *Quantum.*

Flush Joint, Flushing up. See *Pointing.*

Flush Panel. A panel that is flush, or level, with the face of the framing.

Flush Soffit. The soffit of stone stairs when the underneath of the steps is worked to the pitch to form a plain surface. See *Spiral Stairs*.

Fluted Sheet. Blown sheet glass with rolled flutes for ray diffraction.

Flutes, or Flutings. Semicircular sinkings, or hollows, in columns and pilasters, etc. See *Mouldings*.

Flux. Any substance used to cause the flow, or fusion of metals or minerals, or that dissolves impurities during calcination. Limestone, clay, or sand, is used for smelting iron. Lime, potash, soda, common salt, act as fluxes with brick clay. Different metals and different conditions require different fluxes. The following are used in soldering: borax, resin, sal ammoniac, zinc chloride, Venice turpentine, tallow, silver sand, hydrochloric acid.

Fly. The space above and behind the proscenium of a theatre.

Flyaside. A patent sliding-door gear.

Flyer. See *Fliers*.

Flying Bond. See *Monk Bond*.

Flying Buttress. A detached buttress arched over at the top to engage with the wall.

Flying Levels. Levels consisting of backsights and foresights only, to arrive back at the starting point, when no bench marks are available.

Flying Scaffolds. A suspended scaffold. The scaffolding is suspended from an outrigger beam by means of winches, which allow the raising or lowering of the platform by wire ropes.

Flying Screed. An overhead screed cantilevered from the shuttering or mass concrete.

Flying Shores. Shores fixed between two walls without support from the ground. See page 131.

Fly-leaf. A movable leaf to a table.

Fly Nut. 1. A nut for securing the end of a pipe to a lavatory basin, etc. 2. A wing nut, *q.v.*

Fly Rail. A rail drawn out to support a leaf to a table.

Fly Wire. Fine woven wire mesh for covering the joints of wall-boards.

Foam Slag. Treated blast-furnace slag used for light-weight concrete.

FLYING BUTTRESSES

Focus. 1. A point at which a number of light or heat rays meet. 2. A directing point used in the conic sections. See *Ellipse* and *Parabola*.

Fodder. A weight, from 19½ to 22 cwt. according to the district, used for pig lead.

Foils. 1. The inside arcs of tracery work, which is named from the number of foils, i.e. trefoil, quatrefoil, cinquefoil, or multifoil. See *Tracery*. 2. Extremely thin metal.

Folded Flooring. Floor boards sprung into position, to tighten the joints, instead of cramping them.

Folding Doors. 1. Doors consisting of two or more leaves hinged together so that the door can open and fold in a confined space. 2. A pair of doors for a wide frame, with a rebated joint at the meeting stiles. They are also called "single doors hung folding," and are commonly used for entrance doors.

Folding Wedges. A pair of wedges driven in opposite directions and sliding on each other. They are used for increasing the distance between two bearing surfaces, as between a centre and its supports. See *Easing* (2) and *Shuttering*.

Fold Joint. A welt or seam, *q.v.*

Foliated. 1. Tracery work with foils. 2. Leaf-shaped ornamentation. 3. Laminated. 4. Coated with a thin plate.

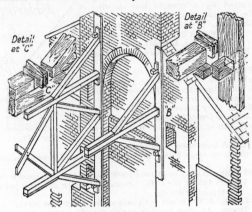

FLYING SHORES

Followers. 1. The slates following the eaves course in stone slating, *q.v.* 2. The bush for the spindle of a lock. 3. A machine part actuated by another part.

Fomerell. A dome or cover to a lantern.

Font. A stone vessel used in churches for holding water for baptisms.

Foot Blocks. See *Plinth Blocks*.

Foot Bolt. A heavy type of tower bolt placed vertically at the bottom of a door.

Foot Bridge. A bridge for foot traffic only.

Footing Beam A tie beam.

Footing Piece. See *Sole Piece*.

Footings. Offsets at the foot of a wall to provide a greater bearing area, where the superincumbent weight, per sq. ft., is greater than the bearing resistance, per sq. ft., of the soil.

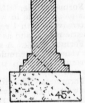

FOOTINGS AND BASE

Foot Irons. See *Step Irons*.

Foot Pace. A landing to a stairs, as quarter- or half-pace. It is more usual to say quarter- or half-*space* landing. See *Pace*.

Foot Plates. 1. Pieces connecting the feet of spars and ashler pieces. They are cogged on the wall plate which is bedded in the brickwork. 2. A hammer beam. See page 132.

Footprints. A pipe-fitter's tool for turning threaded pipes. Two bell-cranked levers are pivoted and the jaws are serrated so that they will grip the pipe securely.

Foot Stall. A plinth or base, or pedestal.

Footstone. The bottom supporting stone of a gable coping. See *Gable*.

f.o.r. Abbreviation for *free on rail*. Carriage charges to the railway station nearest to the destination are included in the price.

Force. That which moves, or tends to move, a body from its condition of equilibrium or state of motion. A force may be represented by a straight line showing magnitude, direction, and sense.

Force Diagram. A scale drawing representing forces, and used in graphical solutions to structural problems. See *Graphic Statics*.

Force Polygon. See *Graphics* and *Vector Polygon*.

Force Pump. One that forces the water to some height above the level of the pump. A pipe, with valve, branches from the barrel, and the piston has no valve.

Force Scale. Any convenient scale for the representation of forces in force diagrams.

Forecourt. An enclosed space in front of a building.

FOOT PLATE

Fore Sight. A term used in surveying when directing the level away from the starting point and towards a changing station.

Forest of Dean. A grey sandstone from Gloucestershire, used for general building. Weight 149 lb. per c. ft. Crushing strength 569 tons per sq. ft.

Foreyn. A cesspool or drain.

Forked Joints. Joints made with indentations like saw teeth.

Forked Tenon. An open mortise sitting astride a tenon not at the end of the material.

Forking Lath. A fillet nailed to a bearer, or forking piece, to receive the notched end of a joist, etc.

Forma. A channel for water. An aqueduct.

Formaldehyde. (HCHO). A colourless gas, soluble in water. It is used extensively with Urea or Phenol for plastic resins, and as an antiseptic.

Former Block. A shaped block to be veneered in circular work, as for a round end step to a stairs. See *Curtail Step*.

Formerets. The ribs lying next to the wall, in groined vaulting; usually half the thickness of the other ribs.

Formica. A widely-used proprietary plastic material. It may be obtained in various colours. A decorative wallboard. It has a paper or cotton fabric foundation for the plastic veneer.

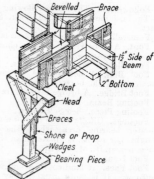

FORMWORK FOR CONCRETE

Form Liner. Material placed on the face of shuttering to produce texture on the face of the concrete.

Forms. Troughs or boxes in which concrete is cast. They may be of wood, metal, or plaster.

Formwork. The temporary timbering or metalwork used for the casting of concrete *in situ*. The correct design and erection of formwork has a great bearing on the cost and efficiency of concrete work. It is also called shuttering or falsework. Several patent kinds of metal shutterings are now extensively used. See *Shuttering*.

Forstner Bit. A patent brace bit for timber, for sinking holes that do not go through the material. It has no centre point and the sharpened circumference fixes the position.

Forum. An open space round which the buildings were grouped in Roman cities. It was used as a market place or place of assembly.

Fosalsil. Proprietary light-weight partition and casing blocks. They are made from diatomaceous earth and are strong and insulating.

Fosse. A moat.

f.o.t. Abbreviation for *free on truck*.

Foul Air Flues. Flues in a ventilating system for extracting vitiated air.

Foundation Bolts. Special bolts for fixing machinery, structural members etc., to concrete or stone foundations. The special types are Lewis, rag, and cottered bolts. See *Masonry Fixings*.

Foundation Forms. Formwork for the concrete for foundations.

Foundation Piers. Piers used instead of a continuous wall for foundations.

Foundations. Applied to the footings and concrete base on which a structure stands, or to the supporting soil. A structure below ground level to distribute the loads evenly and to prevent settlement. Foundations are designed to suit the bearing capacity of the supporting soil and the superincumbent load from the supported structure. They vary from a large reinforced concrete raft on piles to a simple footing course.

Founded. A term applied to a caisson when settled on its bed.

Founder. A person engaged in the casting of metals. See *Smith and Founder*.

Foundry Iron. Cast iron intended for making castings.

Found Stone. Pieces of stone roughly broken by the hammer at the quarry.

Fourcault. A process for drawn sheet glass. Viscous glass is drawn through a slit in a fire-clay block which floats on the molten surface of the glass. The glass is raised vertically between asbestos rollers and annealed in the process.

Four-centred Arch. A Tudor arch, or one struck from four centres. See *Arches*.

Four-leaved Flower. A carved ornament in hollow mouldings, and characteristic of the Decorated style of architecture.

417 Cement. A Portland cement with accelerated setting properties for use during frost.

Fox, or Fox-tailed, Wedges. Wedges in the end of a stub tenon that open out, or split, the end of the tenon in the form of a dovetail, when they are driven in the mortise. See *Stub Tenons*.

Foxiness. Incipient decay in hardwoods, accompanied by a reddish brown stain.

Foyer. A small hall or extended vestibule at the entrance to a public hall or place of amusement; usually includes payboxes, cloakrooms, etc.

fra. Abbreviation for *frame*.

Frame. Usually applied to timbers connected by mortise and tenon joints, and serving as an enclosure or a support. The essential structural elements of a building, etc., that support the remainder of the structure.

Frame Construction. See *Framed Structure*.

Framed and Braced. A framed and ledged door with the addition of diagonal braces to prevent the nose of the door from dropping.

Framed and Ledged. A door consisting of stiles, top rail, ledges, and battens. The ledges are thinner than the stiles, to receive the battens, or boards.

Framed Floor. A timber floor consisting of beams, binders, and common joists. It is so called because the binders are usually framed into the beams.

Framed Grounds. Grounds prepared with stiles and rails and framed together as one piece.

Framed Linings. Jamb and soffit linings for interior doorways, consisting of stiles and rails in the form of three skeleton frames. Thin boards are placed on the face to cover the spaces between the stiles and rails and to act as stops for the door. See *Panelled Linings.*

Framed Partition. Partitions in which members are framed together and braced, making a structural unit supported by the walls only.

Framed Roofs. See *Trussed Roofs* and *Roofs.*

Framed Square. Applied to panelled framing without mouldings round the panels.

Elevation.

SHOWING TYPICAL ARRANGEMENT OF FRAMES IN FRAMED STRUCTURES.

Stringer Beams or Girders.

Piers or Stanchion

Piers or Stanchions

Girders or Beams

Stringer Beams or Girders

Plan.

FRAME CONSTRUCTION

Framed Structure. A building in which the main structural parts are of steel, secured together and braced to form a rigid frame, to carry walls, floors, and roof. A triangulated structure that can resist distortion, as for a roof truss.

Frame Saw. Saws for wood or stone in which a number of blades are secured in a frame which moves in a horizontal or in a vertical direction, making several parallel cuts in one operation.

Fram Floor. A patent fire-resisting floor consisting of hollow fireclay blocks with steel reinforcement.

Framing. 1. The act of constructing a frame. 2. The skeleton or structural parts of a frame.

Framing Timber. Wood suitable for the carcassing of buildings.

Franki. The registered name of a piling system for large structures.

Franking. A joint used in sash stuff in which a spur, or projection, on the mortised member fits in a recess in the tenoned member. It is the reverse to haunching, and does not weaken the mortised member so much.

Frazzi Blocks and Slabs. Patent hollow terra-cotta self-centering blocks, of several types, used in fire-resisting construction for floors, partitions, casings, and roofs.

frd. Abbreviation for *framed.*

Freehold. Land and property over which the owner has full control, subject to the Crown. See *Leasehold.*

Free Moisture. Moisture in the cell cavities of wood, to distinguish from that in the cell walls. See *Moisture Content.*

Freestone. Fine grained and easily cut sandstone or limestone, of close texture.

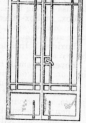

FRENCH CASEMENT (STEEL)

French Arch. See *Dutch Arch.*

French Casement. A large sash hinged and used as door.

French Curve. A drawing appliance of varying curvature for lining or inking in freehand curves.

French Doors. A pair of large glazed sashes in the form of folding doors.

French Fliers. Fliers round a rectangular well.

Frenchman. 1. A sawing stool with a V-shaped top. 2. A knife used by the bricklayer in conjunction with a pointing rule for straightening the edges of mortar joints when pointing.

French Polishing. Finishing the surface of timber with a coat of shellac dissolved in methylated spirits. The pores are usually stopped with a filler and the wood is stained, if necessary, to the required colour. Also see *Fuming* and *Wax Polishing.*

French Spindle. A woodworking moulding machine in which the spindle is slotted to receive the cutters. The machine is specially suitable for circular work of small radius, as the stuff may run on the bare spindle and the cutters are perfectly balanced.

French Stuc. Plaster surfaces imitating stone.

French Tiles. Interlocking tiles of the Belgian *Courtrai* and French *Marseilles* types. Both are deep red in colour and weigh from 6 to 7 cwt. per square. The former requires 210 and the latter 125 per sq.

French, or Belgium, Truss. A steel roof truss for spans up to 60 ft. Various devices are used for overcoming the difficulty at the joint *A*, when obtaining the stresses in the members : "substitution of members," "method of moments," or "method of sections." See *Roofs.*

Fresco. 1. Painting with water-colours on freshly finished plastered surfaces so that the colours incorporate with the plaster and become durable even when exposed to the weather. 2. Distemper in powder form.

Fresh-air Inlet. An inlet to provide a current of fresh air through a drainage system to dispel the foul gases. It is usually placed at the intercepting chamber, and the outlet vent carried above the eaves.

Frestex. A proprietary plastic paint.

FRESH-AIR INLET

Fret. 1. A carved ornamentation having projecting fillets in rectangular formation. A continuous rectangular pattern. 2. The cover before the ash-pan to a register grate. It is usually provided with slots for the draught.

Fretted Lead. Strips of lead of H-section used for leaded lights and stained glass. See *Cames.*

Friary. A monastery for an order of friars.

Friction. The force tending to prevent one surface from sliding on another. This is controlled by the normal reaction between the surfaces and the nature (condition and kind) of the surfaces.

Friction Piles. Piles secured by the friction on the sides of the pile only.

Frieze. 1. A flat or pulvinated member between the architrave and cornice in an entablature. See *Mouldings.* 2. The top part of a wall, between the ceiling cornice and the picture rail.

Frieze Panel. The top panel in a door of five or more panels.

Frieze Rail. The rail immediately below the frieze panel.

Frig-bob Saw. A long hand-saw used in the Bath stone mines.

Frigidarium. The cold bath apartment in an ancient Roman thermae, *q.v.*

Frit. 1. The prepared material used in the glazing of bricks. Smelted enamel glass is quenched in water so that it disintegrates into small pieces ready for grinding before application. 2. Calcined sand and fluxes for glass-making.

Frithstool. A seat near the altar in a church.

Frodingham. A patent steel sheet piling.

Frog. The indentation in the bedding surface of a brick, to reduce the weight and provide a key for the mortar. See *Terra-cotta*.

Frontage Line. A line prescribed by a local authority, beyond which a building must not project.

Front Hearth. The part of the hearth in front of the fireplace jambs.

Fronton. An ornament to an entrance door consisting of consoles, cornice, and pediment.

Frost. Frost seriously affects the setting of limes and cements. Work should be suspended during frost; and protected with straw or sacking if frost is expected before the mortar has set. If it is necessary to proceed with the work, chemicals, such as calcium or sodium chlorides, are mixed with cement, or unslaked lime is used, or rapid-hardening cement. The last two generate sufficient heat to withstand the action of the frost until the mortar has set.

Frost Line. The depth to which the ground may be frozen in an exceptionally cold spell. Foundations for heavy structures should go below the frost line in many soils. The depth in the London area is about 3 ft.

Frowy. A term applied to soft, brittle timber.

Frustum. A solid cut by a plane parallel to the base, i.e. a cone frustum.

F.S. Abbreviation for *factor of safety* and for *flat sweep*.

Fulcrum. The support about which a lever rotates.

Fullering. Caulking riveted joints.

Fumerell. See *Femerell*.

Fuming. Darkening the surface of oak by enclosing the timber in an air-tight chamber and exposing it to the fumes of ammonia, NH_3. It is observed through glass until the required shade is obtained, and afterwards coated with french or wax polish.

FRUSTUM OF PYRAMID

Fungicide. Any preparation that destroys fungi.

Fungus. The mushroom type of plant life that produces mould, mildew, etc. There is a great variety that attack wood and trees, and most of them are destructive. In some cases they increase the decorative value, and in some cases they simply stain the wood. See *Preservation* and *Dry Rot*.

Funicular, or Link, Polygon. The closed polygon, on the linear diagram, the sides of which are drawn parallel to the vectors to the force, or polar, diagram, and which gives the position of the resultant for a system of forces.

Funnels. Small conical vessels used by the plasterer when pouring gelatine for moulds in fibrous plaster work.

Furn. Abbreviation for *furniture*.

Furness. A patent roof glazing.

Furniture. See *Door Furniture*. The term is also applied to fittings of a loose or semi-loose character, and is described according to its special function as school, bank, or church furniture.

Furred. Applied to pipes and boilers that are coated with deposits from the water.

Furrings. See *Firrings*.

Furrowed. Ashlar quoins with draughted margins and projecting face which is furrowed with vertical parallel flutes or grooves, sunk to the level of the margin.

Fuse. 1. To make metal molten. 2. A wire on a porcelain insulator that "blows," or melts, when the safe electric current is exceeded.

Fuse Board. A board carrying fuses for a number of circuits in an electrical installation.

Fust. The ridge of a roof or the shaft of a column.

Fustic. A decorative hardwood from the West Indies, yellow to green in colour. It is very hard, heavy, strong, and durable, and used for superior joinery. Weight about 52 lb. per c. ft.

Fyberstone. A proprietary waterproof wall-board.

G

G. Abbreviation for *gulley*. A symbol for *shear modulus*. **g.** A symbol for *radius of gyration*.

Gab. A pointed tool used when working hard stone.

Gabers, or **Gabbard, Scaffold.** A local name for a scaffold built of square timbers bolted together.

Gabion. A wicker or woven metal cylinder filled with earth.

Gable. The portion of the end wall of a building from the eaves to the ridge. It follows the outline of the roof.

Gable Board. See *Barge Board*.

Gable Post. A short piece to carry the top ends of barge boards.

Gable Roof. A roof open to the rafters and finishing against a gable. See *Roofs*.

GABLE

Gable Shoulder. The springing at the footstone for a gable coping.

Gable Springer. See *Footstone*.

Gable Tiles. Special tiles for the junction between gable and roof. Angle tiles.

Gable Window. A window in a gable ; or one shaped like a gable.

Gablet. A small gable, or a decoration shaped like a gable, often over a niche or small opening.

Gaboon Mahogany. A W. African timber used for plywood, laminated wood, and as a core for veneers. See *Mahogany*.

Gad. An iron wedge for splitting stone.

Gadget. Any small contrivance or fitting to a machine.

Gadroon. A decoration used in a form of nulling on friezes, turnery, etc. It resembles repetitive beads with quirks or large reeds.

Gain. 1. A notch. 2. The lap in lapped timbers. 3. The bevelling of a shoulder.

Gaines. Light metal tubes to enclose the wires in prestressed concrete.

Gaining. An alternative term for *notching*.

Galilee. A small balcony or gallery in a church. Also a porch to a church especially at the west end of a Gothic church.

Gallery. 1. A long platform projecting from the walls of a building and supported by cantilevers or columns. 2. A long narrow apartment. 3. A three-armed support for a globe over a gas burner. 4. A room used for the display of works of art.

Galleting. Inserting small pieces of tile in the hollows of roll-shaped tiles to level up the seating for the ridge tiles or hip tiles. See *Garreting*.

Gallows Bracket. A framed wooden bracket capable of supporting a load at the outer end.

Galmins. Abbreviation for gallons per minute.

Galton Method. A *direct-indirect* method of heating and ventilating. The fresh air is passed through channels behind the fireplace.

Galvanite. Anti-corrosive paint with zinc base in electrolytic fluid. It is an undercoat for metal windows, etc.

Galvanizing. Coating iron with zinc to prevent corrosion.

Galvd. Abbreviation for *galvanized*.

Gamboge. A yellow pigment from Siamese gum resin.

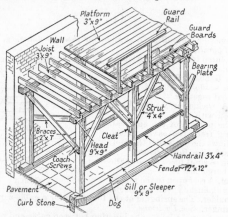

GANTRY OVER PAVEMENT

Gambrel, or Gambril, Roof. A roof hipped for part of its height and terminated by a small gable.

Gambrel Vent. A triangular louvre ventilator.

Gangau. See *Ironwood*.

Gangboarding. Rough boarding used for a gangway; especially applied to one in a roof.

Ganger. A foreman over a number, or gang, of navvies.

Gangmill. A machine in which several reciprocating saws, or gang saws, cut at the same time.

Gang Mortiser. One that cuts a number of mortises in one operation.

Gang Mould. Formwork for casting a number of concrete units at one operation.

Gang Riveting. A process in which a number of rivets can be headed in one operation.

Gangue. Earthy impurities in iron ore.

Ganguillet Formula. Used for calculating the velocity of fluids in drains.

Gangway. 1. A footway formed of rough planks for the passage of workmen. 2. A passage between rows of seats. 3. A plank bridging a gap temporarily.

Ganister. Made from dense siliceous rock containing clay or from a mixture of ground quartz and fireclay. It is a highly refractory material and used for lining furnaces and for fire-bricks.

Gantry. A double staging built of large square timbers or built-up steel members and carrying a travelling crane.

Gantry Girders. Built-up steel girders supported by stanchions, and braced for a gantry.

Garage. A building to house motor-cars. The chief requirements are : fire-resisting, cleaning pit, drainage system, convenience for cleaning and repairs to cars, electric light.

Garchey Sink. A patent sink for household refuse which is emptied by vacuum suction to a central incinerator. It is used for large blocks of flats.

Garden Wall Bond. A modified form of some other brick bond, especially English and Flemish. The bond is used in 9 in. walls faced both sides, and the number of headers is reduced to a minimum because of the difficulty of obtaining flush faces.

Gargoyle. A spout, of grotesque design, to a parapet gutter, to throw the water away from the walls.

Garnet Hinge. See *Cross Garnet*.

Garnet Paper. See *Sandpaper* and *Abrasives*.

Garret. The top rooms of a building immediately under the roof.

Garreting. Inserting small pieces of stone in the joints of rubble and flint walls. See *Galleting*.

Garron. A large wrought nail with flat point and rose head.

Gaselier. An ornamental pendant, for several lights, in gasfitting.

Gas Heating. Applied to buildings, rooms, water, etc, that are heated by gas, by means of boilers, fires, geysers, etc.

Gasket. The hemp packing used in pipe joints. For large pipes the jointing medium, between the flanges, to make the pipes oil- or gas-tight is usually a sheet of thick asbestos between thin sheets of copper

Gaskin. See *Gasket*.

Gas Threads. The standard screw threads used on water, steam, and gas pipes. There are two series used in this country : Whitworth and British Standard Pipe.

Gate. A framework of metal or timber, opening or closing the entrance to an enclosure. Also applied to large batten doors. Gates are named according to their particular use : carriage, crossing, dock, entrance, farm, field, garden, lock, lych, sluice, warehouse, wicket, yard.

Gate Leg. A hinged leg for a falling leaf to a table.

Gate Piers. The hanging and shutting piers for a gate ; usually of brick, stone, or concrete. They are called *gate posts* when made of timber.

Gate Step. An iron block carrying the pivot at the foot of the stile of an iron gate.

Gate Valve. See *Peet's Valve*.

Gatherer. See *Glass*.

Gathering. 1. The patchy appearance in distemper due to irregular absorption. 2. Altering the direction, or contracting the opening, of a flue, by *corbelling over* the bricks. The flue should not have an inclination of less than 45° unless a soot-door is provided.

Gauge. 1. A standard method of denoting the thickness of sheet metal and wire. 2. The distance apart of the bottom edges, or tails, of consecutive rows of slates or tiles. See *Eaves*. 3. A tool for marking parallel lines on timber, or a tool to measure the thickness of sheet metal and wire. 4. To measure the quantities of the

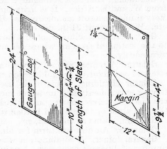

GAUGE FOR SLATES

components of a plastic mix. 5. The distance apart of parallel rails. 6. Dial, or clock, gauges are used to measure pressure, height, etc. 7. See *Preservation*.

Gauged Brickwork. Brickwork with fine joints and in which the bricks have been cut and rubbed to the required shape and size. See *Arches*.

Gauged Mortar. Cement-lime mortar.

Gauged Stuff. Plastering material to which plaster of Paris has been added to hasten the setting and to counteract shrinkage. It is also called "putty and plaster," but any lime mix may be gauged.

Gauge Piles. See *Guide Piles*.

Gauge Pot. A tin can used for pouring cement grout.

Gauge Rod. A rod for setting out the courses in graduated slating.

Gauging. Proportioning plastic materials. Also see *Gauge* (3).

Gauging Board. A platform for mixing the materials for concrete.

Gauging Box. A box used for measuring the correct cubical quantity of dry material for a plastic mix.

Gaul. A defect in plastered surfaces due to faulty trowelling when scouring the setting coat.

Gault. A stiff strata of clay between the layers of Green sand in chalk formations. Used for Gault bricks. See *Suffolk Whites*.

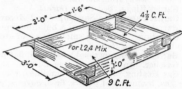

GAUGING BOX (COMPOUND)

Gauze. Perforated metal or woven wire, used for ventilating screens and sieves.

Gavel. See *Gable*.

Gazebo. A summer-house built in the form of a temple, and having a wide prospect.

g.b. Abbreviation for *gas barrel*.

G-cramp. See *Cramp*.

Gd. A. Abbreviation for *gauged arch*.

Gear. Mechanical appliances for controlling moving bodies, as lifts, etc. A train of toothed wheels. Also applied to the equipment or tackle required for a job.

Gedunoha. Budonga Mahogany. A reddish brown, lustrous, hardwood, from W. and E. Africa. It is light and soft, but fairly durable and strong. Very similar to a soft mahogany, *q.v.* Used for superior joinery, shop-fitting, etc. Weight about 31 lb. per c. ft.

Gel. A colloidal solution set to a hard jelly. The term is specially applied to the setting of cement, where gel and crystal formation take place simultaneously.

Gelatin. A refined animal or vegetable glue. It is used for transparent size and as a fixative for gold leaf in water gilding.

Gelatin Moulding. Gelatin is used for moulds for plaster castings having enrichments and undercut mouldings to facilitate the removal of the casting. Glycerine or gum arabic is sometimes added to retard setting. Chrome alum or tannic acid harden it but prevent remelting.

Gelation. Same as curdling, *q.v.*

Gemmels. An obsolete term for hinges.

General Joiner. A combination woodworking machine that performs a number of different processes. A universal woodworker.

Generating Plant. A building equipped for generating electricity.

Geoidal Slope. The gravitational movement of moving bodies, more especially applied to winds.

Geometrical Stairs. Stairs having a continuous string round a semi-circular or elliptical well.

Georgian Glass. Fire-resisting glass reinforced with electrically welded square mesh.

Gerwood. A registered wallboard and hardboard.

Gesso. A plaster surface prepared as a ground for painted decoration.

Getters. Men using pickaxes, or other tools, for loosening the earth in excavations.

Geyser. Any apparatus for the quick heating of water, having water control on the inlet side and a free outlet.

G.G. A trade mark for plate glass suitable for *good glazing*, but not for silvering.

Ghaut. Stairs to a river-side landing stage.

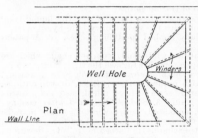

GEOMETRICAL STAIRS (STONE)

GIN WHEEL

Giam. A dark brown hardwood from Malay. It is extremely hard, heavy, strong, and durable. Used where durability and strength are of major importance. Weight about 65 lb. per c. ft.

Gib Door. Same as jib door, *q.v.*

Giblet Check. An external rebate. A rebate for a door opening outwards.

Gibs. Metal clamps to prevent a strap from opening, when driving the cotters, and to provide a seating for the cotters. See *Cottered Joint.*

Gidgee. A decorative hardwood from Australia.

Gig Stick. A *radius rod* to a horsed-up mould, for running circular plaster work.

Gilding. Painted surfaces finished with oil gold-size and then coated with gold leaf. The surface is prepared with a fixative which may be of slow drying oil, a quick drying varnish, or a water soluble jelly, for oil-, japan-, or water-gilding respectively.

Gilflex. A flexible non-metallic conduit for electric wiring.

Gills. Flat plates to a stove to give a greater radiating surface.

Gilsonite. Used as a binder for bitumastic materials, or muckite, *q.v.*

Gimlet. A small screw-pointed boring tool for use by hand.

Gin. A tripod of three poles carrying a gin wheel for hoisting purposes.

Ginnel. A narrow passage.

Gin Wheel. A large single pulley for hoisting materials. It has no mechanical advantage but only serves to change the direction of the effort.

Girder. A large beam of wood or iron forming the main horizontal member of a floor, bridge, etc. The term is specially applied to built-up steel beams.

Girder Casings. Any material used for enclosing the part of a girder projecting below ceiling level. The casing may be concrete, plaster, fireclay, or terra-cotta blocks, or timber. See *Cradling*.

Girdling. A term applied to the operation of removing a strip of cambium layer round the circumference of a tree to prevent growth.

Girt, or Girth. The perimeter of a cross-section. The distance round.

Girts. The steelwork at the sides and ends of a steel frame building. Rails, between head and sill, in a framed wood building. Where they meet on a corner post one is *raised* a little and one *dropped* a little, for better fixing in the post.

Give. 1. The term implies that a structural unit will fail when loaded. 2. Yield or bend without breaking.

gl. Abbreviation for *glass* and *glazing*.

Glacis. A sloping bank.

Glacitex. A registered type of tracing cloth.

Glaisher's Tables. Hygrometrical tables for obtaining the humidity of the atmosphere in conjunction with the wet and dry bulb hygrometer.

Gland. 1. A collar. Specially applied to stuffing-boxes. 2. A patent joint for copper tubes.

Glass. The chief constituents are white sand, soda, and chalk, which are mixed and melted at high temperatures in a furnace 100 ft. × 20 ft. × 5 ft., gas-heated from above. The various processes are drawing, rolling, pressing, blowing. *Types.* 1. Glass with natural surface: sheet, drawn, antique, and crown. 2. Surfaces prepared before cooling: rolled, cast, and pressed; they may be patterned or plain. 3. Surfaces prepared after cooling; ground, polished, enamelled, embossed, and decorated. Glass may be transparent, translucent or obscured, or opaque. *Varieties.* Ambetti; Arctic, or anti-actinic (heat excluding); Cathedral rolled; dewdrop; Fadosan (transparent, but artificial colour protector); Flemish; fluted; glistre; hammered; kaleidoscope; Morocco; muffled; muranese; prismatic; rough cast; safety (wired, or in plies with inside celluloid ply); slab; stippolyte; Tintopal; ultra-violet ray; Vividek; wavene; etc. *Sheet glass* is graded from 11 oz. to 42 oz. per sq. ft., and is produced mechanically. It may be *blown* and flattened in a kiln, but it is usually *drawn*. The Fourcault process is now improved in the Libbey-Owen and Pittsburgh processes. The molten glass is *gathered* by a *bait* through a fire-clay float, and drawn over asbestos rollers through a *lehr* 250 ft. long. The glass may be up to 160 in. by 100 in. *Plate glass* is cast from pots, annealed, ground, and polished. *Crown glass* is prepared by rotating the blow-pipe so that the glass forms a circular disc with a thick central portion called a bull's eye, or bullion. Usual size of square is up to 12 in. × 10 in., and it is used for its antique appearance. *Ground glass* has its surface ground, or powdered glass fused into the surface. *Embossed glass* has the design formed by the action of hydrofluoric acid. *Enamelled* has the design formed by fusing powdered glass into the surface. *Toughened Plate.* Ordinary plate glass heated and suddenly chilled which makes the surfaces in compression and the inner part in tension, hence if the glass is broken it falls into powder. *Coloured glass* is made by adding metallic oxides when melting, or it may be "flashed" by fusing coloured powdered glass on the surface. See *Safety Glass*.

Glass Beads. See *Glazing Beads*.

Glass-crete. See *Ferro-glass*.

Glassed. Marble or granite brought to a smooth finish by means of a felt-covered glossing disc on a disc-polishing machine.

Glass-house. A greenhouse.

Glass-lined Pipes. Iron pipes lined on the inside to resist the corrosive and chemical action of water and other liquids.

Glasspaper. Finely ground glass sprinkled on glued paper and used for finishing the surface of timber. Waterproof glasspaper is used for outside work, cleaning paints, etc. Usual strengths for joinery work: 0; 1; 1½; fine, middle, and strong 2. See *Abrasives, Sandpaper* and *Flourpaper*.

Glass Silk. Molten glass forced through small-bore nozzles and whisked to a rotating drum to give fine flexible fibres, built up to the required thickness of mattress. It is used for heat and sound insulation, Thermolux, etc. It will resist 1000° F., and is immune from acid or insect attack.

Glass Slate. A pane of glass bonded with the slates of a roof.

Glaze. 1. A transparent glass-like coating, or slip, applied to bricks, tiles, terra-cotta, etc. Various materials, chiefly china clay, and colours are used, and fired at a high temperature. 2. Colours used in graining.

Glazed Ware. Stoneware pipes and fittings for drains, etc., glazed by throwing sodium chloride (common salt) into the kiln when burning.

Glazement. A proprietary material, with Portland cement base, for facing brickwork or concrete. It gives a tile-like, jointless, waterproof surface and is in various colours.

Glazeraise. A patent process for producing a tool-raised polished surface on concrete.

Glazier. One who cuts and fixes panes of glass.

Glazing. 1. The cutting and fixing of panes of glass. The glass is usually bedded in putty or wash leather, and is fixed by putty, beads, grooves, clips, cames, or by patent glazing. There are many varieties of patent glazing The usual type consists of a steel bar protected against corrosion, an asbestos cord as a seating for the glass, a condensation groove, and two flanges of lead, or sheaths, attached to the steel bar and which are turned over the glass when fixing the panes. 2. Giving a glassy surface to clayey materials. See *Glaze*.

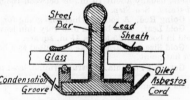

ROOF GLAZING BAR

Glazing Beads. Loose beads, mitred and nailed round rebates to secure panes of glass. See *Bronze Sheathing*.

Glazing Groove. A groove for glass, instead of a rebate. See *Bar Tracery*

Glebe. Land belonging to the Church.

Gliksten. A proprietary wood-fibre wall-board.

Glistre. A decorative obscured glass with sand-blasted or velvet finish. It is made as large or small glistre.

Globe Cock. A tap with globe-like outlet as used for baths.

Gloia Cement. A highly-hydraulic lime.

Gloss Paint. Varnish incorporated in the paint during manufacture.

Glue. An adhesive material made from gelatinous substances, bones, hoofs, hides, and fish refuse. It is made in cakes, varying in size up to 6 in. × 6 in. × ¾ in., and sold as Scotch, English, or French. *Liquid glue* does not require heating above 60° F., and is uniform in quality. *Cold water glue* is in powder form and made from dextrines or caseins. The latter is a milk product and sets like cement. It is waterproof, fireproof, and resists moulds and bacteria, but it is only good for 24 hours after mixing and should be used the same day. *Marine* glue consists of india-rubber, naphtha, and shellac. *Albumen* glues, or cements, are from egg and blood sources and are waterproof. *Film glue* is made from synthetic resin and is in the form of cellophane-like sheets. The film is placed between the wood laminations and the whole is heated in a press which fuses the glue.

Resin glues are synthetic and are superseding other glues where the necessary equipment can be installed. They may be in two parts: a paste-like glue and a liquid hardener. Chemical action takes place when the two are mixed, and the glue sets hard. Different hardeners require different times for setting. They may be obtained in powder form consisting of resin and casein, which must be mixed just before use. Phenol-formaldehyde glues require heat and pressure. Urea-formaldehyde glues can be cold-press or hot-press. See *Plastic Glues*. There are numerous proprietary synthetic glues on the market, but many require expensive equipment in the way of presses and machines. Veneering of circular work is done in a vacuum bag, or "hot-water bottle," etc.

Glyphs. The vertical channels in the triglyph of the entablature in the Doric order, *q.v.*

Gneiss. Stone having similar constituents to granite (quartz, felspar, mica), but stratified owing to the different arrangements of the particles. It can be split along the seams, or planes, of foliation.

g.o. Abbreviation for *get out.*

Go. See *Going.*

Gobbets. Blocks of stone.

G.O.C. A proprietary, tile-red, plastic, waterproofer.

Godfather. A wood or concrete support to a decayed fence post.

Godroon. An ornament in the form of a bead or cable.

Going. The horizontal distance of a tread between two consecutive risers, usually about 10 in., or between the first and last risers in a flight of stairs. See *Step.*

Going Rod. A rod on which the going for a flight of stairs is set out.

Gold Leaf. Gold beaten into very thin leaves and used for gilding. It is sold in books of 25 leaves 4½ in. square.

Gold Size. A varnish-like drier used as a base for gold leaf or as a binder for flat paint.

Gordon's Formula. A strut formula, later modified by Rankine. The formula is sometimes used for brick piers: $\dfrac{P}{A} = \dfrac{f_c}{1 + a\left(\dfrac{l}{d}\right)^2}$. Where $P =$ total load, $A =$ sect. area, $f_c =$ safe crushing strength of short pier, $a = \frac{1}{500}$, $d =$ least thickness, $l =$ height.

Gore. A wedge-shaped piece of material, as the lunes for covering a dome.

Gorge. Same as cavetto, *q.v.*

Gothic. A style of architecture distinguished by the pointed arch. The term is applied to Early English, Decorated, and Perpendicular styles. Thirteenth to sixteenth centuries. See *Mouldings.*

Gothic Arch. Pointed arches: equilateral, drop, and lancet. See *Arches.*

Gothite. A patent retarded hemi-hydrate plaster.

Goufing. Strengthening the foundations of walls.

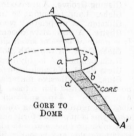

GORE TO DOME

Gouge. A form of chisel with curved cutting edge. There is a big range of sizes and curvatures and they may be inside- or outside-ground. A bent gouge is curved in its length. One with a V-shaped cutting edge is called a parting tool.

Goux Pail. A pail for a conservancy closet, having an absorbent lining of dry refuse. They are relined after emptying.

Grab. A mechanical appliance for picking up and loading loose material from excavations.

Gradient. A slope or inclination. It is stated in terms of 1 unit of vertical height to the number of horizontal units, i.e. 1 ft. rise and 40 ft. level distance = *a gradient of* 1 *in* 40.

Gradin. 1. One of a series of low steps. 2. A tier of seats. 3. A ledge at the back of an altar.

Grading. Arranging in order of size or quality. Specially applicable to the aggregate for concrete, to ensure that all voids are filled when mixed. Most timbers are subject to grading rules for quality, and a few for their strength values. See *Shipping Marks*.

Gradiograph. An instrument for testing the fall in drains. It is placed inside the pipe to ensure the correct gradient.

Graduated Courses. A term used in slating where the gauge of the courses is diminished as they ascend the roof.

Graffito. Ornamentation scratched through plaster applied in layers of different colours.

Grafting Tool. A narrow rectangular spade for digging clay and heavy soils.

Grain. The arrangement of the fibres, or wood elements, in timber. The following terms are used to describe the variations and are now standardised: bastard, chipped, coarse, comb, cross, curly, diagonal, edge, end, even, feather, fine, flat, gnarly, interlocked, interwoven, mottled, narrow-ringed, oblique, plain, quarter, quilted, ribbon, rift, roey, short, silver, slash, spiral, straight, torn, twisted, uneven, vertical, wavey, wide-ringed.

Graining. Painting surfaces to represent the grain of timber. Oak and pitchpine graining is usually done in oil colours, and other timbers are usually represented in water-colours.

Granary. A silo, or storehouse, for threshed grain.

Grandstand. An elevated structure on a race-course or sports ground from which the spectators can obtain a good view.

Grange. A farming establishment. A large house with farm attached.

Granger's. A proprietary waterproofing solution. The firm has several different waterproofing and preservative emulsions.

Granite. Igneous unstratified rock consisting of quartz, felspar, and mica. The variation in the felspar and the mica and the presence of small quantities of other minerals govern the quality and colouring of the granite. *Scotch granites*: Rubislaw, Peterhead, Cairns, Creetown, Sclattie, Dalbeathe, etc. *English*: Colcerrow, De Lank, Lamorna, Penryn, Shap, etc. *Irish*: Castlewellan, Dalkey, Newry, etc. *Swedish*: Bon-accord, Victoria and Deeside grey, Carnation. *Norwegian*: Royal Blue, Emerald Pearl, Standard Grey. Sp.G. 2·5 to 2·8. Granites are very strong and durable, polish well, but offer poor resistance to fire.

Granite Plaster. Quick- and hard-setting plaster with a gypsum base, for interior work.

Granitic Finish. Cement work having a face mix to represent granite.

Granolithic. Floor surfaces consisting of cement, coarse sand, and fine chippings with the addition of a liquid hardener. Carborundum, emery, etc., are often included to make the surface non-slippery.

Granolithic Plates. Iron plates used to facilitate the placing of the face mix when casting concrete in moulds for reconstructed stone.

Graphs. Plotting a series of points, representing two sets of related variable quantities, and joining the points together in one continuous line, forms a graph. The points are plotted on squared paper relative to two axes $x - x$ and $y - y$. Their intersection is the origin o. The distances, or quantities, measured from $x - x$ are *ordinates*, whilst those from $y - y$ are *abscissae*. See *Plotting*.

Graphic Statics. A method of solving problems dealing with forces, moments, stresses, etc., by means of scale drawings, instead of, or in addition to, mathematical calculations. It is based on the triangle, parallelogram, polygon of forces, and the link polygon, with the application of Bow's notation. See *Statics*.

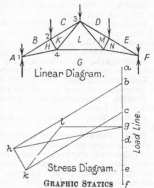

Linear Diagram.

Stress Diagram.

GRAPHIC STATICS

Graphite. A form of carbon. Also called black-lead or plumbago. Used as a lubricant, "oildag" or "aquadag," and in paint. See *Pigments*.

Graphite Paint. Used to protect ironwork against corrosion. It is much more durable than ordinary paint and is unaffected by acids and alkalis.

Grappler. The eye, fixed in the wall, for a bracket scaffold.

Grasshopper Gauge. A joiner's pencil gauge for gauging over obstructions and for double curvature work.

Grass Table. The top of the plinth to Gothic buildings.

Grate. The cast-iron frame and fire bars for a fireplace. There is a great variety, including many patented forms of grates.

Grating. 1. A timber "grillage" for foundations on poor soil. 2. A wood lattice for keeping large gutters clear of snow. 3. A framework of iron bars forming a protection to an opening. See *Grille*. 4. A perforated disc over the outlet to a sink.

Gravel Plank. A board placed on edge along the bottom of a wooden fence to keep the ends of the palings above the ground.

Graving Docks. See *Docks*.

Gravity. The tendency of a body to be attracted towards the centre of the earth. The term is also used when proportioning chemicals in solution for plastic materials, the gravity, or specific gravity, of the solution depending upon conditions and requirements. Also see *Centre of Gravity*.

Gravity System. See *Low-pressure System*. The circulation depends upon gravity, as the density of the water in the flow pipe (hotter water) is less than that in the return pipe (cooler water).

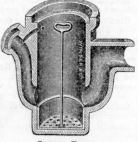

GREASE TRAP

Grease Trap. Large gullies, or traps, to prevent grease from entering and clogging the drains. They are fitted with trays so that the congealed grease may be removed easily.

Grecian. Characteristic of the architecture prevailing in ancient Greece See *Orders*.

Grecian Fret. See *Fret* and *A la Greque*.

Greek Alphabet. Used as symbols in structural and scientific calculations.

Greek Mouldings. See *Mouldings.*

Greek Orders. See *Orders.*

Green. A term applied to unseasoned timber, or to plastic materials before they are set.

Greenheart. *Nectandra rodioei.* Very strong, durable, and resists insect attack. Dark yellowish green in colour. From Guiana. Excellent timber for underwater work. Used where strength and durability are of primary importance : marine structural work, brewer's stagings, piling, etc. Weight about 62 lb per c. ft.

Letter	Name	Letter	Name
α	alpha	λ	lambda
β	beta	μ	mu
γ	gamma	ξ	xi
δ	delta	π	pi
Δ	Delta	Σ	Sigma
ε	epsilon	φ	phi
η	eta	ψ	psi
θ	theta	ω	omega
κ	kappa	Ω	Omega

SYMBOLS FROM GREEK ALPHABET

Greenhouse. A building consisting of glass and, usually, wood bars except for about 3 ft. from the ground. It is used for the cultivation of plants.

Green Pigments. Brunswick Green (Prussian blue and chrome, reduced with barytes). Zinc green (zinc chrome and Chinese blue).

Green Room. A rest room for artists in a theatre or concert hall.

Greenstone. See *Diorite.*

Greenwood Airvac. The registered name for several patent ventilators.

Grevak. A patent deep-seal anti-siphon trap.

Grey Stocks. Sound bricks but irregular in colour. See *London Stocks.*

Greystone Lime. See *Lime.*

Greywood. See *Chuglum.*

Gribble. Marine insects very destructive to wood.

Grid. 1. Timber framework built above low-water level to receive vessels floated into position at high tide for repairs. 2. Perforated metal or a framework of iron to prevent the passage of solid matter, or for a ventilator. 3. A room, with open floor, over a theatre stage for controlling the scenery and lighting. 4. The *plan lay-out* in steel frame construction.

Grid Formation. The arrangement of the steelwork in a steel-framed building so that the members cross each other in rectangular formation in plan and elevation, but modified to suit the site. See *Frame Construction.*

Grill. 1. A room in a restaurant where meat is grilled and served. Usually applied to a room where meals are served *à la carte.* 2. See *Grillage.* 3. See *Wicket Gate* and *Grille.*

Grillage. A foundation consisting of two or more layers of rolled steel joists (B.S.Bs.), embedded in concrete. The layers of joists, or grills, are at right angles to each other. Sometimes heavy timbers are used in the same way, instead of steel joists.

Grille. A metal open-work guard, or grating, to protect windows or openings. They are of wrought iron, cast-iron, or bronze, and form a decorative feature.

Grinning. A defect in distempered surfaces on old wallpaper due to the colours showing through the distemper.

Grip. 1. A long vertical bar serving as a handle to a swing door. 2. A temporary channel for conveying water. 3. The distance between the heads of a rivet.

Grip-a-block. A proprietary bituminous adhesive for fixing wood blocks to floors by cold-process application.

Gripfast. A patent fixing plate for wall-boards.

Grip Length. The distance that a bar is embedded in concrete for anchorage. For tension it is 40 × diam. and for compression 25 × diam.

Grippon. A proprietary sealer and mastic, for fixing glass, Vitrolite, etc., to cement or hard plaster surfaces.

Grips. Small trenches to convey rain-water away from foundations during erection.

Grit Stone. Sand stone used for grindstones.

Grizzles. Stocks chilled in burning. They are grey in colour and not durable for exterior work. Also called *samel*, or *place*, bricks.

Grog. Ground bricks or waste from clay works. It is mixed with clay in brickmaking.

Groin. The salient angle formed at the intersection of two vaults crossing each other. See *Vaulting*.

Groined Roof. A vaulted ceiling having groins.

Groove. A long shallow recess or furrow. It may be vee, rectangular, or semicircular in shape. See *Tongue and Groove*.

Gross Features. A term used in the identification of woods. It includes the features that can be distinguished without laboratory help: colour, odour, weight, taste, grain, rays, resin, etc.

Grotesque. 1. Unnatural ornamentation. Caricatures of human and animal forms. 2. Artificial grotto, or rock work.

Ground Air. Carbon dioxide and other gases expelled from the earth. The accumulation of these gases in confined spaces assists decay in floor timbers. For this reason, and for health, building sites are covered with a layer of concrete.

Grounded Work. Joinery work fixed to grounds.

Ground Floor. The floor of a building near to ground level.

Ground Glass. See *Glass*.

Ground Line. 1. The intersection of the picture plane with the ground plane in perspective drawing. The xy line in orthographic projection. 2. A line marked on the feet of framing denoting ground level.

GROOVES

Ground Mould. A mould for the invert of a tunnel or drain.

Ground Plan. A plan of the first story of a building.

Ground Plane. 1. Horizontal plane. 2. The level plane on which the object is supposed to be resting in a perspective drawing.

Ground Plate. The lowest horizontal member of a timber framed building, and into which the vertical members are framed.

Grounds. The timber fixings, usually unwrought, plugged to the wall to receive joinery. If they are required over 6 in. wide they should be framed. They are usually fixed flush with the floating coat of plaster. See *Finishings*. 2. Backings, or cores, for veneer.

Ground Table. See *Earth-table*.

Ground Water. The water held in the sub-soil.

Grout. Cement mortar in a fluid state for filling crevices, or for forming a key for joining together new and old cement work.

Grouting. Filling the crevices of brick and stonework with grout.

Groutnicks. Special grooves formed in masonry joints for pouring in grout.

Groynes. A low wall of concrete, stone, piles, or timber, projecting into flowing water, or the sea, to retain the shingle and to prevent scour.

Grub Saw. A small saw used with sand and water for hard stone.

Grub Screw. A set screw having a small head with slot for screwdriver.

Grummet or **Grummit.** A mixture of hemp and red lead putty, to make joints water-tight.

Grypterete. Patent fibrous plaster ceiling slabs, used for coil, or panel heating.

Gtg. Abbreviation for *grating*.

Guambo. A dark yellowish-brown hardwood from C. America. Fairly hard and heavy, durable, with rather coarse texture but smooth and easily wrought. Useful for general purposes. Weight about 32 lb. per c. ft.

Guanacaste. A variegated brown hardwood from C. America. Variable in weight and hardness. Better qualities resemble walnut and polish well. Useful for interior fittings. Weight from 24 to 36 lb. per c. ft.

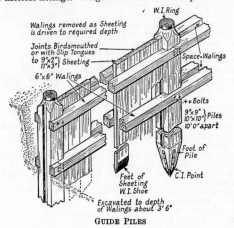

GUIDE PILES

Guard. 1. An appliance on a machine to protect the machinist from the cutters or saw. 2. A protection to an opening or on a scaffold.

Guard Bead. See *Inner Bead* and *Sash and Frame.*

Guard Boards. See *Gantry.*

Guard Rail. A handrail to a landing or balcony. A rail that protects from machinery, falling from scaffold, etc.

Guatacre. A light-brown hardwood from Brazil. Very hard, heavy, durable, and strong. Used for structural work, piling, etc. Weight about 64 lb. per c. ft.

Gudgeon. See *Band and Gudgeon.*

Guichet. An enquiry opening to an office. A small shutter, grating, door, window or opening.

Guide Bead. See *Inner Bead.*

Guide Piles. Large piles at regular intervals apart and connected by horizontal runners, or walings. The latter guide the sheet piles used to fill the spaces between the guide piles.

Guijo. A brownish-red hardwood from the Philippines. It is hard, heavy, and strong but not durable. Polishes well and not difficult to work. Used for superior joinery, flooring, etc. Weight about 48 lb. per c. ft.

Guilloche. Two intertwining bands carved as an ornamentation on classic mouldings.

GUILLOCHE

Guillotine. See *Trimming Machine*.

Gullet. The root, or hollow, of a machine saw tooth.

Gully or **Gulley.** A channel for conveying water.

Gully Grating. The cast-iron perforated top to a gully trap.

Gully, or **Yard, Trap.** A glazed earthenware receptacle to trap
the drain where rain and waste water is
collected before entering the drain. The
circular reversible type is the best.

Gum. 1. The name given to a number of
species of Eucalyptus, *q.v.* They are chiefly
excellent woods, very hard, heavy, and
durable and used for most purposes, and
especially where durability and strength are
of primary importance. Figured woods are
used for superior joinery, etc. Blue Gum
is probably the best. Weight varies from
52 to 64 lb. per c. ft.

Gum Animé. See *Copal*.

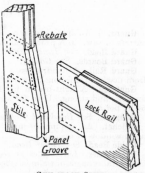

GULLY TRAP

Gumhar or **Gamari.** A brownish-yellow
variegated hardwood from India and Burma. Fairly light and soft, but
durable. Like some softer mahoganies. Used for superior joinery, etc.
Weight about 36 lb. per c. ft.

Gun. 1. An instrument for spraying cement. The ingredients are
hydrated as they emerge from the nozzle. Another type is used for spraying
paint. See *Cement Gun*, and *Spraying*. 2. A bricklayer's templet for the
skewback of an arch.

Gunite. A method of constructing walls and finishing wall surfaces by
means of a gun. Steel wire mesh forms
the tensile reinforcement for the walls.

Gun Metal. A bronze alloy, copper
and tin. Used extensively for fittings,
door furniture, etc.

Gun Stock. See *Diminished Stile*.

Gun Template. A triangular template
built up of three laths, and used for
describing circular arcs with inaccessible
centre of curvature.

Gunter's Chain. A chain used for
measuring distances when surveying.
It is 66 ft. long and divided into 100
links of 7·92 in. each. Varied shaped
brass tags are attached at every tenth
link to facilitate reading. See *Chain*.

Gurjun or **Gurgan.** A reddish-brown
hardwood from India, Burma, and
Andaman Islands. It is fairly hard,
heavy, and strong with handsome
grain. It is used for flooring, superior
joinery, etc. Weight about 45 lb. per c. ft.

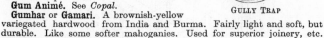

GUN-STOCK STILE

Gusset. Steel packing pieces, or plates, to which the various members,
meeting at a joint, are riveted or bolted, as in steel girders, trusses, etc.

Guttae. Drop-like ornaments under the triglyph and mutule in the Doric
entablature. See *Doric Order*.

Gutter. A channel for conveying water, at the sides of a road or on a
roof. The types used on roofs are: box, eaves, parapet, saddleback, secret,
tapering, trough, valley, and V. They may be of wood, asphalt, lead, zinc.
copper, or iron.

Gutter Bearers. Short pieces of about 2 × 2 in. carrying the boards which in turn carry the lead for a roof gutter.

Gutter Bed. A sheet of lead passing over the tilting fillet and behind the eaves gutter to prevent any overflow from penetrating the wall.

Gutter Boards. 1. See *Snow Boards*. 2. See *Gutter Bearers*.

Gutter Bolts. Small bolts used in C.I. gutters for bolting the spigot and faucet ends together. The joint is bedded in red lead to make it watertight.

Gutter Plate. 1. A beam under a lead gutter. 2. The sides of a box gutter that carries the feet of the rafters.

Guy. A rope stay to a mast or shear legs. See *Mast*.

Guy Derrick. A crane with an upright mast secured by guys.

Gymnasium. A building used for physical exercises and gymnastics.

Gymnosperms. Trees bearing naked fruit, or seeds. Conifers, or softwoods.

Gyn. Same as shear-legs, *q.v.*

Gypklith. A proprietary wood-wool building slab.

Tapering Gutter

Box Gutter

GUTTERS

Gyproc. A trade name for numerous building products : plaster boards, partitions, ceilings, etc.

Gypsum. A compound of sulphate of lime and water, calcined for plaster of Paris. It is greyish white with pearly lustre. Hydrous calcium sulphate.

Gypsy. A proprietary patching plaster for repairs to walls, ceilings, etc. It may be used as a filler for woodwork.

Gyration. See *Radius of Gyration*.

H

Habitacle. A niche for a statue.

Hachures. See *Hatchings*.

Hacienda. A S. American farm or farmhouse.

Hack. To roughen a wall to form a key for cement or plaster. See *Hacks*.

Hack Caps. Small roof-shaped wooden structures to cover the hacks.

Hacking. 1. A course of masonry rubble walling in which part is of single stones and part with two stones in the height. Each course is a mixture of *coursed rubble* and *rubble built to courses*. 2. Removing old putty from the rebates of sashes, to replace the glass. 3. See *Hack*. 4. Cutting material carelessly.

Hacks. Stacks of bricks arranged for drying after moulding. The hacks are allowed to stand for a few weeks before burning.

Hacksaw. A saw for cutting metal. It is shaped somewhat like a coping saw or fretsaw, with an iron frame.

Hadang. An olive-brown hardwood, with darker streaks, from India. It is lustrous, smooth, hard, durable, strong. Rather difficult to work. Suitable for superior joinery. Weight about 42 lb. per c. ft.

Hade. The angle with the vertical that a fault plane makes in strata, due to faulting, *q.v.*

Haematite. Iron ore imported to improve the quality of the metal from British ores.

Haffit. 1. The fixed part of a cover, or lid, to which the opening part is hinged. 2. An end to a church seat, or pew.

Haft. The shaft, or handle, of striking tools.

Hagioscope. See *Squint.*

Ha-Ha. A sunken fence.

Hailes. Two varieties of sandstone, pink and white, from Edinburgh. Both are used for general building purposes. Weights 142 and 144 lb. per c. ft. Crushing strengths 511 and 523 tons per sq. ft.

Hair. Used in coarse stuff for plastering, about 1 lb. to 3 cub. ft. It is usually long and clean cow hair, and is beaten to separate it. Ropemaker's waste is used as a cheap substitute.

Hake. A *hack* for drying tiles.

Haldu. A yellowish-brown hardwood from India. Used for superior joinery, etc. Fairly hard, heavy, durable, and strong, and resists insects. Stains and polishes well and not difficult to work. Weight about 39 lb. per c. ft.

Half Bat. Half of a brick, cut crosswise.

Half Blind Dovetail. A *Lap Dovetail.*

Half Header. A closer for forming the bond in brickwork.

Halford Window. A registered metal window for buildings of Gothic design.

Half Pace. A raised floor to a fireplace or bay window. See *Half Space.*

Half Plain Work. The labours to beds and joints of ashlar.

Half Slating. Open slating.

Half Socket Pipe. A drain-pipe with socket to lower half only.

Half Space. A landing the full width of a staircase, extending across two flights.

Half Timbering. 1. Framed timber buildings with the spaces filled with other building materials. The timber should be of oak and not too seasoned. 2. More correctly applied to a building of framed timbers that do not go through the full thickness of the wall, but are backed by an inside lining of bricks or stones. This lining adds to the stability of the structure. See *Timber Framed.*

Half Tuck. Bastard tuck pointing. See *Pointing.*

Hall. 1. A vestibule or entrance lobby. 2. A large room for public meetings. 3. The chief apartment in a medieval house or in a college.

Halving. Cross joints in which half of the material in each piece is cut away to give flush surfaces. There are many variations, and combinations, of halving joints.

Ham Hill. Two varieties of Limestone, yellow and grey, from Somerset. Both are used for general building purposes. Weights 136 and 146 lb. per c. ft. respectively. Crushing strengths 207 and 260 tons per sq. ft.

Hammer. A striking tool, of which there are many varieties and weights : ballpane, Canterbury, claw, cross-pane, Exeter, framing or heavy, sledge, pin, Warrington, etc.

Hammer Beams. The short cantilevers at the springings of a hammer-beam roof truss. See *Roofs.*

Hammered Cathedral. A rolled glass with raised rounded pattern.

Hammer Faced, or Blocked. Stones roughly squared in the quarry.

Hammer Headed. Applied to mason's chisels with which the hammer is used, and not the mallet.

Hammer-head Key. A hardwood key, enlarged at the ends to form a seating for wedges. It is used to tighten and secure an end-to-end joint, as between stile and head in a circular headed frame.

Hammering. 1. A term used in veneering. The hammer (which is a piece of sheet brass, 5 in. long in a wooden handle) is worked in a zig-zag manner from the centre outwards to squeeze out the surplus glue. 2. Noises in boilers and pipes, due to loose scale, air-lock, etc. 3. Correcting a buckled hand-saw.

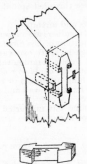

HAMMER-HEAD KEY

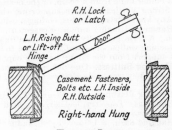

HAND OF DOOR

Hammer Post. The post springing from the hammer beam in a hammer-beam truss.

Hance. Applied in early architecture to the small arcs at the springing of 3- or 4-centred arches.

Hand. A term used to distinguish on which side a door or casement is hinged. When the door is viewed from the side showing the knuckles of the hinges it is termed *right-* or *left-*hand *hung*, according to the position of the hinges. There is more confusion when referring to the locks and furniture. If the door opens away from one, with the lock on the right and the latch bevel facing one, it is termed a right-hand lock, otherwise it is a left-hand lock, and the furniture is named accordingly.

Hand Barrow. A small platform with handles, for conveying goods by two men.

Handcraft. Registered asbestos-cement products.

Handcraft Roofing. Asbestos-cement sheets with widely spaced corrugations. The sheets have two-way reinforcement and are intended for low-pitched roofs. The sheets, 10 ft. long, are given 9 in. overlap, and weigh 5 lb. per sq. ft.

Handrail. A guide for the hand in a staircase. The top member of a balustrade.

Handrail Bolt. A double-ended bolt for joining together two pieces of handrail lengthways. See *Bolts*.

HANDRAIL SCROLL

Handrailing. The art of constructing handrails to stairs, especially when entailing wreaths. Various methods are used : tangent, normal sections, falling line, bevel cut, etc.

Handrail Scroll. A carved spiral ornament terminating a handrail. It consists of circular arcs of increasing radii forming a continuous curve.

Hand Saw. Usually applied to the carpenter's cross-cut saw. Other saws used by hand include: block, bow, coping, dovetail, fret, keyhole, pad, rip, tenon saw, etc.

Hand Screw. A wooden cramp operated by two threaded spindles, or shafts.

Hangar. A building or shed to house aeroplanes and airships.

Hangers. 1. Iron shoes or straps by which cross members are suspended, as a binder hung from a beam, or the bearers for shuttering. See *Shuttering*. 2. Strong vertical timbers supporting the walings in shaft excavations. 3. Any kind of strap for suspending a ceiling joist from a common rafter.

Hanging. Hinging a door or casement. Attaching the cords to a sliding sash. A term generally applied when fixing anything by hinges or cords. Attaching wallpaper to walls. Applied to fitments fixed to a wall and without support from the floor: hanging cupboard, etc.

Hanging Buttress. One not rising from the ground but supported on a corbel. They were a decorative feature of the Perpendicular style of architecture.

Hanging Flashings. See *Stepped Flashings*.

Hangings. Tapestry, leather, etc., used as linings to walls before paper-hanging was introduced.

Hanging Sash. A balanced sash, sliding vertically. See *Sash and Frame*.

Hanging Steps. Masonry cantilever steps. They are securely built in the wall at one end and free at the outer end. See *Treads*.

Hanging Stile. The hinged stile of a door or casement.

Hardboard. Wall-boards of wood fibre, etc., compressed to produce a smooth, hard, weatherproof board. Pressure up to 2000 tons is applied. The boards are easily cut and may be bent for covings, etc.

Hard Core. Broken bricks or stone consolidated as a foundation for concrete in solid floors.

Hard Finish. A gritty setting coat in plastering due to a greater proportion than usual of fine sand. Usually intended for distempering.

Hard Gloss. See *Gloss Paint*.

Hardrow. A registered concrete tile.

Hard Wall Plaster. Calcium-sulphate powders used with water for the final skim on plastered surfaces. They must not be used with lime putty, but with whiting. Used for stopping cracks in plaster.

Hard Water. Water containing a mineral substance such as lime. Lathers badly with soap. Temporary hardness is removed by boiling.

Hardwood. Timber from broad-leaved, or deciduous, trees. The name applies to the structure of the wood, which is more complex than in softwoods. Generally the woods are harder than softwoods, but some hardwoods are the softest and lightest known woods. See *Timber*.

Hardwood Flour. Fibrous powdered wood for use with composition and cement for special types of concrete work.

Harewood. Figured sycamore or maple dyed to a silver grey colour.

Harl or **Harling.** A local term for *rough cast*.

Hartham Park. A deep-cream Bath stone from Wiltshire. It has fine grain and is used for interior carving. Specially selected for exterior work. Weight 123 lb. per c. ft. Crushing strength 123 tons per sq. ft.

Harwich Cement. See *Cement*.

Haskinizing. A proprietary method of treating wood with a preservative.

Hasp. A hinged and slotted plate, or link, to fit over a staple, so as to secure a door by means of a padlock.

Hassall's Joint. A patent double-seal joint for drain pipes. It is a bituminous joint, with a space to receive liquid cement to seal the joint.

Hassock. Sand and soft stone interstratified with Kentish ragstone.

Hatch. 1. A sliding trap door. 2. A cover to an opening in a floor or roof 3. A dwarf door with an opening above. A half door. 4. See *Serving Hatch*.

Hatchet. A small axe with short haft.

Hatching. Section lines on drawings. See *Sectioning*.

Hatching and Grinning. Applied to an irregular face of brickwork. Not flush.

Hatchment. An escutcheon, *q.v.* A coat-of-arms, of a deceased person, placed on a building.

Haulage. Same as *cartage*.

Haunch. A small projection left after reducing the width of a tenon. It is prepared so that the tenon may be wedged, when the mortise is near the end of a stile. See *Mortise and Tenon* and *Haunches*.

Haunched Concrete. The concrete carried up from the base over the footings, to provide a bearing for the surface concrete.

Haunches. 1. The parts of an arch about half way between springing and crown. See *Arch*. 2. The bracket-like projections at the top of a column to give a greater bearing surface to the beam and to help the shear resistance. See *Secondary Beam*.

Haunching. A recess formed to receive the haunch, or haunchion, of a tenon.

Hawk. A rectangular board, with a handle beneath, for holding mortar or plaster when pointing or plastering.

Hawksley's Formula. Used to find the thickness of C.I. pipes, $t = \cdot 18 \sqrt{d}$, where d = diam. in inches.

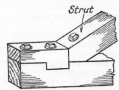

HAWK

Haydite. Clay or shale fused at high temperatures to produce a cellular material for light-weight concrete.

Hayton Floor. A patent, glass, dance floor on springs.

Hayward. A patent roof glazing.

Hazards. The danger points, or risks, that may give rise to fire. If they are from the building or contents they are *internal hazards*, if from adjacent buildings they are *external*.

H.B. Abbreviation for *half brick*.

H.B.S. Abbreviation for *herring bone strutting*.

Head. 1. The top horizontal member of a frame or opening. See *Centering*. 2. See *Head of Pressure*. 3. A cutter block for a spindle machine or four-cutter. 4. The highest part or top of anything, as head of stairs. 5. The foremost part of anything, as pier head.

Head Block. A block bolted to a timber to take the thrust from another timber, as from a strut. The block may be keyed or tabled.

Head Board. 1. A board near a manger to which an animal is secured. 2. The board forming the top, or head, of a bed.

Header. 1. A brick with the end showing on the face of the wall. 2. A nog in a stud partition for the horizontal edges of wallboards.

Head Flashing. The flashing along the top edge of any projection through a roof. It is usually in the form of a small gutter.

HEAD BLOCK

Head Gate. A lock gate that admits water.

Head Guards. Lead gutters formed over openings in cavity walls to protect any timber from moisture, such as lintols, frames, etc.

Heading. A small underground passage, formed to examine the strata and to give passage to the workmen, when tunnelling.

Heading Bond. Brick bond formed by all headers. It is used for walls circular in plan to give a more even curvature.

Heading Course. A course of headers as used in English bond and in foundations.

Heading Joint. An end-to-end joint between two pieces of material.

Head Nailing. The nailing of slates 1 in. from the head instead of at the centre. This method requires more slates per square, repairs are more difficult, and it is not so good against strong winds, but the nails are covered by two slates which is a big advantage. See *Gauge* and *Centre Nailing*.

Head of Drain. The highest point of a drainage system.

Head of Pressure. The pressure of water at any point due to the height or depth of water. The pressure is proportional to the vertical distance from the source of supply, and is not influenced by other factors such as bore or arrangement of pipes. If the supply cistern is 20 ft. above the level of the boiler then the pressure on the boiler is 20 × ·434 lb. per sq. in., or 20 × 62·5 lb. per sq. ft.

Head of Stack. The top of a stack of rain-water down pipes. It collects the water from the gutters, and is usually of an ornamental character.

Head Piece. The capping piece of a quartered partition.

Head Post. The post near the manger, in a partition between two stalls in a stable.

Head Room or **Headway.** The clearance above a flight of stairs. It is measured from the top of a tread, vertically with the face of a riser, and should not be less than 6 ft. 6 in. to any obstruction.

Head Tree. A short horizontal timber on a post to increase the bearing surface.

Head Wall. The wall in the same plane as the face of the arch for a bridge.

Healing. Covering a roof with slates, tiles, lead, etc.

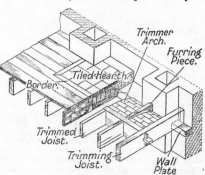

HEAD TREE

Heart Bond. Bonding in the middle of a wall in the direction of its thickness.

Hearth. The fireproof part of a floor adjacent to a fireplace. The *front hearth* extends 6 in. beyond the grate on each side and 1 ft. 6 in. in front and is usually tiled. The *back hearth* is immediately beneath the fire and behind the face of the breasts.

Hearting. Filling the interior of a wall between the facings.

Heart Shake. Shakes radiating from the pith in the heart wood. Usually due to over-maturity and the medullary rays failing to convey the sap to the inner part of the trunk. Also due to uneven seasoning. See *Timber*.

HEARTH DETAILS

Heart Wood. The inner part of a tree trunk, or duramen, as distinguished from the sap wood, or alburnum. The timber is harder and denser and offers a greater resistance to decay and disease. See *Timber*.

Heat. A form of energy. The degree of hotness of a body is stated as its *temperature*, which is usually measured by Fahrenheit or centigrade scales. The former is graduated so that boiling water is 212° and freezing water 32°. The same points on the centigrade are 100° and 0°. The British method of measuring quantities of heat is in British Thermal Units (B.Th.U.), in which one unit is the quantity required to raise 1 lb. of water through 1° F. The scientific unit is the *calorie*. Heat may be transmitted by radiation, conduction, or convection.

Heaters. See *Geyser* and *Immersion Heater*. Also see *Heating*.

Heathman's Scaffold. A patent scaffolding consisting of extension ladders braced together.

Heating. Buildings may be heated by coal or coke fires or stoves, gas radiators, electric convectors, hot water or steam, or by hot air (plenum system). Large buildings are invariably heated by hot water or steam, either by *low pressure* or *high pressure* systems. See *Central Heating*.

Heating Panels. See *Coil Heating*.

Heave. The relative horizontal displacement of strata due to faulting, *q.v.*

Heck. A local term for a door latch.

Heel. 1. Sometimes applied to the cyma reversa moulding. 2. The bottom of the hinging stile of a door. 3. The back end of a plane.

Heel, or Kicking, Post. A post in a stable partition at the opposite end to the head post.

Heel Stone. One taking the foot of the hanging stile of an iron gate.

Heel Strap. A W.I. strap used at the foot of the principal rafter of a roof truss. The strap passes over the back of the rafter and two eyes engage with a bolt passing through the tie beam. It is intended to take the thrust from the rafter but the shrinkage of the timber often defeats this object.

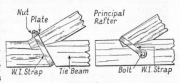

HEEL STRAPS

Heft. A handle.

Height of Buildings. Controlled by the local authority. In London the maximum is 80 ft. from the footway to the parapet or base of gable for ordinary buildings, or, in narrow streets, no higher than the width of the street, except by the consent of the Council. The revision of the maximum height is now under consideration. See *Skyscraper*.

Helical, or Helicoidal. Applied to twisted surfaces in "ramp and twist" work, skew arches, and in double-curvature work generally. The centre line for a handrail wreath. Having the form of a spiral, or helix.

Helical Hinges. A special form of hinge for swing doors opening both ways.

Helical Surface. A surface generated by a straight line moving along and revolving round a fixed straight line, with the angle between the two lines constant.

Heliodon. A device for tracing the path of a ray of light, or sunshine, through any required period.

Helioscene. An outside blind consisting of small hood frames together and secured by stays. It allows for light and air but provides shade.

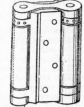

HELICAL HINGE

Helix. 1. A spiral or volute. 2. A line round a cylinder such that the increase of rise is constant with the increase of angular advance.

Helliwell. A patent roof glazing.

Helmet. 1. A cast-iron protection on the top of a reinforced concrete pile. 2. A reinforced concrete housing for the end of a main beam.

Iron Helmet to protect Pile whilst being driven.

Sand Cushion.

Helying. See *Healing.*

Hem. The rim of the Ionic volute.

Hemihydrate Plaster. See *Plasters.*

Hemlock. Two species: Eastern and Western Hemlock from N. America and Canada. Only the latter is satisfactory for carpentry work. It is coarse in texture and resembles pine in many respects, non-resinous, light, soft, strong, and durable if kept wet or dry, and is in increasing demand for construction, joinery, flooring, form-work, stable fittings, etc. Weight about 28 lb. per c. ft.

Hemmingway. A preservation process for stone that precipitates calcium silicate.

Hench. The narrow side of a chimney shaft.

HELMET TO PILE

Hendecagon. See *Endecagon.*

Hennibique. A system of reinforced concrete in which the steel reinforcement consists of round steel bars held in position by flat stirrups which act as shear members. The tensile members in the beams are bent upwards at the points of contraflexure to take the reversed bending moment.

Heptagon. A polygon with seven sides. Area = sq. of side $\times$ 3·634, for a regular figure.

Heraklith. A proprietary insulating material, up to 3 in. thick.

Hermes. A pillar with a finial in the form of a bust.

Hermetex. A waterproof plastic asbestos compound for roof repairs.

Herring-bone. Timbers or other materials placed obliquely in opposite directions.

Herring-bone Bond. Used to give longitudinal bond to thick walls. About every fifth course the interior of the wall is filled with bricks placed diagonally and converging to the middle of the wall. Also used for ornamental brick panels.

Herring-bone Strutting. Used to stiffen floor joists. Small pieces of timber are cut diagonally and fixed in pairs between each pair of joists, across the floor. See *Floor Strutting.*

HERRING-BONE BOND

Herse. A portcullis, *q.v.*

Hessbit. A proprietary bituminous underlining, or sarking, felt.

Hewn. Implies *dressed* in masonry, and *axed* in woodwork.

Hexagon. A polygon with six sides. Area = sq. of side $\times$ 2·598, for a regular figure.

Hexahedron. A cube.

Hexastyle. A portico with six columns, in classic architecture. See *Temple.*

Heywood. A patent roof glazing.

H-Hinge. See *Parliament hinge.*

H.I.B. Abbreviation for *hoop-iron bond.*

Hick Joint. A raked joint pointed with a "flat joint" with fine mortar. See *Pointing.*

Hickory. A tough, flexible North American wood used for shafts bentwork, sports equipment, etc.

Hico. A registered centering or falsework system.

High-pressure System. The Perkin's or "small-bore," system of heating.

The system consists of strong 1 in., W.I. pipes, with a furnace coil acting as boiler. The water is heated in tanks, in suitable positions, by the flow of water through the coils. The system is not healthy, due to the *parching* of the atmosphere, and dangerous if a joint fails.

Hill Adams. A patent sliding-door gear.

Hinge Bound. Applied to a door difficult to close because the hinges are sunk too deeply. No clearance between the hanging stile and frame.

Hinged Truss. A steel roof truss fixed at one end only, to allow for expansion and contraction with change of temperature.

Hinges. Types in common use : butt, rising butt, helical, pew, parliament, back-flap, counter-flap, spring, pin, strap, garnet, box, Bommer, centre, H, continuous, cross garnet, dolphin, floor spring, kneed, knuckle, screen, T. Hinges may be of various metals and with different finishes. See *Door Furniture.*

Hinge Stone. The stone to which the pin for a band hinge is fixed when there is no door frame. The hinge pin, or gudgeon, may be forked and embedded in a joint of the pier or it may be in the form of a rag bolt let into a dovetailed hole and caulked with lead.

Hip. The salient angle formed by the intersection of two inclined roof surfaces.

Hip Bath. See *Sitz Bath* and *Bidet.*

Hip Hook, or Iron. A hook fixed at the bottom of a hip to prevent the slipping of the hip tiles.

Hip Knob. A finial at the end of the ridge in a hipped roof.

Hip Rafter. The rafter forming the hip of a roof.

Hipped Roof. A roof inclined at the end as well as the sides. It is without gable. See *Roofs.*

Hippodrome. Literally implies a racecourse for horses and chariots, a circus, but now applied to a place for variety entertainment.

Hip Roll. A circular timber, with flat V underneath, to cover the intersection of the roofing materials at the hip.

Hip Tile. The saddle tiles fitting over the intersection of the roof tiles at the hip. There are varied shapes such as bonnet, cone, half-round, and angular.

Hit and Miss Ventilator. A fixed slotted plate with a similar sliding plate behind so that the slots may be open or closed as required.

Hitch. 1. A particular form of knot used with rope lashings, for hoisting timbers, etc. 2. An angle connection for structural steelwork. It is usually a short piece of standard angle iron.

H.M.D. Abbreviation for *hydraulic mean depth.*

H.N.W. Abbreviation for *head, nut, and washer.*

Hoarding. A close boarded fence to enclose a building during erection or alterations, or for advertising purposes.

HITCH

Hob. A horizontal plate to a fire grate to hold pans, etc.

Hod. A box-shaped tray with long handle for carrying on the shoulder. It will hold 12 bricks, or nearly half a bushel of mortar.

Hoffman Kiln. A continuous kiln for brickmaking. It contains from 12 to 16 chambers which, by the arrangement of the dampers, are at varying temperatures, ascending from the chamber being loaded to the chamber where they are burnt, and then descending to the one being unloaded.

Hog Back Girder. A built up girder with the top chord slightly segmental, i.e. curved from the centre downwards to the abutments. See *Roofs.*

Hoggery. A pigsty or piggery.

Hoggin. Coarse sand or fine ballast.

Hogging. Cambering, or raising the middle of a horizontal structural member.

Hog's Back Ridge. A ridge tile with rounded top.

Hoist. A lift, or elevator, for goods rather than passengers.

Hoisting. The act of raising materials by mechanical appliances.

Holbric. Patent hollow terra-cotta blocks, suitable for external walls with facings.

Holderbat. The best type of fixing for pipes. It is of metal and has a dovetail lug, for fixing in a joint or running in with lead. The front is removed to insert the pipe.

Holdfast. 1. A metal fastening driven into the joints of walls, to secure framing by means of a screw through the head. See *Fastenings*. 2. An appliance for clamping material to the bench whilst it is being shaped or carved.

Holing Machine. A drill for making the nail holes in slates.

Hollocast. Patent hollow beams of reinforced concrete with metal core, used for fire-resisting floors. The ducts can be used for ventilation, wires, pipes, etc.

Hollock. A greyish-brown hardwood, with dark striations, from India. Fairly hard, heavy, strong, durable, and stable. Rather difficult to work. Used for joinery, construction, plywood. Weight about 42 lb. per c. ft.

Hollong. A reddish-brown hardwood from India. Fairly hard, heavy, and strong. Should be treated for exterior work. Used for structural and constructional work, plywood, etc. Weight about 42 lb. per c. ft.

Hollow. 1. A concave moulding. 2. A joiner's plane to form a round, or convex, surface.

Hollow-backed. Hardwood floorboards hollowed on the underside to provide a better seating or ventilation or reduced weight.

Hollow Beams. Pre-cast reinforced-concrete hollow beams resting on the secondary beams for fire-resisting floors. The units are placed side by side and are self-centering for the *in situ* concrete.

Fixed Block. 2 sheaves

Movable Block. One sheave

Fall

P

HOISTING WITH TACKLE

Hollow Bed. Masonry joints riding at the outer edges due to the surfaces not being perfectly plane. The pressure on the small areas flushes the edges of the stone.

Hollow Blocks. The many types of blocks carried by filler joists in fire-resisting floors. See *Armoured Tubular*.

Slots to cover Flange

HOLLOW FLOOR BLOCKS

Hollow Drill. See *Drill*.

Hollow Newel. The well-hole of a winding stair. See *Geometrical Stairs*.

Hollow Partitions. Those formed of clay, terra-cotta, or breeze, hollow blocks for lightness and for sound insulation. They may also be formed by staggering the studs. A double partition with space between for a sliding door.

Hollow Roll. A lead roll dressed hollow over 2 in. wide copper tingles, placed at about 2 ft. intervals.

Hollow Tiles. The many patented types of hollow blocks used in fire-resisting floor construction. They are usually of fireclay or terra-cotta, and reinforced with steel and concrete. They are laid in panels on shuttering until the *in situ* concrete has set.

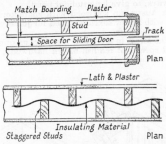

HOLLOW PARTITIONS

Hollow Walls. See *Cavity Walls.* Also walls constructed of hollow concrete blocks. See *Hollow Partitions.*

Hollow Wood Construction. Deal-framed cores faced with plywood.

Holophane. The name of a firm specializing in lighting, patent glass, reflectors, etc. Their specialist productions are marketed as *Holophane.*

Homan and Rodger's Floors. 1. A patent T-shaped reinforcement for concrete, having a horizontal flange and wavy web. 2. Hollow triangular fireclay blocks or lintols, with dovetailed grooves on all sides to provide a key for plaster or concrete. The blocks rest on the bottom flange of filler joists and are self-centering. See *Lintel.*

Homasote. A prefabricated wood house made on the panel system.

Honeycomb Slating. Asbestos slates laid diagonally (corners vertical). The bottom corner is removed otherwise it is similar to "drop-point" slating.

Honeycomb Walls. Sleeper walls supporting ground floor joists. Occasional bricks are omitted for ventilation purposes.

Hood. A cover or canopy. It is usually supported on cantilever brackets and provides shelter to a doorway.

Hood Moulding. Same as *Label Mould,* but usually applied to a moulding over an internal opening. A projecting moulding at the top of a casement window to protect against the weather.

Hood's Rule. A formula used by heating engineers for obtaining the length of pipe required to heat a room. It has been superseded by more scientific formulae.

Hook. 1. See *Saw Teeth.* 2. The projection of the cutting iron through the sole of a plane. 3. See *Band and Gudgeon.*

Hook and Eye. 1. A fastening consisting of a pivoted hook and a fixed staple, or eye. 2. A gate strap-hinge with a hook to engage in a gudgeon in the form of a circular eye. It allows for easy removal of the gate.

Hooke's Law. "In any elastic body the strain is proportional to the stress, within the elastic limit," i.e. if a piece of material lengthens ·1 in. under a pull of 1 ton it will lengthen ·2 in. under a pull of 2 tons, etc. See *Modulus of Elasticity.*

Hook Rebate. An air-tight S-shaped rebate used for show cases, and sometimes for the meeting stiles of casements.

Hook Ring. The eye for a cabin or cupboard hook. See *Hook and Eye*.

Hoop. Wood or metal bent into circular form. A circle. See *Jacket*.

Hooping. Thin spiral bars for circular reinforced concrete work.

Hoop Iron Bond. Brickwork with tarred and sanded or galvanized hoop iron laid lengthways, to give longitudinal bond.

Hoop Pine. An Australian softwood with characteristics of white pine but stronger and with firmer texture. Used for joinery, construction, etc. Weight about 34 lb. per c. ft.

Hoop Stress. The stress in a circular member, or ring.

Hopea. Several species of reddish-brown hardwood, with purplish cast and white lines, from India. It is hard, heavy, and very durable and strong.

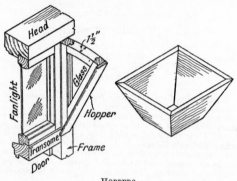

HOPPERS

Polishes well, but difficult to work. An excellent wood where durability and strength are of primary importance. Weight about 48 lb. per c. ft.

Hope's Destroyer. A proprietary preservative for wood.

Hope's Patent Glazing. Specialized forms of windows and roof glazing, involving several patents.

Hopper. 1. A draught preventer at the side of a sash hinged at the bottom. 2. A large wood, steel, or concrete funnel to hold granular material, as grain, coal, etc. See *Bunker* and *Concrete Mixer*.

Hopper Head. A head to a stack of rain-water pipes in the form of a funnel, to collect the water. It is also used to form a break in a long stack, for the escape of gases, or to receive the discharge from more than one pipe.

Hopper Light. A sash hinged or rotating on a *rocker*, on the bottom rail, and opening hopper-wise. See *Hopper Window*.

Hopper W.C. A long conical pan with a syphon trap. An obsolete type of wash-down, or pedestal closet, consisting of basin and trap.

Hopper Window. A window formed of a series of horizontal sashes hinged at the bottom and having hoppers at the sides.

Hopton Wood. Cream, grey, or brown limestones from Derbyshire. Polish well and used for monumental and decorative work. Wrongly called marble. Weight 150 lb. per c. ft. Crushing strength 800 tons per sq. ft.

Horizontal Plane. 1. An imaginary level plane used in orthographic projection. Ground level. See *Planes of Projection*. 2. A plane parallel to the horizon, through the level of the eye, and at right angles to the picture plane, in perspective drawing.

Horizontal Shores. See *Flying Shores.*

Horn. A projecting end of a framed member left as a protection until the framing is fitted in position, or to assist in fixing, or to strengthen the

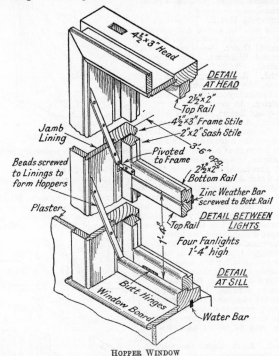

4½"×3" Head

DETAIL AT HEAD

2½"×2" Top Rail

4½"×3" Frame Stile

2"×2" Sash Stile

Jamb Lining

3'-6" opg.

Pivoted to Frame

Beads screwed to Linings to form Hoppers

2½"×2" Bottom Rail

Zinc Weather Bar screwed to Bott. Rail

DETAIL BETWEEN LIGHTS

Plaster

Top Rail

1'-4"

Four Fanlights 1'-4" high

DETAIL AT SILL

Butt Hinges

Window Board

Water Bar

HOPPER WINDOW

frame. See *Joggle.* 2. The bevelled shoulder to a tusk tenon. 3. The projection on a W.C. basin for connection with the service pipe.

Hornbeam. *Carpinus betulas.* A European hardwood. Close grain, tough, and very hard when seasoned. Yellowish white. Weight about 47 lb. per c. ft.

Hornblende. A dark green to blackish crystalline mineral, formed of silica, lime, and magnesia; and found in igneous rocks, such as syenites and greenstone. Very durable.

Horse. 1. An appliance, or thwacking frame, for bending pantiles, before burning. 2. A perforated board for fixing the damper rod in a brick kiln. 3. The slipper or board carrying the various parts of a mould for running plaster cornices and mouldings; or the complete mould. 4. A trestle. 5. Any form of framing used as an intermittent or temporary support.

Horse Chestnut. *Æsculus hippocastanum.* A European hardwood. The timber is light, soft, and not durable. White. Weight about 31 lb. per c. ft.

Horsed Joint. See *Saddle Joint.*

Horse Pot. A trap for the drainage of a stable before it enters the drain.

Horseshoe Arch. See *Moorish Arch.*

Horseshoe Drains. U-shaped pipes for agricultural drainage.

Horsing-up Mould. Preparing, or building up, the mould for running plaster or cement mouldings, etc.

Hospital Doors. See *Flush Doors.*

Hospital Windows. See *Hopper Windows.*

Hostel. A lodging house for students. An inn for young tourists, generally run on non-commercial lines.

Hostelry. An inn.

Hot-press Resins. Thermo-setting, urea-formaldehyde, resins, that require "cooking" in an autoclave. They have the advantage of allowing

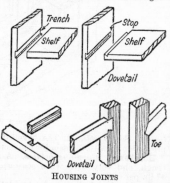

HORSED MOULD

a time interval of several days before setting which is necessary in large-scale assembly work. See *Plastic Glues.*

Hot Short. Applied to iron that cracks when bent at red heat.

Hot-water Apparatus. Applied to any of the various systems of heating water for domestic purposes. The usual method is to heat the water by means of a boiler behind the kitchen fire. See *Low Pressure.* When the system is used for heating the building it is usually called *Heating Apparatus* and may be by low pressure or high pressure systems.

Housebreaker. One that specializes in the demolition of old buildings.

Housed String. See *Close String.*

Housemaid's Sink. A slop sink.

Housing. 1. A sinking in one timber to allow for the insertion of the end of another timber. 2. A niche for a statue. See *Tabernacle.*

Housing Scheme. Usually applied to the cheaper type of dwelling houses erected by local authorities to overcome the shortage of houses, and to be let at a low rental.

Hovel. 1. A mean dwelling. 2. See *Tabernacle.*

Hovelling. Carrying up two sides of a chimney stack above the level of the top of the flue to allow the smoke to escape at a lower level when strong currents are blowing over the chimney.

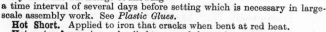

HOUSING JOINTS

Howe's Formula. Used to obtain a preliminary estimate of the weight of a steel roof truss : weight of truss in lb. per sq. ft. of horizontal projection of roof $= \frac{3}{4}\left(1 + \frac{\text{span}}{10}\right)$.

Howe Truss. A wood roof truss for large spans up to 80 ft. It has king, queen, and princess posts and auxiliary posts for the larger spans. See *Roofs*

Howley Park. A brown sandstone from Yorkshire. It is used for general building, dressings, steps, and landings. Weight 140 lb. per c. ft. Crushing strength 467 tons per sq. ft.

H.P. Abbreviation for *horse-power*, or *high pressure*, or *horizontal plane*.

H.r. Abbreviation for *half-round*.

Humidifiers. Mechanical appliances for keeping the air in the correct condition of humidity.

Humidity. A measurement of the amount of moisture present in the atmosphere.

Hungry Spots. Defects, or small holes, in the surface of new concrete.

Hung Sash. See *Hanging Sash*.

Hung Slating or Tiling. Slating or tiling on walls.

Hunting Box. A small country house generally only used during the hunting season.

Huntland Rail. A patent curtain rail.

Huntonit. A registered wood-fibre wall-board.

Huon Pine. *Dacrydium Franklinii.* A Tasmanian resinous timber. Fine grain, lustrous, and beautifully marked. Yellowish-brown colour. Polishes well. Used for joinery, fittings, etc. Weight about 33 lb. per c. ft.

Hurlhinge. A patent self-aligning butt hinge.

Hurter. A block of stone, concrete, cast-iron, or wood used to protect a quoin from vehicular traffic.

Hut. A light building for workmen, soldiers, etc. A small cabin or shelter.

Hutch. 1. A chest for sacred utensils. 2. A trough used in ore-dressing. 3. A low wagon used in mines. 4. A rabbit coop.

Hutton's Formula. Used to obtain the wind pressure on a sloping surface, θ, in terms of one on a vertical surface. $P_\theta = P_V \sin \theta^{1\cdot84 \cos \theta - 1}$.

Hyatts. A patent iron-framed pavement light with plain or prismatic lights and inset tiles to cover the iron frame.

Hydralime. A proprietary hydrated lime.

Hydramel. A proprietary oil-bound washable distemper.

Hydrant. A plug, or stop valve, to a water pipe to control the supply from the main.

Hydrated Lime. Lime slaked, dried, and ground in powder form. It is mixed dry with sand and then wetted to make it plastic, in the same way as cement. It should, however, be prepared the day before using.

Hydraulic. Applied to lime or cement that will set under water.

Hydraulic Jack. See *Jack*.

Hydraulic Mean Depth. If the area of the cross-section of a flowing liquid $= a$ sq. in., and the wetted perimeter (the length of the section of the pipe or channel submerged) $= p$, then $\dfrac{a}{p} =$ h.m.d.

Hydraulic Mortar. Made with hydraulic lime for wet situations.

Hydraulics. The science dealing with the motion of liquids, and the application of water to machinery.

Hydrofuge. Waterproof.

Hydrol. A proprietary waterproofer.

Hydrostatic Arch. A linear arch designed to sustain normal pressure at each point proportional to the depth below a given horizontal plane.

Hydrostatic Test. The water test for drains. The outlet is plugged and the drain filled with water and examined later to see if there is any fall of level due to leakage. It is a severe test and should only be applied to new drains. A head of 7 ft. exerts a pressure of about 3 lb. per sq. in. and this should not be exceeded. See *Drain Testing*.

Hydrous. Containing water.

Hydrox. A proprietary waterproofer for cement, in powder form.

Hygeian Rock. A waterproof composition for vertical or horizontal D.P.C.'s. It is poured in a liquid state and sets hard.

Hygrometer. An instrument used to measure the humidity of the atmosphere.

Hygroscopic. Applied to materials that absorb moisture from the atmosphere.

Hypæthral. A building or temple without roof, in classic architecture.

Hyperbola. See *Conic Sections*.

Hypertherium. A projecting cornice over the architrave of a door.

Hypocaust. A space below the floor of ancient Roman buildings where the heat from the furnace accumulated for heating the bath or warming the rooms of the house.

Hypostyle. A hall of which the roof rests on columns characteristic of Egyptian temples.

Hypotrachelium. The junction of the shaft with the capital of a column. See *Neck*.

Hyrib. A patent metal lathing. It is bent at intervals to form a stiff rib so that it is practically self-supporting. It is used for solid partitions and for reinforcing concrete.

Hysteresis. A variation from Hooke's law when testing materials.

I

I. Abbreviation for *moment of inertia*. See page 366.

I$_e$. A symbol for *equivalent moment of inertia*.

I.B.A. Abbreviation for *International Bath Association*.

I-beam. See *British Standard Sections*.

Ibeco. A proprietary waterproof Kraft paper.

Ibo. 1. An attractive reddish-brown, lustrous, hardwood from Nigeria. Suitable for interior fittings. 2. A prefabricated wood house.

Ibus Board. A proprietary wall-board.

I.C. Abbreviation for *intercepting chamber*.

Ice Concrete. Crushed ice is mixed with the concrete. The melting of the ice leaves cavities for light-weight concrete.

Ichnograph. A ground plan or horizontal section of a building.

Icon. A sacred image or representation.

Iconography. The study of ancient mosaic work, frescoes, statues, etc.

Icosahedron. A solid having 20 equal faces. Surface area = sq. of linear side × 8·66. Vol. = cube of linear side × 2·182.

Idigbo. A West African hardwood used for superior joinery, etc. It is light-oak in colour with brownish stripes, moderately hard, heavy, durable, and stable. Weight about 37 lb. per c. ft.

Idler Pulley. A loose pulley to a machine.

Igneous. Rocks of volcanic origin such as granites, syenites, and traps. The rocks were originally in a semi-fused or molten condition.

Illumination. Artificial lighting. See *Light*.

Imbricate. To overlap, like roof tiles.

Imbrices. Half-round tiles to cover the joints of the tegulae in Old Roman tiling.

Imbuia. A yellowish to chocolate-brown hardwood from Brazil. It is fairly hard, heavy, durable, but not strong; fragrant and lustrous. It is very decorative, polishes well, and easily wrought. Used for superior joinery, interior fittings, etc. Weight about 36 lb. per c. ft.

Immersion Heater. An electric heater for a hot-water cistern.

Impact Load. A suddenly applied load, as from reciprocating machinery, rolling loads, etc.

Imperfect Frames. See *Indeterminate Frames*.

Imperial Paper. See *Drawing Paper*.

Imperial Slates. Size 30 in. × 24 in., requiring 45 to the square at 3 in. lap.

Impervion. A proprietary plastic and liquid waterproofer.

Impervious. A registered metal-casement window.

Impluvium. A walled tank built in the compluvium of ancient Roman buildings to catch the rain-water.

Impost. A projecting springing stone. The top, or cap, of a pier or pilaster from which an arch springs.

Improved Wood. A name applied to numerous proprietary materials consisting of wood treated with synthetic resins, heat, and pressure, to give a much stronger and more stable wood. Some of them are also called "man-made wood," because other materials are incorporated, such as paper, sawdust, etc. Jicwood, Weyroc, Benalite, Compreg, Hydroxylin, Hydulignum, Impreg, Permali, Plyscol, Uraloy, are some of the registered names.

In. See *Eng*.

Inband. A jamb stone used as a header in a wall and recessed for a frame

Inbow. To vault or arch.

Incandescent. Light obtained from glowing bodies subject to high temperatures, as gas mantles and electric filaments.

Incidence, Angle of. The angle at which a ray of light, heat, etc., strikes a surface.

Incinerators. Specially constructed furnaces to burn refuse.

Incised. Cut or carved. Applied to carvings that are below the surface of the material.

Incised Tiles. Partially or wholly glazed tiles with sunk lines which are filled in with Portland cement when the tiles are laid.

Incrustation. A hard coat or covering. The term is specially applied to the coating of hot-water pipes due to the presence of lime in *hard waters*.

Indene Resins. Produced from coal tar, naptha and sulphuric acid and used for varnishes.

Indented Bar. Round or square steel reinforcement for concrete. The bars are regularly indented along the length to bond with concrete.

Indented Joint. A fished joint in which beam and plate are indented to prevent one surface from sliding on the other. The

INDENTED JOINT

sinkings in one piece of a scarfed joint to receive the projections on the other piece are *indents*.

Indeterminate Frames. Structural frames that present difficulty in determining statically the stresses in the members, due to the presence of redundant members or lack of triangulation. They are called imperfect frames.

Indexed Plan. A plan in which the heights of the points above the horizontal plane are indicated by numbers, instead of by an elevation.

Indian. Architecture characteristic of India. There are many styles, but usually with the common feature of rich carved ornamentation and inlays of great delicacy.

Indian Red. An earth pigment for paint. The colour is due to the presence of iron.

Indian Stones. Oilstones made from corundum and alundum.

Indigo. A blue pigment obtained from the indigo plant.

Indirect Heating. The heating of rooms by convection currents. The source of the heat is some distance away, and the heat is conveyed to the rooms as hot water, or air, or steam.

Indurated Clays. Shales that have to be crushed to be made plastic.

Induroleum. A proprietary jointless flooring.

Inert Fillers. Adulterants or bases used in paint to regulate weight and cost: barytes, calcium carbonate, charcoal, gypsum, magnesia, silica, silicates of alumina and magnesia, whiting, etc. Different fillers are required for different purposes and materials.

Inertia. The property of matter by which it retains its state of rest or motion in a straight line.

Inertia Areas. The selected portions of a section of a structural member that are resisting the bending moments. If we consider this sectional area as built up of a large number of very small areas then the *moment of inertia* (I) is the sum of the second moments of these small areas about the neutral axis. If a_1, a_2, etc., are the areas and d_1, d_2, etc., their distances from the axis, then $I = a_1 d_1^2 + a_2 d_2^2 + \ldots = \Sigma a d^2$. See page 366.

Infilling. Thin stone panels, called webs or cells, between the framework of ribs, in rib and panel vaulting. Panel walls between stanchions, etc.

Influence Lines. A curve showing the shear, bending moment, or other required function, at any section of a beam for different positions of a rolling load.

Ingate. The hole for pouring the fluid material into the mould when moulding.

Ingle Nook. An enclosed seat near a fireplace.

Ingoing. A jamb, or the reveal to a jamb.

Ingot. A mass of unwrought metal.

Ingyin. Also called Burma Sal. A hard, heavy, strong and durable hardwood from India and E. Indies. It is the colour of mahogany. Used for structural and constructional work. Weight about 51 lb. per c. ft.

Initial Set. The first stage of the setting of cement, just as it loses its plasticity.

Injector Hot-water Lifter. Patent hot-water heaters that automatically raise the water to a higher level.

Inlay. Ornamentation formed by cutting out of a surface and substituting a different material.

Inlet. Specially applied to the entrance of a drain into a manhole.

Inn. 1. See *Hostel*, as *Inns of Court*. 2. A tavern, or public-house, providing accommodation for travellers.

Inner Bead. The beads mitred round the inside of a *sash and frame*, q.v., to control, or guide, the sliding sash.

Inner Lining. The inside lining forming the box for the weights in a *sash and frame*.

Inner Plate. The inside wall plate when two are used.

Inodorous Felt. An asphaltic flax felt used as underlining, etc.

Inscribed Circle. A circle drawn inside a rectilinear figure so that the sides of the figure are in contact with the circle.

Inserted Tenons. Hardwood tenons inserted in the rails of oblique and circular work because the cross grain in the solid tenon would be too weak. Also loose tenons to prevent lateral movement.

In situ. Applied to concrete and other materials that are moulded in their required position in a building.

Insley Plant. A mechanical contrivance for lifting and placing concrete when moulded *in situ*. See *Placing Plant*.

Inspection. The periodical inspection by the local building inspector of building operations. His approval is usually necessary at the following stages: excavations, foundations, damp-proof courses, drains, and completion of building.

Inspection Certificate. A certificate from a local authority that the drains are satisfactorily completed and the building fit for occupation.

Inspection Chamber. A brick chamber, or manhole, for collecting several drains together. It is large enough for a man to descend and clean the pipes when necessary.

Inspection Eye. A small chamber at the change of direction of a drain pipe or between manholes. It is sufficiently large to lower a lamp that can be seen from the manholes to locate an obstruction.

Inspection Pit. An *Inspection Chamber.*

Inspection Pockets. A flat panelled cover to cast-iron pipes that can be removed for cleaning purposes in the event of a stoppage.

Installation. The complete equipment of a building with electric light and power.

Insulating. 1. Protecting electric wires from contact, by tape, lead sheathing, vulcanized rubber, or other non-conducting material. 2. Applied to materials that prevent the passage of heat, sound, or vibration of any kind. The insulation is usually obtained by means of trapped air in porous materials or by scientific construction.

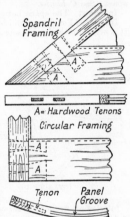

Spandril Framing

A = Hardwood Tenons

Circular Framing

Tenon   Panel Groove

INSERTED TENONS

Insulating Boards. Applied to the numerous registered wall-boards, that are specially made, usually of wood fibre, to resist sound.

Insulating Bricks. Very porous bricks used for heat insulation in kilns, etc., to reduce consumption of fuel, and for sound absorption in sound-proof work. The clay is mixed with combustible material that burns away in the kiln, leaving a porous brick.

Insulating Tape. Adhesive tape impregnated with insulating compounds.

Insulcrete. Proprietary insulating partition blocks made of pumice.

Insulex. A mineral, gypsum, insulating material for ceilings, partitions, etc.

Insulite. A proprietary wood-fibre wall-board.

Insulwood. A proprietary wood-fibre wall-board.

Insurance. The various insurances included in the primary charges when obtaining labour costs for estimating. They include National Insurance, Workman's Compensation, Unemployment, Guaranteed week, Holidays with pay, Third party, Fire, etc.

Int. Abbreviation for *interval.*

Intagliated. Carved on the surface. Sunk ornamentation.

Intaglio. Tiles with sunk patterns in relief, formed by pressing the tiles in a mould with a sunk die.

Intake. The alternative to offset, *q.v.* The practical result is the same in both cases.

Integral Process. Incorporating powder, liquid, or paste, with concrete, as a void filler or fattener, to increase its waterproof properties.

Intensity of Stress. Unit stress. The force acting on unit area at a section of a structural member. See *Stress.*

Intercepting Chamber A manhole with interceptor.

Interceptor. A trap placed between a drain and a sewer to prevent the passage of sewer gas and vermin.

Intercolumniation. The spacing of columns in classic architecture, pycnostyle 1½, systyle 2, enstyle 2¼, diastyle 3, araeostyle 3½ to 4 diameters apart. See *Columns*.

Interjoist. The space between two joists.

Interlaced Fencing. Solid fencing formed by interlacing very thin boards. It is also described as *wovenboard* and *interwoven fencing*.

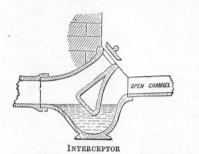

INTERCEPTOR

INTERLACING ARCHES

Interlacing Arches. A series of arches springing from alternate columns so that they cross each other. They are used in arcades, chiefly as an ornamental feature.

Interlocked Grain. A spiral arrangement of the grain during growth that makes the wood difficult to work smooth, because of the ribbon grain running in opposite directions.

Interlocking Tiles. The various patterns of roofing tiles that engage with each other, or lock, along the edges. Popular types are: Bridgwater, Marseilles, Mulden, Boulet, Du Nord.

Intermediate Rib. The shorter ribs in a rib and panel vault between the diagonal and the wall, or transverse, ribs.

Internal Dormer. Vertical casements set back within the plane of a sloping roof, which requires a lead flat in front of the casement.

Interstice. A crevice, or an intervening space.

Interstitium. The central space under the tower of a church.

Intertie. A horizontal member in structural framing between head and sill. The member just above the doorways in a trussed partition. See *Partition*.

Interwoven Fencing. A close fencing consisting of thin material woven, or interlaced, to make a large sheet. It is generally sold under a trade name.

Intrados. The underside, or soffit, of an arch or vault. See *Arch*.

Inundator. A closed drum for gauging the water in sand for concrete.

Invert. The bottom of a brick sewer. An invert block is one that forms the bottom of the sewer and is generally of concrete or terra-cotta.

Inverted Arch. A segmental arch reversed and placed at the bottom of piers to distribute the pressure over the foundations. See *Arches*.

Invincible. A patent glazing bar for roof lights.

Invisible Glazing. A patent non-reflecting shop window. The glass is curved inwards and a black base trough makes the glass invisible.

Involute. The curve formed by unrolling the perimeter of any plane figure. The plane figure is the eye, or evolute.

Ioco. A patented rubber, jointless flooring material.

Ionic. One of the orders of architecture the chief characteristic of which is the scroll, or voluted, capital. See *Architectural Orders*.

Ionic Volute. See *Volute*.

Ipil. A yellowish to dark-brown hardwood from the Philippines. It has black spots due to a secretion and is hard, heavy, strong, and durable. Rather difficult to work. Used for furniture, fittings, and where strength and durability are of primary importance. Weight about 54 lb. per c. ft.

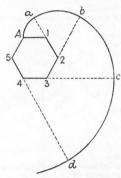

INVOLUTE OF HEXAGON

IONIC

Iroko. A West African hardwood used for superior joinery, etc. It is brown with dull streaks, moderately hard, heavy and stable, but very strong and durable and used as a substitute for teak. Weight about 42 lb. per c. ft.

Iron. Fe. Extracted from ores : magnetic, red haematite, brown haematite, spathic. The processes for extraction are : crushing, weathering, calcination, smelting. Different ores require different treatment. See *Cast Iron*, *Wrought Iron*, and *Steel*.

Ironbark. The name applied to several species of Eucalypti. They are pale to dark brown, and very hard, heavy, strong, durable, and excellent woods for structural work, piling, flooring, etc. Weight about 63 lb. per c. ft.

Ironbox. The name applied to several species of Eucalypti. They are too hard for ordinary purposes. Used for structural work, piling, and as a substitute for lignum-vitae. Weight about 65 lb. per c. ft.

Iron Cement. A proprietary material for stopping leaks in boilers, pipes, etc., and for plugging walls to receive bolts, etc.

Iron Core. A continuous metal plate tying together the tops of the balusters to a stairs. The handrail covers the core.

Ironite. A surface waterproofer for brickwork and concrete. It is mixed with cement and brushed on in the form of a slurry. Ironite No. 2 is for smooth surfaces and is mixed with water only.

Ironmongery. A general term for articles made of iron, etc. Hardware. Metal fastenings and fittings used on doors, windows, etc.

Iron Oxide. Used in cheaper paints, especially for iron and steelwork, but not so protective as red lead. It is produced from iron ore. See *Pigments*.

Iron Pyrites. Brass-like crystals found in slates. Bisulphide of iron, FeS_2.

Irons. The cutters of planes or machines. See *Plough*.

Ironwood. The name applied to nearly 100 tropical and semi-tropical woods. They are chiefly species of acacia and are all very hard, heavy, and difficult to work. The commonest of them are known by other names: Billian, Ekki, Gangau, Kajoe, Lapacho, Melkhout, Merban, Mesquite, Muhimbi, Olivewood, Penago, Pyinkado, Quebracho, Ru, Snakewood, Tembusu, Tempinis, etc.

Island Station Roof. A truss supported on a central column only, as used on station platforms. See *Roofs.*

Isoflex. A proprietary plastic insulating material.

Isolament. A proprietary casehardener for cement surfaces to prevent dusting.

Isolator. See *Floating Floor.* A term used in acoustics for materials that prevent vibration from being transferred from one part of a structure to another part.

Isometric. A form of pictorial projection in which measurements are made along three axes that are arranged at an angle of 120° to each other, with one vertical.

Isosceles. A triangle having two sides and two angles equal.

Isteg. A patent twisted steel bar for reinforcing concrete. Now called Tentor.

I.T. Abbreviation for *intercepting trap.*

Itako. A light-yellow lustrous hardwood from Nigeria. It is hard, heavy and with slightly interlocked grain. Polishes well and suitable for superior joinery, etc. Weight about 51 lb. per c. ft.

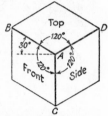

ISOMETRIC PROJECTION

Italian. A comprehensive term embracing the architectural styles of Florence, Rome, and Venice.

Italian, or **Roman, Mosaic.** Laid on concrete and formed of ¾ to 1 in. square regular pieces. The design is prepared and attached to paper until bedded on the cement. When set, the paper is removed and the mosaic rubbed down.

Italian Tiles. Consist of a flat tapered *under-tile,* or tegula, with flanges, and a half-round tapered *over-tile* that covers the flanges of adjacent under-tiles.

Italic Tiles. Patent tiles similar to pantiles but with water bar underneath.

Ivory Black. See *Black Pigments.*

I.W. Abbreviation for *iron weights.*

J

Jack. An appliance for raising heavy loads, usually by means of a screw thread and a lever handle. Other forms are the ratchet and hydraulic jacks.

Jack Arches. Small arches used to support the road over the spandrels of bridges. This cellular construction is used instead of earth filling to lessen the dead load.

Jacket. An insulating cover to protect a cistern and to retain the heat. It protects the insulating material. The term is also applied to the facing veneers of laminboard.

Jack Plane. A joiner's plane, about 17 in. long, used for preliminary and rough work.

Jack Pump. A lift pump with open barrel discharging directly into the spout.

Jack Rafters. The short rafters between the hip or valley rafter and wall plate, and in the same plane as the common rafters.

Jack Rib. 1. A curved jack rafter, as in a domical turret roof. Also called *cripple* or *crippling.* 2. A lierne rib, *q.v.*

Jack Timbers. Jack rafters. See *Roofs.*

Jacobean. A transitional style of architecture prevailing from A.D. 1603 to 1649. Gothic features with Renaissance detail were characteristic, with twisted or banded columns.

Jagged. Cut roughly, uneven, or torn. Shaped with indentations.

Jagger Table. A patent machine for vibrating concrete to make it compact.

Jali Panel. A richly carved, or perforated, panel characteristic of Indian architecture. The carving consists of a network of intersecting projecting mouldings, similar to vermiculated work. Also called Jolli panels.

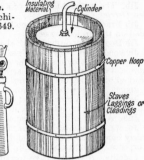

JACK   JACKET FOR CYLINDER

Jalousie. An outside blind framed with slats like louvres that open and close. A Venetian blind.

Jaman. A brownish-grey hardwood, with darker streaks from India. Fairly hard, heavy, durable, and strong. Interlocked and wavy grain. Used for structural and constructional work, piling, etc. Selected wood used for superior joinery and fittings. Weight about 42 lb. per c. ft.

Jamb Linings. The linings covering the jambs of wall openings. See *Finishings.*

Jambs. 1. The vertical sides of wall openings. 2. The vertical posts of framing fixed to the jambs of openings. 3. The sides to a fireplace.

Japan Gold Size. A quick-drying varnish used as a mordant for gilding and for flat paints.

Japanned. Ironwork coated with a hard black varnish to prevent corrosion.

Jarrah. *Eucalyptus marginata.* A hard, tough, durable timber. Reddish brown colour. Rather coarse grain. Works smoothly and polishes well. Used for all purposes. Weight about 54 lb. per c. ft. See *Eucalyptus.*

Jasil Water Bar. A patent metal bar to check the entry of water at the bottom rail of a casement opening inwards.

Jaw Box. A wood sink with lead lining.

Jelly Moulds. See *Gelatine Moulding.*

Jenny. A travelling winch.

Jenny Wheel. See *Gin Wheel.*

Jequitiba. A deep red-brown hardwood with darker streaks, also called Brazilian or Columbian mahogany. Very variable in characteristics. Used for carpentry, superior joinery, veneers, etc. Weight from 30 to 42 lb. per c. ft.

Jerkin Head. A roof that is hipped towards the top of a gable. The feet of the hips are usually about half-way up the triangular gable.

Jerry. A term applied to inferior work.

Jesse. A particular type of church candlestick or sculpture with branches representing the "Tree of Jesse."

Jesting Beam. A beam introduced °r appearance and not for utility.

Jettied Construction. Half-timb " buildings with projecting upper story.

Jetty. 1. A projecting structure built on piles forming a landing stage at the water's edge. 2. A projecting or overhanging part of a building.

Jhil Mill. A mesh reinforcement for concrete floors and partitions. It is self-centering if necessary.

Jib. The inclined member of a crane from which the load is suspended. See *Derrick Crane.*

Jib Bracket. A framed wood bracket with brace so that it can carry a load on the projecting end.

Jib Door. A door flush with the wall so as to be inconspicuous.

Jicwood. See *Improved Wood.*

Jig. A guide or templet for controlling tool operations for shaping the material on a machine. An arrangement to hold small stuff whilst being worked on a machine. See *Building Jig.*

Jigger. A reciprocating trough feeder for brickmaking.

Jigger Saw. A machine fret saw. A short saw blade moves up and down and is used for circular work in moderately thin stuff.

Job. A piece of work. A small contract.

Jocote. A yellowish-brown, lustrous, hardwood from C. America. It is fairly hard, heavy, durable, and strong. Used for construction and general purposes. Weight about 42 lb. per c. ft.

Jodelite. A preservative for timber. It is applied hot and is effective against disease and insect attack.

Joggle. 1. A small projection left on a framed member to strengthen an angle joint. It is usually ornamental in character as on the stile of a sash. See *Sash and Frame.* 2. A projection on one stone to fit in a corresponding recess in another stone; or adjacent recesses filled with cement grout. See *Masonry Fixings.*

Joggled Lintel. A flat arch of stone or terra-cotta formed of small blocks jointed together with joggle joints.

Joggled Piece. A post having a shoulder as an abutment for a strut.

John Bull. A proprietary oil-bound washable distemper.

Joinery. The art of preparing and fixing the wood finishings of buildings.

Joint. A prepared connection between two or more pieces of material. WOODWORK JOINTS. These are classified as (a) *Angle joints*: halving, mortise and tenon, dovetailing, housing, bridling, cogging, keying, scribing, etc. (b) *Lengthening joints*: scarfing, lapping, halving, fishing, tabling, etc. (c) *For increasing the width*: butt, tongued and grooved, rebated, etc. (d)*Hinging and shutting joints.* In the first three classes the joints are often strengthened by glue, nails, screws, dowels, bolts or other metal fastenings. MASONRY JOINTS. A joint in masonry implies one not subject to much pressure, as distinct from a bed joint. The usual types are: butt, rebated, joggled, or tabled, and they may be secured by dowels, cramps or plugs. See *Masonry Fixings.* METAL JOINTS. Riveted, bolted, brazed, welded, soldered. See *Pointing.*

Joint Bed. Applied to stone in which the natural bed is vertical and at right angles to the face of the wall, as used for projecting courses, such as cornices, etc.

Jointer. 1. A woodworking plane larger than the try-plane for straightening long edges. 2. A bent steel tool used for forming brickwork joints when pointing. A jointer saw is one for sawing stone.

Joint Fastener. A corrugated metal fastener, driven in at the ends, to secure and tighten glued butt joints.

Jointless Floors. The surface of floors formed of composition applied in a plastic condition and trowelled. The composition has a magnesia cement base compounded with waste materials such as cork, sawdust, rubber, asbestos, etc., and is of various colours. It sets hard but with the qualities of wood or linoleum. See *Sorel Cement*. The term may also be applied to cement, granolithic, and bituminous finishes.

Joint Mould. See *Section Mould*.

Joint Rule. A straight-edge used with the jointer when pointing brickwork.

Joist Hanger. A pressed-steel or cast-iron shoe to carry the end of a joist.

Joists. Horizontal timbers to carry floors or ceilings. Also small steel beams to carry floor slabs. See *Filler Joist*.

Jolli Panel. See *Jali Panel*.

Joule. A unit of heat. The heat produced by 1 ampere through 1 ohm in 1 second.

Journal. The portion of a shaft in the bearing.

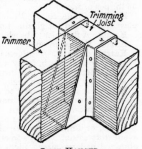

JOIST HANGER

Jt. Abbreviation for *joint*.

Jtg. Abbreviation for *jointing*.

Jube. 1. A choir screen. 2. See *Rood Loft*.

Judas. A peep-hole in a door.

Juglans. See *Walnut*.

Jumbo Bolt. The bolt of a rim lock operated by the thumb instead of a key.

Jume. A patent slate clip.

Jump. A step in a brick or masonry course or in foundations.

Jumper. 1. A long octagonal or round steel chisel used by the mason, chiefly for drilling holes in hard stone. 2. The seating for the washer in a water tap. 3. A riser in snecked rubble masonry.

Jump Joint. A butt joint in riveted steel plates.

Junction. A specially prepared drain pipe with a socket at the side to receive a branch drain.

Junction Block. The joint between drain and sewer. See *Drainage System*.

Junk. A collection of comparatively useless things.

Jury. An adjective implying "temporary."

Jutahy. A reddish to dark-brown hardwood from C. America. It is very hard, heavy, strong, and durable, and rather difficult to work. Used for structural work, piling, etc. Weight about 54 lb. per c. ft.

Jute Fibre. Used in plaster as a substitute for hair.

Jutty, or Jetty. A projecting or overhanging part of a building.

K

K. A symbol for *radius of gyration* and *bulk modulus*, and abbreviation for *knot*. **k.** A symbol for *constant*.

Kahikatea. New Zealand white pine. Characteristics of yellow pine and used for the same purposes. Weight about 29 lb. per c. ft.

Kahn System. A system of reinforcing concrete. The trussed bar has a diamond-shaped section for the core with wings on each side. These wings are cut and bent to resist the shear where required. The Kahn rib has projecting lugs for providing a mechanical bond with the concrete.

Kaim. A greyish-brown hardwood from India. Fairly hard, heavy, but not durable. Polishes well, rather difficult to work. Used for construction, joinery, etc. Weight about 38 lb. per c. ft.

Kalamein Sheathing. Sheet bronze drawn on a hardwood core. Used for shop fronts and fittings, etc. See *Bronze Sheathing*.

Kaleidoscope. A figured rolled glass giving obscurity with high light diffusion.

Kalsomine. A brush having stiff scoured bristles, used for distempering.

Kalumpit. A yellowish to greyish-brown lustrous hardwood from the Philippines. It is fairly hard, heavy, and strong, but not durable. Somewhat like a white Lauan. Used for joinery, flooring, etc. Weight about 36 lb. per c. ft.

Kaolin. China clays. Primary, or pure, clays. They are deficient in plasticity. Used for glazing bricks, etc.

Kapok. A flocculent absorbent filler used for acoustical purposes.

Kapor. Camphor wood (*q.v.*).

Karrah, or **Karri.** *Eucalyptus diversicolor.* Similar to Jarrah but stronger. Very dense and tough, fairly durable, and fire-resisting.

Kata Thermometer. An instrument for measuring the rate of cooling. It is used by the heating engineer for obtaining the comfort index, *q.v.*

Katon. A greyish-brown to mahogany-red hardwood from Burma and Siam. Interlocked grain, dark-brown flecks, white deposit, variable characteristics. Polishes well. Suitable for superior joinery, fittings, etc. Weight from 30 to 40 lb. per c. ft.

Kauri, or **Cowri.** *Agathis Australis.* New Zealand pine. Close and even, silky grain. Warps freely and shrinks endways. Works easily, and obtained in large sizes. Light yellowish-brown colour. Used for joinery, fittings, etc. Weight about 33 lb. per c. ft.

Kaye's Patent. A method of fixing door knobs to the spindle.

K. Ct. Abbreviation for *Keene's cement.*

Keaki. A golden-brown, lustrous, hardwood, of the Elm family, from Japan. It is excellent wood and very decorative, and easily wrought. Highly esteemed for cabinet and decorative work, etc. Plain wood is used for structural work because of its strength and durability. Weight about 38 lb. per c. ft.

Keedon. A patent reinforcement for concrete.

Keel. The projecting fillet that forms a roll or scroll.

Keel Blocks. The blocks on the bottom of a dry dock on which a boat rests whilst undergoing repairs.

Keel Moulding. A Gothic moulding formed of two ogees meeting with a sharp edge like the keel of a ship.

Keene's Cement. A "baked" plaster. Gypsum is twice calcined at high temperatures and soaked in alum between the burnings to act as an accelerator. It is used for arrises, reveals, etc.

Keep. A stronghold in an ancient castle.

Keeper. 1. A recessed piece of metal fixed over the fall bar of a Norfolk latch to limit the movement. Also see *Tower Bolt.* 2. The socket fitted on door jambs to house the bolt of the lock when in the shut position.

Keeper Plate. A striking plate, *q.v.*

Keeping the Perps, or Perpends. Making the vertical joints of alternate courses of brickwork perpendicularly over each other. The term is also used for slating or other bonded materials.

Ke-it. Small tapered dovetailed strips of rubber. They are wrapped in paper and laid on the formwork before the concrete is poured. They are easily removed to form a key for plastering.

Keith's Rule. Used to find the required radiation surface of pipes for hot-water heating.

Kemnay. A silver-grey granite with black specks of mica, from Aberdeen. Used for constructional work. Weight 161 lb. per c. ft. Crushing strength 1212 tons per sq. ft.

Kenmore. A registered wood-fibre wall-board.

Kentish Rag. A hard compact limestone from the Greensand formation.

It is used chiefly for polygonal rag work and rubble walling, as it is very difficult to dress.

Kentledge. Heavy material placed on top of a monolith to help it to sink into the ground.

Kenton. A light-brown sandstone from Northumberland. Used for general building, dressings, etc. Weight 142 lb. per c. ft. Crushing strength 467 tons per sq. ft.

Kenyon Interceptor. A patent interceptor for a manhole, in which a vertical valve operated by a chain can empty the manhole through the cleaning arm when there is a stoppage in the trap.

Kerb. A stone edging to a raised footpath. Also see *Curb*.

Kerf. A saw cut only partly through the material.

Kernel. Same as Crenelle, *q.v.*

Keruing. See *Gurjun*.

Kestner Cement. Refractory hydraulic cement. It will stand temperatures up to 1,300° C.

Ketton. Varied coloured limestones, grey, cream, brown, from Rutlandshire. Used for general building and dressings. Weight 156 lb. per c. ft. Crushing strength varies.

Ketton Cement. A Portland cement manufactured in the Midlands.

Kevil, or Kevel. 1. Used in the Portland stone quarries for dressing blocks of stone. It is a combined pick and spall hammer and weighs about 7 lb. 2. A cleat with two pegs for securing the free end of a rope.

Key. 1. A piece of hardwood in a joint to prevent sliding. See *Fish Plates*. 2. A wedge used in carpentry. 3. A dovetailed batten to prevent the warping of wide boards. 4. Roughening a surface to receive a coat of plastic material; or the spaces between laths, or in wire mesh, to provide a grip for the plaster. 5. An appliance for shooting the bolt of a lock. 6. A spanner for turning a nut. 7. Same as *cotter*. 8. A tool for jointing brickwork. 9. Applied to many things that secure one thing to another, as a pulley to a shaft, railway keys, etc.

Key Block. The last block in a series to be placed in position and that gives stability to the whole. A *key-stone*. See *Arch* and *Bull's Eye*.

Keyed. Fixed or secured by a key, or wedge.

Keyed Joint. See *Pointing* and *Key* (1) and (8).

Keyed Strutting. Strong pieces tenoned through floor joists and wedged like a tusk tenon to provide stiffness.

Keyhole Saw. See *Pad Saw*.

Key Pile. The last pile to be driven in sheet piling to complete the bay.

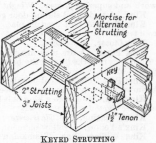

KEYED STRUTTING

Key Plan. A small-scale plan showing the relative positions of a number of units constituting a whole, as the plan of an estate or a survey.

Key Plate. A metal plate with keyhole. It is screwed over the keyhole in the door, to prevent the wood from being worn by constant insertion of the key.

Key Post. Same as *Puncheon* (1).

Key-stone. The central voussoir at the crown of an arch, or the top stones of a vault, or the boss of a dome. See *Key Block*.

Keystone. A proprietary paint for new cement work.

Kibble. 1. See *Kevil*. 2. An iron bucket for hoisting loose materials.

Kick. 1. The projection on the stock board to form the frog in bricks. 2. Recoil, or backlash.

Kick Back. Applied to the surplus antiseptic released from the wood when the pressure is released in preservation.

Kicker. A rail to prevent a drawer from dropping at the front when open.

Kicking Plate. A wide metal plate to protect the bottom rail of a door.

Kicking Strap. A heel-strap that only ties the timbers together and does not take any thrust.

Kieselguhr. See *Moler*.

Killed Spirits. Zinc chloride. Zinc dissolved in hydrochloric acid, or spirits of salts.

Killesse. 1. See *Coulisse*. 2. A local term for a dormer or a hipped roof.

Killing Knots. Covering knots with some form of *knotting* to prevent the resin from penetrating the paint.

Kiln Dry. Applied to kiln-dried wood with less than 12 per cent moisture content.

Kilns. Chambers in which great heat is generated for the manufacture of various materials such as bricks, cement, lime, etc., and for the seasoning of timber. There are many types: continuous, intermittent, Scotch, Hoffman, Manchester, English, etc. The kilns for seasoning wood are much cooler and there are numerous types, up to 50 ft. long. The heat, humidity, and circulation of air are automatically controlled. See *Progressive Kiln*.

Kilowatt. Unit of electric power, equal to 1000 watts.

Kilt. The weathering to stone steps.

Kimoloboard. A proprietary fire-resisting wallboard of mineral composition but easily sawn. Kimolo is the trade name for numerous diatomaceous earth products.

Kindal. A greyish-brown to rich-brown hardwood from India. Moderately hard, heavy, strong, and durable. Somewhat like black walnut and a good substitute for teak. Used for constructional and structural work and superior joinery. Weight about 45 lb. per c. ft.

King. A registered sliding-door gear.

King Bolt. A bolt used in place of a post in a composite or king *bolt* roof truss.

King Closer. See *Closer*.

King Floor. A fire-resisting self-centering floor in which hollow blocks, or lintels, are forked on to the bottom flanges of the filler joists, which are placed at 3 ft. centres.

King Post. The vertical member from the apex to the middle of the tie beam in a king-post roof truss. See *Roofs*.

King Rod. Similar to a king bolt but in an all-steel king truss, hence the ends are riveted.

King's Glazing. See *Glass-crete*.

King's Yellow. A yellow pigment, chiefly tri-sulphide of arsenic.

King William Pine. Tasmanian pine. Pinkish, soft, light, but fairly strong and durable. Used for joinery and cheap cabinet work, etc. Weight about 22 lb. per c. ft.

Kinked. Undulating or buckled.

Kino. An astringent gum from Eucalyptus.

Kiosk. An open pavilion, supported by pillars. A small wood structure with serving hatch for the sale of popular commodities.

Kips. Abbreviation for kilo-pounds, or 1000 lb.

Kisol. A proprietary vermiculite concrete for light-weight, insulation and fire resistance.

Kiss Marks. Due to unequal burning of bricks where they have been in contact in the kiln.

Kit. The equipment required by a craftsman to do his work. An outfit or set of tools.

Kitchen. A room set apart for the preparation of food.

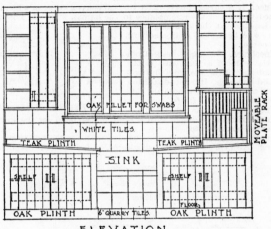

ELEVATION

PLAN

KITCHEN FITMENTS

Kitchener. A kitchen range.

Kitchenette. A combined kitchen and scullery.

Kitchen Fittings or Fitments. Cupboards, dressers, shelves, tables, racks, etc., suitable for the storing and preparation of food. There are many registered designs and patents, involving labour-saving devices, for kitchen fitments.

Kite Winder. The centre winder of three, forming quarter-space winders. It is so called because of its resemblance to a kite, or trapezium.

Kleine Blocks. Self-centering hollow fireclay blocks for fire-resisting floors. Steel reinforcement and concrete are used in conjunction with the blocks.

K.L. Storage. A soundness test for cement by alternate storage in water and air.

Knapen System. A patented method of curing damp walls by trapping the flow of moisture and aerating the interior of the walls.

Knapped Flints. Snapped flints roughly squared on the facing end to show a gauged, or squared flint in walls.

Knee. 1. A curved brace. 2. A vertical convex curve in a handrail. So named because of the resemblance to the human knee 3. The return of the dripstone at the springing of an arch. 4. A bent piece of iron serving as a corbel.

Kneebrace. An angle brace from the tie beam of a roof truss to the column supporting the truss, in steel-framed buildings. It resists distortion due to the wind pressure on the side of the building. See *Roofs.*

Kneed Bolt. A cranked bolt, *q.v.*

Kneed Hinges. A hinge with a right-angled bend for the knuckle for skylights, box lids, etc.

KNEES TO HANDRAILS

Kneeler, or **Kneestone.** A stone bonded with a gable wall to support the coping. Also called *bondstone, skew-corbe,* and *club-skew.* See *Gable.*

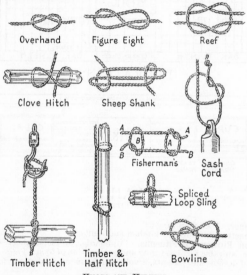

KNOTS AND HITCHES

Kneeling Boards. Sloping boards in church pews for kneeling upon during prayers.

Knitting. Joining together. Especially applied to plastic materials.

Knitting Layer. A layer of cement grout or fine concrete between old and new concrete.

Knobbing. The operation of breaking off pieces of stone roughly with a hammer.

Knobs. See *Door Knobs*.

Knocker. A hinged piece of metal striking on a metal plate and fixed to an entrance door.

Knocking. The noise in water pipes due to trapped air or a faulty washer.

Knocking Up. Working up plastic materials partly set.

Knot. 1. A section through a branch where it penetrates the trunk of a tree. The following names are standardized. *Branched*: two or more springing from a common centre. *Dead*: not joined to the surrounding wood. *Decayed*: infected by fungus. *Encased*: a dead knot surrounded by pitch or bark. *Enclosed*: not seen on face of wood. *Large*: over $1\frac{1}{2}$ in. diameter. *Live*: firmly joined to surrounding wood. *Loose*: a shrunken dead knot. *Oval*: obliquely cut knot. *Pin*: less than $\frac{1}{2}$ in. diameter. *Pith*: a sound knot with pith hole. *Rotten*: advanced stage of decay. *Round*: cut at right angles to axis. *Spike* and *Splay*: sawn lengthwise. *Sound*: same as live. *Standard*: less than $1\frac{1}{2}$ in. diameter. *Tight*: sound and securely fixed. 2. A length of sash cord of 12 yd. 3. Looping a cord on itself as a fastening in scaffolding, etc.

Knot Brush. Painter's brushes with the bristles arranged in round or oval bunches. A single knot is used for painting, but two- or three-knot brushes are used for distemper. See *Flat Brush*.

Knotted Over. Joinery prepared for priming by coating the knots with knotting.

Knotting. Various compounds, usually shellac and methylated spirits, used to cover knots in timber before the initial coat of paint. Shellac solution is also used on ironwork that has been coated with Angus Smith composition, or other tar products, to prevent the tar from showing through the paint.

Knott's Clay. Found in the Oxford clay around Peterborough. It contains carbonaceous matter and so requires little fuel for the manufacture of Fletton bricks.

Knowle's Free Flow. A patent joint for drain-pipes making them self-centering and ensuring alignment.

Knuckle. That part of a hinge containing the pin, or pivot.

Knuckle Joint. 1. Used for lengthening steel rods. The end of one rod is forked to receive the end of the other rod, and a pin, bolt, or rivet is passed through to secure them. 2. A joint for a falling leaf to a table. See *Rule Joint*.

KNUCKLE JOINT

Knuckle Soldered Joint. A plumber's joint at the junction of two lead pipes forming a right-angled bend.

Kokko. An Indian dark-brown hardwood with irregular darker markings. Fairly hard, heavy, and durable. Used for superior joinery, flooring, etc. Nigerian kokko is similar. Weight about 40 lb. per c. ft.

Kontite. A patent joint for copper tubes.

Korkoid. Registered decorative composition floor.

Korkoleum. A proprietary composition for floors.

K.P.S. Abreviation for *knot, prime, stop*.

Kraft Paper. A thin tough paper used or Cabot's quilt, sheathing paper, etc.

Kraftwood. A registered decorative type of plywood for interior wall surfaces.

Ku Bricks .A patent type of engineering brick in the form of a folding wedge for pinning new concrete beams to old work.

Kudampro. A proprietary copper damp-course strip.

Kulm. A patent, light, fire-resisting, partition block, made of crushed pumice and fire-resisting cement.

Kurrajong. Several species of a cream-coloured hardwood from Queensland. It is light, soft, not durable or strong. Used for interior fittings. Weight about 25 lb. per c. ft.

Kursaal. An amusement centre, public hall, etc., at a health resort.

Kusam. A reddish-brown hardwood from India. It is very hard, heavy, and strong, but not durable. Difficult to work smooth. Used chiefly for heavy constructional work. Weight about 55 lb per c. ft.

KW. Abbreviation for *kilowatt*.

Kwikform. A patent shuttering for concrete.

Kyanizing. Impregnating timber with corrosive sublimate, as a preservative.

Kyljack. A plastic material to protect water pipes against frost.

Kynal. A proprietary aluminium non-slip tread.

L

l. Abbreviation for *linear*.

Lab. Abbreviation for *labours*.

Label or **Label Mould.** A hood moulding. A moulding terminated by corbels, over a window or doorway. A drip stone.

Labours. A term used in estimating for the labour expended on the materials.

Labradorite. A Swedish stone belonging to the augitesyenites. Used for decorative and monumental work.

Laburnum. A very hard European timber used for ornamental work. Yellow sapwood and brownish-green heartwood.

Lac. See *Shellac.*

Lace. A brace or tie. A timber tying, or lacing other timbers together.

Laced Column. A square column consisting of four heavy constructional timbers held together with diagonal braces on all four sides. It is similar to a slender derrick tower, and is used in temporary buildings, as for exhibitions.

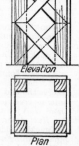

Elevation

Plan

LACED COLUMN

Laced Valley. A valley formed by tiles without a valley gutter. A "tile and a half" tile is used at the angle between the roofs, on a wide board laid in the valley.

Lacet. A winding roadway forming the approach to a building on the top of a sloping site.

Lacing Course. Courses of bricks, stones, or tiles, used to level up and strengthen walls built of irregular materials such as rubble, flints, etc. A radial course of stretchers in a large arch built of half-brick rings.

Lacings. Vertical timbers tying together the horizontal struts in a coffer dam.

Lacotile. An imitation tile made in large sheets with waterproofed compressed-wood backing.

Lacquer. 1. A varnish for brass, usually shellac dissolved in alcohol, to prevent the metal from tarnishing, or oxidizing. 2. Japanese lacquer is a black resinous varnish.

Lacrinoid. A registered name for numerous plastic fittings, etc.

Lacunar. A coffer. A sunk panel in a ceiling.

Lacunaria. Masonry ceilings in ancient Greek architecture.

Lade. A mill-race or water-course.

Ladies. Medium-sized slates, 16 in. × 8 in. usually, and requiring about 280 to the square at 3 in. lap.

Ladkin. A boxwood tool for opening the cames in lead lights.

Ladle. A bowl-shaped vessel with long handle used by the plumber for pouring molten lead.

Lady Chapel. A chapel dedicated to the Virgin Mary. Cathedrals and large churches usually have a Lady Chapel for special services with small congregations.

Laggings, or Lags. 1. Narrow strips of wood, as for tying together the two frames, or felloes, of a center and carrying the voussoirs. See *Centering*. 2. See *Snow Boards*. 3. Narrow strips for jacketing boilers and cylinders. See *Jacket*. 4. Lagging implies insulating pipes and cisterns against frost.

Lag Screw. Similar to a coach screw but with conical point instead of gimlet point. Used in heavy timber construction. See *Bolts*.

Laing's Floor. A fireproof floor formed of ferro-brick blocks. The blocks are placed end to end with cement joints, and reinforcing rods are placed in grooves on each side and secured by cement mortar. The completed beam is then in effect a pre-cast beam, and is self-centering.

Lairages. Temporary quarters for live-stock landed from ships.

Laitance. Scum or slurry formed on the surface of concrete due to the separation of gypsum from the cement, or to dirty materials.

Lakeland Walling. A dry boundary walling common in Westmorland.

Lake Pigments. Pigments produced by fixing organic dyes on inorganic bases. They are crimson, purple, scarlet, carmine, madder, and black, lakes; Dutch pink and rose pink. They possess little obscuring properties.

Lalique. A registered moulded, decorative, heat-resisting glass.

Lamb's Tongue. A flat ogee moulding commonly used on sash stuff. See *Mouldings*.

Lamella. A patent steel construction. It is a light triangulated pressed steel for roof construction, to dispense with trusses for large spans. The principle is also applied to a network of wood.

Laminated. A member built up to the required size by several thin layers. The method is useful for building up circular work.

Laminated Glass. A safety glass consisting of outer layers of glass with celluloid-like core.

Laminated Wood, or Laminwood. A development of plywood. It consists of a core of strips, with reversed grain, glued together, and faced on both sides with a thick veneer. The core is usually of deal or Oregon pine and the veneers of Gaboon mahogany. Sizes up to 15 ft. × 5 ft. and from $\frac{3}{16}$ in. to $1\frac{1}{2}$ in. or more in thickness.

Laminating. Peeling away in thin flakes, as in defective tiles.

Lamorna. A greenish-grey Granite from Cornwall. Used for constructional work. Weight 168 lb. per c. ft. Crushing strength 1100 tons per sq. ft.

Lamp Black. See *Black Pigments*.

Lamp Holes. See *Inspection Eyes*.

Lamsom Tubes. Pneumatic tubes for conveying cash from counter to cashier's office in stores.

Lancashire Bricks. Dense, impervious, and uniform in size and bright red colour. They are excellent facing bricks and used extensively in sanitary engineering work. Also called *Accrington Bricks*.

Lancet Arch. A Gothic, or pointed, arch with the radius of curvature greater than the span. See *Arches*.

Land Drain. These are usually formed of unsocketed earthenware pipes laid end to end in a trench and covered with broken stones.

l & h. Abbreviation for *lime and hair*.

Landing. 1. A wide resting place at the top of a flight of stairs. See *Half* and *Quarter Space*. 2. A large paving stone, over 1 sq. yd. area.

Landing Stage. A platform, often floating, at the waterside for embarking passengers, etc. A projecting platform to a scaffold for landing the materials.

L. & P. Abbreviation for *lath and plaster*.

Landscape Panel. Same as lay panel, *q.v.*

Land Tie. A tension bar secured to the back of a retaining wall and anchored in the earth to stone or concrete or by means of a plate.

Langite. A proprietary cork material for damping vibrations.

Lantern Light. A projecting erection above a roof, to provide light. The outline of the plan may be of any form.

Lap. 1. The amount that one slate covers the next but one below it. See *Gauge*. Also see *Passings*. 2. Similar to onyx and used for decorative purposes. 3. A rotating disc for polishing hard stones or metals.

LANTERN LIGHT

Lap Dovetail. A dovetailed joint for drawer fronts. The end grain is seen on one side only. See *Dovetail*.

Lapidolith. A proprietary, liquid, chemical hardener for concrete.

Lapis Lazuli. 1. A semi-precious stone used with marble as a decorative feature. 2. A silicate, with sulphur, from which a bright blue pigment is obtained.

Lap Joint. A joint in which the two pieces overlap each other.

Laplage. A proprietary white glazed tile of specially large dimensions, 8 in. × 8 in.

Lapping. See *Grummet*.

Lap Scarf. A joint used for joining wooden gutters lengthways. The lower part of one projects into a corresponding space removed from the other. The joint is secured and made good by paint, white lead, and screws.

Lap Sidings. Weatherboards that overlap each other. There is no preparation of the joints. See *Sidings*.

Larana. A Nigerian hardwood somewhat like sycamore.

Larch. *Larix Europœa.* The toughest and most durable coniferous timber. Straight grain, free from knots, but shrinks and warps badly. Excellent for posts, piles, and sleepers. Weight about 37 lb. per c. ft.

Larder. A domestic place for storing food.

Large Knot. See *Knots*.

Larmier. The corona or drip in an entablature or cornice.

Larried Up. Lime and sand mixed together with the larry.

Larry. 1. An implement with long handle for mixing lime, sand, and hair for plaster. 2. Semi-liquid mortar.

LARRY

Larrying. Bedding bricks in semi-liquid mortar. A sliding motion flushes up the mortar in the heading joints and the pouring of the mortar for the next course fills the crevices. The method is adopted for filling between the facings of thick walls in inferior work.

Larssen Piling. A registered steel sheet piling, trough-shaped and with interlocking joints.

Lashings. Ropes for tying together scaffold timbers. See *Whips.*

Latch. A pivoted bar engaging with a catch to keep a door closed. See *Norfolk Latch* and *Night Latch.*

Latchet. A *tingle.*

Latent Heat. The heat required to change the state of a substance without its temperature changing.

Lateral Escape. The *squeezing out* of soft soils from under the foundations of heavy structures, which is prevented by sheet piling.

Lateral Wind Bracing. Diagonal bracing lying in the plane of the roof.

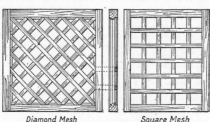

Diamond Mesh   Square Mesh

LATTICES

The connections are made on the backs of the principal rafters. Any type of bracing to prevent lateral movement due to the wind.

Lath. See *Laths.*

Lath Bricks. Special bricks, 22 in. × 6 in., used in hop-kilns.

Lathe. A machine for turning wood, metal, stone, etc., into circular form.

Lathing. The nailing of laths or perforated metal on joists and studs to receive plaster. Laths should be butt jointed and arranged, in foot widths, to break joint on alternate joists, and about ⅜ in. apart to provide a key for the plaster. Various patent types of metal lathing are used for plaster work. The spacing of slater's laths, or battens, depends upon the *gauge.*

Laths. Strips of wood of small section for carrying plaster work, or slates and tiles. Plasterer's laths are sawn or riven from Baltic fir, and 3 to 4 ft. long. Single laths are 1 in. × ³⁄₁₆ in., lath and a half 1 in. × ¼ in., and double laths 1 in × ⅜ in. Slate and tile laths, or battens, are about 1½ in. to 2 in. × ⅝ in. to 1 in. The laths, or battens, for shingles are stronger and depend upon the spacing of the rafters. See *Pantiling,* and *Venetian Blinds.*

Lath Scratcher. Several short pieces of laths pointed and held together by laths nailed on the side. It is used for scratching plaster surfaces to provide a key for a succeeding coat.

Latona. A patent portable oil radiator.

Latrines. A number of water closets connected with each other and under one roof. They may be in the form of a continuous trough.

Lattens. Sheet iron less than No. 24 B.G.

Lattice Girder. A built-up girder braced by diagonal braces crossing each other between top and bottom chords. The stresses are obtained graphically by superposition, *q.v.* See *Roofs.*

Lattices. 1. Any arrangement in wood or metal in which narrow strips, or laths, cross each other to form an open network. 2. Skeleton construction in steel work.

Lattice Roof. One built up in lattice formation like a Belfast Truss, *q.v.*

Lattice Window. One in which the spaces are filled with panels of lattice work, or in which the glass is bedded in bars crossing each other diagonally.

Lauan. Philippine mahogany. About 14 species of *Dipterocarpus* are imported under the names of red and white lauan, hence there is great variation of the woods. They are similar to mahogany and used for similar purposes.

Launder. A local name for eaves gutter.

Laundry. A place where clothes are washed and pressed.

Laurel. There are numerous species but they are of little importance in building. The most important at present is Indian laurelwood, which is not a true laurel. It is light brown to brownish-black with darker markings. It is an excellent wood and highly esteemed for decorative work, electric casings, etc. Californian laurel is also called myrtle, *q.v.*

Lavatory Basin. A wash bowl.

Lavatory Waste. The pipe emptying the lavatory basin. It may be of lead, copper, or alloy, and it is trapped just below the bowl.

Lay. To apply the pricking-up coat in plastering.

Layer, or **Lear, Boards.** 1. The boards carrying lead gutters. 2. A board to level up to the battens and over which the lead is dressed for valleys and gutters.

Plaster Ceiling   Sash   Ceiling Joist

Section

Laying Trowel. A bricklayer's trowel. It may be left- or right-handed and from 10 in. to 13 in. long.

Lay Light. See *Ceiling Light.*

Laymat. A proprietary floor surfacing. It is supplied in powder form and mixed with water. It can be applied to any base and is hard wearing and insulating.

Lay-out. Arrangement of a building and its parts on a building site.

Lay Panels. Panels in doors, etc., with the grain running horizontally.

Lazaretto. A hospital for the poor, especially those suffering from infectious disease.

L.C. Abbreviation for *lead covered cable*, in electric wiring.

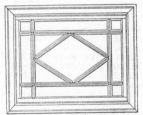

Inverted Plan

LAY LIGHT

L.C.C. Regulations. The various Acts under which the London County Council control the erection of buildings. See *London Building Acts.*

Lead. *Pb.* A heavy, soft, malleable, and easily fused metal, obtained from lead ores (galena). It is used extensively in building operations for pipes and roof coverings. Sheet lead is rolled from slabs of cast lead in thicknesses to correspond with a standard weight of from 3 to 10 lb. per sq. ft. Pipes are solid-drawn in a hydraulic press by means of a steel core and die and a powerful piston. See *Red* and *White Lead.*

Lead Acetate. Ground in oil and used as a driers for light coloured paints. It is a white crystalline solid, $Pb2C_2H_3O_2$, obtained by dissolving litharge, or lead, in acetic acid.

Lead Chromate. See *Chromes* and *Pigments.*

Lead Damp Course. Sheet lead used for damp-proof courses. It is the best material for the purpose but too costly for general use.

Lead Dot, Button, or Rivet. Used to prevent the wind from lifting the free edge of sheet lead on the tops of cornices, etc. A conical hole is drilled in the stone and a corresponding hole in the lead. Molten lead is poured in a mould over the hole which forms a dovetailed plug with a rivet head and secures the sheet lead to the surface.

Leaders. The slides for the drop hammer of a pile-driver.

Lead Flats. Nearly level roofs covered with sheet lead. Flat roofs.

Lead Glazing. Pottery glazed by dipping in some form of lead solution before burning.

Lead Lights. See *Glazing, Fretted Lead*, and *Cames*.

Lead-lined Pipes. Iron pipes lined with lead, for sanitary work.

Lead Oxide. A yellow powder obtained by heating lead in air, so that it absorbs oxygen. Also called Massicot or Letharge.

Lead Plugs. Circular plugs formed by rolling a piece of sheet lead. They form a good fixing in damp walls and for fixing grates, as timber is unsuitable in both positions.

Lead Safe. A shallow lead-lined tray with waste pipe, placed under cisterns, baths, W.C.s, etc., to catch any leakage or overflow.

Lead Shoot. A trough passing through a wall to convey the water from a flat roof to a rain-water head.

Lead Tack. See *Tingle*.

Lead Wire. Used instead of lead wool.

Lead Wool. Thin strands of lead used for caulking the joints of C.I. pipes instead of using molten lead.

Leaf. 1. A movable part to a table top. See *Knuckle Joint*. 2. One piece of a folding screen or folding doors. 3. See *Gold Leaf*. 4. See *Slate Ridge*.

Lean. A term applied to plastic materials having a low proportion of the essential ingredients, as cement in concrete.

Lean Clay. Clay having a high percentage of silica.

Lean-to. A roof with only one slope. It is often formed against the side of a building so that the tops of the rafters are carried by the wall of the building. See *Roofs*.

Lear Boards. See *Layer Boards*.

Leasehold. Applied to property that reverts to owner, or ground landlord, after a stated period.

Leca. A proprietary light-weight expanded clay aggregate for concrete.

Leclanché Battery. A primary battery for generating the current for electric bells. It consists of a glass jar containing a zinc and a carbon pole immersed in a saturated solution of sal-ammoniac. The carbon pole, or plate, is in a porous pot containing a depolarizing mixture.

Lectern. A reading desk in a church.

Ledbit. A bituminous damp-proof course with a lead core.

Ledge. 1. A narrow shelf. The top surface of a projecting member, as the top of a string course. 2. The strong horizontal members carrying the battens in a ledged door.

Ledged Doors. Doors consisting of vertical boards, or battens, and horizontal ledges, or bars. Also called batten or barred doors. The door is strengthened by the addition of braces. See *Framed and Braced*, or *Framed and Ledged*.

Ledgers. Horizontal timbers in a scaffold. See *Scaffolding* and *Ribbon*.

Ledgment Table. A string course or plinth.

Ledkore. A proprietary bituminous damp-proof course with a lead core.

Ledloc. A patent roof glazing.

Leeds Bricks. Those made in the neighbourhood of Leeds, but more especially applied to the large variety of special shapes of glazed and enamelled bricks made by the Leeds Fireclay Co., Ltd.

Left Hand. See *Hand*.

Lehr. A cooling chamber for annealing glass. The sheet is drawn and bent horizontally over a roller and passed through a roller lehr.

Lenscrete. A registered glass and reinforced concrete construction used as walls and partitions, etc. See *Nevada*.

Leucomb. A projecting cover to a gin wheel, crane, etc.

Leucon. A proprietary plastic material impervious to acids and alkalis and insoluble in water. It may be formed into any kind of article and cuts easily and machines well. It is a good insulator, transparent or opaque, and is made in various colours.

Levecel. See *Penthouse*.

Level. An instrument for detecting any variation in a horizontal line or surface, or for obtaining the relative positions of points above the plane of the horizon. There are many types from the mechanic's level to the elaborate types used by the surveyor : dumpy, Wye level, etc.

Level Book. The specially ruled book in which the surveyor records his observations when levelling.

Level Crossing. A road crossing railway lines at the same level. The road is opened or closed as required by crossing gates.

Levellers. The thin stones that build up the unit in snecked rubble.

Levelling. 1. Surveying a site to obtain the contour and irregularities. See *Datum*. 2. Removing irregularities to form a plane surface or a horizontal line. 3. Setting out a horizontal line from which depths may be measured to obtain a regular slope.

Levelling Pegs. Pointed wooden pegs driven into the ground and made level at the top. They are used as a guide for obtaining a plane surface or for measuring the depths of excavations.

Levelling Staff. See *Sopwith*.

Level Ridge Vaults. Intersecting vaults with both ridges in the same level.

Lever. A rigid rod rotating about a fixed point (fulcrum), by means of which a resistance at one end is overcome by an effort applied at the other end. There are three types of levers, according to the relative positions of fulcrum and resistance. See *Machines*.

Lever Arm. The perpendicular distance between two equal and parallel forces. See *Couple*.

Lever Boards. Like louvres but the boards are pivoted and can be operated like a venetian blind.

Lever Lock. A lock in which several levers have to be raised by the key to shoot the bolt.

Lewis. An appliance consisting of three pieces of mild steel secured together by a shackle and steel pin. The outer pieces, *A*, are dovetailed in shape to fit in a corresponding recess in stone or concrete for hoisting the stone into position.

Lewis Bolt. A bolt with one side of one end dovetailed to engage in a corresponding recess. A plug or feather on the vertical side of the bolt presses the dovetailed surfaces together so that the bolt resists withdrawal.

Lewis Sheeting. See *Dovetail Sheeting*.

Lews. Hack caps, *q.v.*

Leza. A greyish-brown hardwood from India. It is hard, heavy, strong, lustrous, and smooth. Used for superior joinery, fittings, flooring, etc. Weight about 39 lb. per c. ft.

Lias Lime. See *Blue Lias Lime*.

Liassic Limestone. Argillaceous limestone used for hydraulic limes.

Libby-Owens. A patent process for making drawn glass.

LEWISES

Library. A room, or rooms, for the keeping of books, for *reading, reference*, or *lending*.

Lich Gate. See *Lych Gate.*

Lierne Ribs. Ribs, in vaulting, not springing from an abacus. Short ribs between the intermediate groin ribs to strengthen them.

Lierne Vaults. Elaborate vaulting, sometimes called "stellar" vaulting, from the star shapes or geometrical patterns of the panels formed by the lierne ribs. See *Fan Vaulting.*

Lift. A hoist or elevator. The usual type is operated by electricity, but hydraulic lifts are sometimes used. For small lifts an endless hauling rope worked by hand is often used.

Lifting Jack. See *Jack.*

Lifting Shutters. Window shutters sliding vertically, like sashes. When not in use they are hidden behind the panelled window back.

Lifting Tackle. Appliances used for hoisting materials ; but more especially applied to pulley blocks and ropes. See *Hoisting, Mast,* and *Pulley Tackle.*

Lift Shaft. The enclosed chamber in which a hoist, or elevator, runs.

Ligger. A plasterer's mixing board.

Liggers. Hazel binders used in reed thatching.

Light. That which stimulates the organs of sight and renders the object, from which it proceeds, visible. Light travels in straight lines at a speed of 186,000 miles per second. The intensity of illumination is inversely proportional to the *square of the distance* from the source of the light, i.e. if the distance of the source of light is doubled the illuminating power is reduced to one quarter. See *Reflection* and *Refraction.*

Lighting. The artificial lighting of buildings may be by candle, oil, gas (coal, acetylene, or air gas), or electricity.

Lightmeter. A photometer. An instrument for measuring the illuminating power of an artificial source of light, or the intensity of illumination at any point.

Lightning. A proprietary high alumina cement.

Lightning Conductor. A rod of copper acting as a conductor from the clouds to the earth. It projects above the highest part of a building to protect the structure against lightning.

Lights. The subdivisions of a window. The glazed portions of a window, or the openings between mullions. See *Two-light Frame* and *Windows.*

Light-weight Concrete. Concrete with light-weight aggregates : tuff, pumice stone, slags foamed while hot, vermiculite, or mixed with gas-producing powders that leave voids.

Lignum Vitae. A very hard, heavy timber, used for decorative purposes. It is similar to ebony, but not so black or flexible.

Lillington's Metallic Liquid. A proprietary colourless waterproofer.

Lime. Produced by burning calcium carbonate, $CaCO_3$, or limestone. The calcining leaves oxide of calcium, CaO, or quicklime, which is used for mortars and plasters. The quality of the quicklime depends upon the constituents of the limestone. Pure lime, from chalk, is used for plasterer's putty, whitewashing, etc., as it has poor setting qualities. Grey stone lime is used for mortar for general building purposes. Blue lias, or hydraulic, lime is used for damp situations, as it has the properties for setting under water, and some qualities approach very nearly to cement. The slaking of lime causes it to generate great heat and to expand, hence limes, except hydraulic, should be slaked some time before required for use.

Lime Blue. See *Pigments.*

Limed Oak. A finish given to oak furniture, by rubbing a paste of chloride of lime and water into the grain. Other materials may be used for the same effect. It is then given a wax polish or egg-shell finish.

Lime Plaster. Lime, sand, and hair, mixed together for plastering.

Lime Putty. Plasterer's putty. Slaked lime sieved to remove lumps and grit, and used for finishing plastered surfaces. See *Setting Coat*.

Lime-Sand Bricks. Bricks made from damp sand with about 8 per cent slaked lime. They are pressed under steam pressure for several hours in a steaming chamber. See *Midhurst Whites*.

Limestones. Calcium carbonate combined with other substances. They are sedimentary, aqueous rocks, formed of deposits of decomposed organisms, and are known as shell, coral, crinoidal, according to the type of organic deposits. Limestones are classified as : siliceous, calcareous, argillaceous, magnesian, dolomitic, ferruginous. Oolites consist of egg-shaped grains cemented together with carbonate of lime. The limestones in general use are : Portland, Bath, Ancaster, Anston, Bolsover, Chilmark, Clipsham, Ham Hill, Hopton Wood, Ketton, Mansfield, Purbeck, Roche Abbey, etc. Also see *Marbles*.

Lime Tree. *Tilia.* The linden tree. A white, soft, close-grained timber, excellent for carving. Weight about 37 lb. per c. ft.

Limewashing. A temporary protection of stone, by the formation of calcium sulphate.

Limewash Brush. See *Turk's Head*.

Limewhiting. Whitewashing walls with freshly-slaked chalk lime, applied whilst hot, for hygienic reasons.

Limiting Friction. The point when friction is just being overcome.

Limits. The amount of error allowed, when preparing machine parts, etc., for satisfactory assembling.

Limmer. See *Asphaltic Rock*.

Limnoria Terebrans. A marine insect, very destructive to timber. See *Gribble*.

Limpet Washer. Conical washers for use with bolts for securing corrugated sheets.

Lincrusta. An embossed wall-paper.

Lincrusta-Walton. An embossed wall covering, made from cork dust and oxidized linseed oil.

Lindapter. A patent fixing or bolt adapter for steelwork to eliminate drilling.

Lindoco. A registered name for mass-produced doors.

Lindsay's Surface Fabric. Patent steel strips for reinforcing cement surfaces subject to hard wear.

Lineal. See *Linear Measure*.

Linear Arch. The link polygon, for an arch, drawn for the loading on the arch with a polar distance, on the polar diagram, equal to the horizontal thrust in the arch.

Linear, or **Line, Diagram.** A scale drawing, showing the centre lines only, of the different members of a structure. See *Graphic Statics*.

Linear Measure. A measurement of length only.

Linear Scale. The scale to which a linear diagram is drawn.

Linen-fold Panel. See *Drapery Panel*.

Line of Chords. See *Scale of Chords*.

Line of Nosings. See *Nosing Line*.

Line of Pressure. *Linear Arch*.

LINEN-FOLD PANEL

Line of Thrust, or Resistance. A line containing the resultant thrusts on the horizontal joints in a section of a retaining wall, to see if the "middle third rule" applies to all the bed joints.

Line Pins. Steel pins with long blade-like shank and large head. They are used to insert in brick joints and to carry a line which serves as a guide when laying the bricks.

Liner. A registered name for concrete machinery: breakers, mixers, vibrators. A registered machine for brick and block-making.

Lining. Running a line of colour to distinguish between two merging colours, or as a simple kind of decoration. Special brushes are used: fitches, riggers, or lining pencils.

Lining In. Finishing a drawing by thickening the preliminary draughting lines.

Lining Out. Marking out the stuff preparatory to sawing.

Lining Paper. Plain paper used to hide defects in plastered surfaces before distempering.

Linings. Thin boards, or other materials, used as finishings for covering rough surfaces, cavities, etc. Also see *Sash and Frame*, and *Finishings*.

Lining Up. Planting a narrow piece under the edge of thin material to make it stronger and stiffer.

Link Dormer. A dormer between the roof and a chimney or some other part of the structure. The lights are on the sides.

Link Polygon. See *Funicular*.

Linotile. A patent jointless composition floor.

Linseed Oil. Obtained by crushing and pressing the seed of the flax plant, which produces raw oil. Boiled oil is raw oil heated with a drier, such as red lead or litharge, which makes it dark coloured and quicker drying. Both are used as vehicles for paint.

HOMAN AND RODGERS LINTEL

Lintel, or Lintol. A horizontal beam across an opening, usually carrying the wall above. Also see *Straight Arch, Hollow Blocks, Relieving Arch*, and *Armoured Tubular*.

Lintile. A patent floor covering.

Linville Truss. A trussed girder with N-shaped bays. The vertical members are in compression and the diagonals in tension. See *Roofs*.

Liotex. A proprietary magnesite flooring, etc.

Lip. See *Lip Piece*.

Lip Joint. A joint for terra-cotta blocks in which a flange on one block engages with a rebate in the adjacent block.

Lipped. *Buttered.* Mortar placed round the edges only, in a brick or masonry joint.

Lip Piece. A short timber nailed on the joint between waling and strut to secure the joint in trench timbering.

Lipping. Facing the edge of inferior wood with superior wood, or a joint between two pieces of different thicknesses so as to form a rebate.

Lip Trap. An obsolete form of trap to a drain, in which a projecting lip dips into the water.

Liqualino. Used with cement and sand as a binding agent to hard impervious surfaces to carry cement rendering.

List or Listel. A square fillet moulding. See *Mouldings*.

Listed. Applied to boards in which the waney edges have been sawn off.

Listing. Pointing at the junction between a roof surface and projecting brickwork.

Litany Desk. A low desk at which the priest kneels whilst reciting the Litany.

Lithalum. A proprietary wood-wool building slab.

Litharge. Oxide of lead, PbO. It is used as a driers for oil mixtures, paints, mastics, etc., as it sets very hard.

Lithocrete A proprietary asphalt flooring. It is damp and vermin-proof, quick-setting and hard-wearing.

Lithofalt Paving. Proprietary asphalt paving blocks for roads for heavy traffic.

Lithopone. Zinc sulphide and barium sulphate precipitated to form a white pigment for paint. It is only suitable for inside work. See *Pigments*.

Little Joiners. Small inlays to remedy defects in timber, such as dead knots, etc.

Live Knot. A sound and firm knot.

Live Load. A variable force or moving load on a structure. See *Superimposed Load*.

Livering. Same as curdling, *q.v.*

Liver Rock. Good sound stone from the best part of a quarry. Also applied to *Freestone*.

Lloyd. A registered wood-fibre wall-board.

L.M. Abbreviation for *lime mortar*.

Load. The total weight acting on a structure. See *Dead Load, Live*

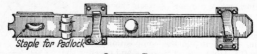

LOCKING BAR

Load and *Superimposed Load*. A *load of softwood* is 50 c. ft. of squared timber or 40 c. ft. hardwood, or 600 ft. super of 1-in. boards.

Load Bearing. Applied to structural members that carry loads.

Loading Concrete. The layer of concrete over the asphalt in a basement floor. In addition to forming the floor surface the weight prevents the upward pressure of water from disturbing the asphalt.

Loading Shed or **Loading Way.** A covered yard, or vanway, to a warehouse where vehicles are loaded and unloaded.

Loading Slab. See *Floating Floor*.

Load Line. A line representing the sum of several forces drawn to scale. See *Graphic Statics*.

Loams. Clays containing sand or gravel. They are known as mild clays, and produce bricks less liable to warp and shrink than strong clays.

Lobby. A passage, small hall, or waiting room used as a means of communication between different apartments.

Lobe. Same as *foil*, *q.v.*

Lock. 1. A metal fastening actuated by a key and usedf or securing doors, drawers, etc. The various kinds are: box, cupboard, desk, door, drawer, wardrobe, and they are distinguished as barrel, beak, cut, cylinder fall, dead, latch, lever, link plate, mortise, night latch, pad, rim. sliding door, stock, straight, cupboard, two-bolt, Yale, etc. 2. A basin in a canal by which the height of the water may be controlled by means of gates. 3. See *Air Lock*.

Lock Block. A block in the framing of a flush door (*q.v.*) to provide a fixing for the lock.

Locker. A small cupboard.

Lock Gate. The gate to a canal lock.

Locking Bar. An iron bar for gates. The bar is pivoted at the centre and engages with a hook plate at one end and a staple at the other end, and is secured by a padlock.

Lock Light. A patent fire-resisting window consisting of built-up copper sections. The patent is applied to stainless steel external windows to protect against "smash and grab" raiders.

Lock Nut. A nut screwed into position on the bolt and fixed, or locked, so that it will not work loose with vibration. There is a great variety, including many patents.

Lock Rail. The middle rail of a door, and usually carrying the lock.

Locor. A force vector, in graphic statics.

Locus of Points. If a series of points are plotted under certain determinate conditions, then the line so formed is the locus of the points.

Locust. An orange to dark-brown hardwood from C. America. It has darker streaks and is hard, heavy, strong, and fairly durable. Difficult to work and polish. Used for construction, superior joinery and fittings. Weight about 50 lb. per c. ft. Also see *Robinia*.

Lode. A watercourse or open drain.

Lodge. 1. A small house, in a park, subordinate to a larger one. 2. A hut or place for temporary residence. 3. A meeting place for freemasons.

Loft. 1. A space immediately beneath the roof, used for storage purposes. 2. The gallery of a church.

Log. A felled tree with bark and branches removed.

Logarithms. The exponent of the power to which a number (base) must be raised to equal a given number. The use of logs. simplifies calculating, and tables may be obtained giving the logs for all numbers when the base is 10.

Log Frame. A machine saw with a number of vertical or horizontal blades for converting, or breaking down, a log to smaller stuff.

Loggia. A covered gallery or path on the outside of a building.

Log Saw. A saw for converting logs. It may be a band saw or a frame saw. See *Log Frame*.

Lok'd Bar. Hope's patent steel casements in which the intersection of the bars is secured by a patented method.

London Building Acts. Building Act, 1894; Amendments, 1905, 1932, 1937, 1939; General Powers Acts, 1902, 1908, 1909, 1930. Also see *Acts*.

London Clay. A fluviatile clay from the London basin. It is a strong clay and requires treatment for good bricks. On the outcrops the clay is more loamy and is used for London stocks.

London Standard. A timber measurement of 270 c. ft.

London Stocks. Bricks made on the outskirts of London from river-deposited clay containing chalk. The bricks vary considerably. They are usually buff coloured with rough surfaces and irregular arrises, but they weather well in the London atmosphere, and are chiefly hand-made. They are classified as malms, malmed, and common, according to the quality.

Long and Short Work. Applied to masonry quoins in which alternate quoin stones are placed on end.

Long Arm. A long pole with hook for operating fanlights, etc.

Long Headers. Bricks laid lengthwise across a thick wall. Headers continued through the wall.

Longitudinal Bond. Used in thick walls. The bricks in occasional courses are all laid stretcher-wise to strengthen the wall lengthways.

Long-leaf Pine. The best quality of pitch pine marketed in this country.

Lookum. A projecting cover to a wall crane or gin wheel.

Loop Hole. A narrow opening in a wall.

Loop-hole Door. A trap door.

Looping Out. The usual method of connecting up the lamps to the fuses in electric wiring.

Loose Box. A large stall in a stable in which the horse is allowed free movement.

Loose Ground. Applied to irregular ground, soft in places, when excavating.

Loose Moulds. Glazing mouldings, or *beads*.

Lorymer. See *Larmier*.

Lounge. A room provided with easy chairs, etc., for relaxation.

Louvre. A turret with a series of inclined surfaces of wood, glass, slate, or metal, at the sides; the horizontal inclined surfaces are called *louvres* and admit light and air but exclude rain. A lantern with louvres instead of glass.

Louvre Door. A door with louvres instead of ordinary panels.

Louvre Window. A window with louvres instead of glass.

Low-pressure System. The usual method of heating water for small buildings. The essentials are: boiler, supply tank, hot-water cylinder or cistern, circulating and expansion pipes. The water is heated in the boiler and rises by convection; whilst the pressure, due to the head of water, and the difference in density of the water in the flow and return pipes, causes a circulatory motion. It is sometimes called a gravity system.

Low-side Window. A small window or opening in the south chancel wall of old churches.

Lozenge. Small diamond-shaped panels. Ornamental mouldings consisting of diamond-shaped carvings characteristic of the Norman style of architecture.

L.P. Abbreviation for *low pressure*.

L. P. F. & S. Abbreviation for *lath, plaster, float, and set*.

Lucarne. A vertical window in a parapet wall. It is similar to a dormer window but the latter stands back from the face of the building.

Ludlow. A registered interlocking roof tile.

LOW-PRESSURE HOT-WATER SYSTEM

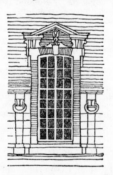

LUCARNES

Luffer Boards. *Louvres*.

Luffing. The horizontal or vertical movement of the jib of a crane.

Lug. An ear, or small projection, on a casting for fixing purposes.

Lumbayau. A light to dark-red, lustrous, hardwood from the Philippines. Fairly hard, heavy, durable, but not strong. Not difficult to work and polishes well. Used for superior joinery, and as a substitute for Mahogany. Weight about 42 lb. per c. ft.

Lumber. An American term for sawn or cleft timber and to felled logs for the lumber or saw-mill.

Lumens. A unit of illumination of foot-candles per sq. ft.

Luminous Paint. Sulphide or oxy-sulphate of calcium ground in a colourless varnish or other vehicle.

Lundberg Adapter. A patent combined socket and plug electrical connector.

Lunes. Shaped like a crescent. The gores of a developed spherical dome are half-lunes. See *Gore*.

Lunette. A curved opening in a concave ceiling to admit light. The space formed by a wall intersecting a vault, especially when containing a window.

Lunumidella. Also called Ceylon Mahogany or Cedar. Similar to Toon, *q.v.*

Luxfer. A patent bar for roof glazing.

L.W. 1. Abbreviation for *lime white* or *lavatory waste*. 2. A proprietary wood-fibre wall-board.

Lych Gate. A covered entrance to a churchyard.

Lyctidae. Includes about twenty species of wood beetles, or worms, very destructive to new timber; especially to

LYCH GATE

oak, ash, chestnut, and walnut. The cure is to spray petrol, paraffin, or peroxide of hydrogen, etc., into the holes, and then seal the holes with bees wax. See *Death Watch Beetle*.

Lye-byes. See *Skews*.

Lying Light. See *Ceiling Light*.

Lying Panel. See *Lay Panel*.

Lysis. A step or plinth over the cornice of a podium.

M

M. Symbol for *Bending Moment, Moment of a Force,* and *Mechanical Advantage.* **m.** A symbol for a *constant,* and an abbreviation for *matched* and *thousand.*

M.A. Abbreviation for *mechanical advantage.*

Macaboard. A proprietary wood-fibre wall-board.

Macadam. Broken stone levelled and consolidated by a heavy roller. The surface may be bound by various materials (water, bitumen, or tar), according to the quality of road required. The interstices between the stones are filled with gravel.

Macasfelt. A patent flat roofing system. It consists of three layers of water-proof sheeting, with plastic bitumen between, and makes a pliable, durable roof.

Macassar Ebony. A decorative hardwood with varied black and dark brown grain. It polishes well and is used as veneer on plywood for better class furniture and fittings.

Mace's Patent. A method of fixing door knobs to the spindle.

Mac-Girling. Patent building units, consisting of cast-stone hollow blocks.

Machicolations. Apertures in parapets of ancient castles, through which missiles were hurled at attacking enemies.

Machilus. Four useful species of grey to light reddish-brown hardwood from India. Fairly light, soft, strong, and durable; similar to chestnut. Easy to work and polishes well. Used for construction and joinery. Weight from 30 to 40 lb. per c. ft.

Machine. A contrivance that produces or changes motion, or that serves to regulate the effect of given forces. Any arrangement that converts work of one kind into work of a required kind. See *Mechanical Powers.*

Machine Guards. Appliances designed to protect operatives from moving parts of machinery. There are numerous variations and patents that comply with Home Office regulations.

Machines. Labour-saving devices used in the manufacture of materials and the preparation of finished products. There is a great number applied to the building trades, and there are scores of specialist machines for woodworking alone. Modern machines are motorized and large machines may have twenty motors for the various operations. Improvements are constantly taking place, especially in automatic and portable machines.

Maclast. A proprietary emulsion for cold setting of wood-block floors.

Maco-block. A registered concrete block-making machine.

Maco Gauge. An adjustable templet for obtaining irregular contours of mouldings, etc.

Made Ground. Built-up ground formed by filling pits with loose material and refuse.

Maf. A registered name for numerous wood connectors, anchors, etc.

Maftex. A proprietary wood-fibre wall-board.

Magbestic. A magnesite, jointless, flooring composition.

Magnesia. MgO. Oxide of magnesium. In brick earth it gives the bricks a yellow tint. Magnesium is a white malleable metal.

Magnesian Limestones. Those composed chiefly of carbonates of lime and magnesium.

Magnesite Cement. Sorel cement. Magnesia paste, MgO, and magnesium chloride, $MgCl_2$. It is a very strong cement used as a binding agent in most jointless floorings. Magnesium chloride corrodes metals and the composition must be on a dry inert surface. Specialist firms usually have their own registered mix, coloured as required.

Magnet-Blaton. A system of prestressed concrete using high-tensile steel wires.

Magnetite. Magnetic iron ore, Fe_3O_4, from which Swedish iron is obtained.

Magnolia. Imported with canary wood and basswood as American whitewood.

Mahabon Teak. Borneo camphor-wood.

Mahogany. The most important timber used in good-class joinery. Over sixty kinds of wood are sold under this trade name. The best quality is Cuban, followed closely by Honduras. The African timber is mostly inferior. Highly figured stuff (crotch, curl, fiddleback, and mottle) is used as veneer. Cuban is a rich brown red with a chalk-like substance in the pores, and is hard, heavy, and durable. S.G. ·8. Honduras is golden red, and softer, lighter, and less durable than Cuban. S.G. ·6. Mahogany is obtained in large sizes. It varies greatly for working, but polishes well, and is the most stable of timbers when once seasoned. W. African mahogany forms the bulk of the supplies in this country. Owing to the popularity of mahogany many other woods are often wrongly marketed as mahogany.

They are usually excellent woods and include: Coromandel, Colombian Mahogany, Crabwood, Dhup, Espavé, Gaboon, Jequitiba, Lauan, Makore, Meranti, Okoume, Rhodesian Mahogany, Sapele, Seraya, Surinam, Thitka, Toon, and several species of Eucalyptus.

Mahtal. A proprietary laminwood.

Maidou. A very ornamental hardwood from Indo China used as veneer for panels in good-class fixtures and furniture.

Main. The principal gas or water pipe from which branch pipes draw supplies for the consumers.

Main Beams. The principal girders in a floor. They transmit the loads from the secondary beams to the walls and columns.

Maire, Black. A deep-brown hardwood with black streaks from New Zealand. It is very hard, heavy, durable, and strong. Difficult to work. Used where strength and durability are of primary importance. Weight about 62 lb. per c. ft.

Maisonnettes. Self-contained residential flats.

Maizewood. A proprietary insulating material.

Majastic. A figured rolled glass, translucent but with high light transmission.

Majolica Glaze. A glaze used for bricks and tiles that have been fired before glazing. The bricks, when glazed, are fired at a low temperature in a muffle kiln.

Making Good. Repairing plaster, woodwork, etc., after alterations to any part of a building.

Makore. Cherry mahogany. A Nigerian reddish-brown hardwood used for superior joinery, veneers, etc.

Malachite. Carbonate of copper found in solid masses. It is a beautiful green colour and used for ornamental purposes.

Malenite. A decorative plywood with "wovenwood" face, for panels, etc.

Mall, or Maul. A large mallet used for driving small piles, levelling sets and pavements, etc.

Malleable. Applied to a material that can be beaten or rolled into plates.

Malleable Iron. Cast iron annealed and given the properties of wrought iron.

Mallet Head Chisel. A mason's chisel with rounded head for use with mallet.

Malm. Artificial marl, obtained by adding chalk to clay and mixing in a wash mill, for brick making. Also a rich chalky clay, or marl.

Malm Bricks. Bricks made from marl or clay with added chalk. They vary from red, through buff, to white, in colour. The best qualities are nearly white. They are graded as cutters, best seconds, mean seconds, pale seconds, brown and hard paviors, shippers, bright stocks, grizzles, and place bricks.

Manbarklac. A variegated olive to reddish-brown hardwood from C. America. Very hard, heavy, strong, and durable, and resists marine insects. A good substitute for greenheart. Weight about 65 lb. per c. ft.

Mandril, or Mandrel. 1. A cylindrical piece of hardwood used by the plumber for straightening lead pipes. 2. The axis of a lathe to which the work is fixed for turning.

Manganese (Mn.). A grey metal occurring in different forms, which are used in numerous alloys, both ferrous and non-ferrous and in glass.

Manganese-bronze. The trade name for several special high tensile brasses used for forgings, castings, or as rolled sections.

Manganese Dioxide. Added to sand for facing multicoloured bricks.

Manganese Oxychloride. A cement for jointless floors.

Manganin. A copper-manganese alloy with high electrical resistance.

Manger. The box, or trough, from which horses or cattle feed.

Manger's Soap. A solvent or cleanser for paints.

Manhole. A chamber into which branch drains are collected. It is used or inspection and cleaning purposes.

Manhole Cover. An air-tight cast-iron cover fitting into a frame and used for covering manholes. Some types for street use are not air-tight, but form a ventilator for the sewer.

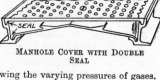

Manilla Fibre. Used in plaster as a substitute for hair.

Manlid. A manhole cover.

Man Lock. See *Air Lock.*

Man-made Wood. See *Weyroc* and *Improved Wood.*

Manometer. An instrument for showing the varying pressures of gases.

Mansard Roof. A roof of double pitch on both sides. See *Roofs.*

Manse. A house belonging to a church, for the minister.

Mansfield. Red and cream sandstones from Notts. They are both used for general building purposes, but not very good for polluted atmospheres. Weight 143 and 140 lb. per c. ft. respectively. Crushing strengths 591 and 462 tons per sq. ft.

Mansfield Woodhouse. A rich yellow limestone from Notts. Used for dressings and selected for general building. Weight 145 lb. per c. ft. Crushing strength 577 tons per sq. ft.

Mansions. Often applied to large blocks of self-contained residential flats.

Mansonia. A Nigerian greyish-brown, variegated hardwood. Resembles black walnut and is used for similar purposes, superior joinery, etc.

Mantel, or Mantelpiece. The ornamental front to a fireplace, usually consisting of pilasters, shelf, and mirror or panelling above the shelf. It may be of marble, polished hardwood, slate, brick, tiles, or cast iron.

Mantel Tree. The lintel of a fireplace.

Mantlet. A protecting movable screen in fortifications.

Manu-marble. A proprietary synthetic marble.

Maple. *Acer spp.* A hard, compact timber that works to a lustrous finish and does not splinter readily. Used for hard-wearing floors and kitchen fitments. Ornamental varieties (bird's eye and fiddle back) used as veneers. Weight 47 lb. per c. ft.

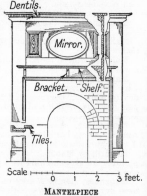

MANTELPIECE

Maple Silkwood. A Queensland hardwood more like mahogany than maple. It polishes and fumes well and is used for superior joinery and fittings.

Marb-l-cote. A plastic paint for textural finishes. It is in powder form, which is mixed with water, and sets like marble in 24 hours.

Marbleboard. Proprietary wall-boards with surface imitating marble.

Marble Facings. Thin layers of marble used as a veneer on walls, etc. They are usually secured by small brass wire cramps built into the joints of the wall. A large amount is imitation, or synthetic, marble, in modern buildings.

Marbles. A compact calcareous stone of beautiful appearance. A metamorphic limestone capable of taking a high polish. There is a great variety, with a large range of colouring and veining due to minerals in combination. The following are amongst the best known used in building: Bardillo, Black and Gold, Bleu Belge, Breccia, Caroline, Corona, Dove, Fleur de Peche, Florido de Barchito, Jaune, Levanto, Meruil, Pavonazzo, Petitor, Pentelikon, Porphyry, Purbeck, Rosarea and Violette, Rouge, San Stefano, Sicilian, Sienna, Skyros ,Subiaco, Tinos, Travertine. There are many that are not true marbles but used for similar purposes: alabaster, onyx, serpentine, soapstone, travertine, tufa. There are also numerous imitation marbles, many of them with proprietary names.

Marbling. The art of painting wood, plaster, etc., to imitate marble.

Marbolike. Insulating sheets finished like marble and used extensively for lining refrigerators.

Marbolith. A proprietary jointless flooring consisting of magnesium oxy-chloride ingredients mixed with asphalt emulsion.

Marcasite. White iron pyrites. A mineral easily decomposed. See *Pyrites*.

Marchioness. Slates 22 in. × 11 in., or 12 in. Requires about 138 per square at 3 in. lap.

Marezzo. Artificial marble. Threads of silk are dipped in a coloured slip of Keene's cement and laid on a sheet of glass to form the veins. A thin slip of body colour is poured over the veins and the silk threads removed. Dry Keene's cement is sprinkled to absorb the moisture and a canvas backing applied. Cement is applied to the canvas to the required thickness. The panel is removed from the glass and polished.

Margin. 1. The narrow strip, or border, mitred round the hearth to a fireplace. See *Hearth*. 2. The projection of a string above the nosing line. 3. The exposed area of a slate when fixed. See *Gauge*. 4. The faces of stiles and rails exposed in panelled framing. 5. A variation, above or below, from a stated amount in a contract.

Marginal Mould. A moulded fascia, or lining, scribed to the wall, to cover the space between frame and wall, and leaving a parallel margin of the frame exposed.

Margin Draught. A chiselled border round the edges of stone.

Margin Lights. Narrow panes of glass around a large pane due to introducing bars.

Margin Trowel. A small box-shaped trowel used by the plasterer for finishing internal angles and margins.

MARIGOLD WINDOW

Marigold Window. A rose, or Catherine wheel, window.

Marine Borers. Molluscs and crustaceans that attack wood in salt waters. The most destructive are the shipworm and the gribble. Few timbers can resist these borers, so timber should be treated and sheathed in infected areas. Timbers with silica content are most resistant. Greenheart, Australian turpentine, Jarrah, Opepe, Manbarklac, are very resistant.

Market Terms. The following terms are used in the marketing of wood:

balk, batten, board, casewood, deals, die square, ends, firewood, flitch, log, mast, pitwood, planchette, plank, planking, pole, quartering, scantling, sets, thick stuff. Wood is sold by standard, cubic measure, board measure, float, load, fathom, square, or by weight.

Marking Knife. A woodworker's tool for marking out the stuff preparatory to sawing.

Marl. A rich calcareous clay earth.

Marley. The registered name for numerous concrete products, roofing tiles, and light-weight roofing.

Marlith. A proprietary wood-wool building slab.

Marmorene. A proprietary polished sheet glass for facing walls, etc., in over forty colours and in several thicknesses. It does not craze or disintegrate.

Marmorite. A proprietary opaque, cast and polished plate glass. It is in sixteen colours and several thicknesses and used for external or internal facings.

Marquetry. Inlaid work of thin, differently-coloured woods.

Marquise. A large canopy, or shelter, at the entrance to a theatre, shop, etc.

Marseilles Tiles. A French interlocking tile used extensively in this country.

Martin's Cement. Gypsum and carbonate of potash. Used for the same purposes as Keene's.

Mascolite. A proprietary shock-absorbing and insulating material.

Mash Hammer. A mason's hammer with square head.

Mask. A carving of grotesque form placed on panels, friezes, keystones, etc.

Masonite. An insulating wallboard used extensively for lining walls and ceilings.

Mason Master. A patent power drill.

Masonry. The preparation and fixing of stone. The term also includes heavy engineering work in brickwork, concrete, and stone, such as harbour and dock walls.

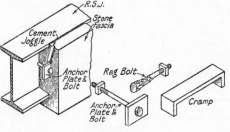

MASONRY FIXINGS

Masonry Fixings. The various metal fastenings used with stonework.

Masonry Nails. Screw nails that can be driven into bricks, etc., with a hammer. For harder materials a pilot hole is necessary. They are up to $2\frac{1}{2}$ in. long and usually cadmium plated.

Mason's Joint. Pointing in the form of a triangular projection.

Mason's Mitre. An internal mitre for a moulding carved in the solid. It is the only practicable method for stone and terra-cotta, and it is generally used on oak framing for church work and small rounded arrises. See *Bar.*

Mason's Putty. A mixture of lime putty and stone dust used for fine joints in wrought stonework.

Mason's Scaffold. A scaffold with two rows of standards so that it is independent of the wall.

Mason's Stop. See *Mason's Mitre* and *Mitre.*

Mason's Trap. See *Dipstone Trap.*

Mass Concrete. See *Bulk Concrete.*

Massicot. Chemically the same as litharge, but yellow. Used as a pigment.

Mast. The vertical post in a crane, or a guyed pole carrying pulley tackle. See *Derrick Crane.*

Master Key. A special key that will operate all the locks to a building.

Mastic. A pointing material in which linseed oil is an ingredient. Various mineral or rubber mixtures are used, but usually litharge is the chief ingredient. Powdered brick or stone, litharge, red lead, and linseed oil are a common mastic. Sand, litharge, gypsum, and boiled linseed oil are used for glazing in

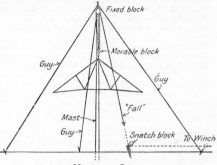

MAST AND GUYS

stone. Putty, white lead, and varnish are used for bedding in metal.

Mastic Asphalt. A mixture of asphaltic bitumen with clean sharp sand. It is moulded into blocks which are heated and worked in position with hot irons for roof coverings, etc. See *Asphalt.*

Mastic Cement. A mixture of boiled linseed oil, red lead, litharge, and fine sand, mixed into a paste for pointing between wood frames and the brickwork.

Masticon. A proprietary mastic for glazing metal windows.

Mastic Process. Applying bituminous or asphaltic materials, in a plastic state, as a surface waterproofer.

Mastobas. Small structures used as tombs in ancient Egypt.

Mat. A dull surface as produced by a *flat* coat of paint. Frosted surfaces on glass.

Matai. New Zealand black pine. A species of Podo, *q.v.* The wood is light brown, fairly hard, strong, and durable, with close grain and texture. It is used for all kinds of joinery, flooring, etc.

Match Boarding, or Matching. Tongued and grooved boards with veed or beaded edges to hide shrinkage.

Matched Ceiling. A ceiling formed of tongued and grooved boarding.

Match Planes. Pairs of planes for preparing the tongues and grooves on matchboarding.

Material Lock. See *Air Lock.*

Matobar. A patent welded, fabric reinforcement for concrete.

Matrix. 1. The cement or lime mortar, in concrete, in which the aggregate is embedded. 2. A mould for casting plastic or fluid materials.

Matt. See *Mat.*

Matt Bitu. A proprietary, bituminous, washable water-paint, made in several colours.

Mattock. A pick with adze-shaped points.

Mattress. Ready-made steel mesh for reinforcing concrete roads.

Maul. 1. See *Mall.* 2. A hand hammer, 3 to 4½ lb., used in granite dressing.

Mausoleum. An elaborate tomb or sepulchral monument.

Mavitta. A registered drafting machine. A drawing board with several mechanical devices.

Maximum Daylight. A patent rolled prismatic glass for deflecting light into basements.

Maxlite. A patent system of sliding windows and partitions.

Maxtrip. A patent system of concealed lighting.

Maxweld. A metal fabric used as a key for concrete, etc.

Mayapis. A species of lauan, *q.v.*

Maycoustic. A proprietary pre-cast stone and tile with acoustic properties.

Meander. A complicated fret carving, similar to Greek fret.

Mean, Geometrical. The square root of the product of two quantities.

Measuring Frame. A box-like frame, without bottom, of known cubical capacity, for measuring the aggregate for concrete.

Measuring, Standard Method. An agreed method of measuring building work, for purposes of estimating, published by the Surveyors' Institution.

Mechanical Advantage. The ratio of resistance to effort, in a machine.

If resistance $= W$, and effort $= P$, then $M = \dfrac{W}{P}$.

Mechanical Powers. Simple machines : lever ; wheel and axle ; pulley ; inclined plane ; wedge ; screw. These are illustrated in the following : crowbar ; winch ; lifting tackle, gin wheel, differential pulley ; an inclined run for rolling up heavy materials ; plug and feathers, folding wedges, cutting tools ; screw jack, bolts, screws, etc.

Mechanical Shovel. A machine for excavating. It is usually fitted with scoops, buckets, grab, back filler blades for replacing and spreading, and crane blocks.

Mechanical Water Bar. See *Adam's Water Bar.*

Medallion. A circular or elliptical raised tablet or panel, with carved or inscribed surface.

Medang. A name including several different woods of the Lauraceae family from Malaya. There is considerable variation in properties and characteristics. Yellow to olive, feels greasy, soft to fairly hard, good qualities are fairly durable and strong, and stable. Used for construction, furniture, planking, etc. Weight from 33 to 44 lb. per c. ft.

Median. A line from the apex to the middle of the opposite side, bisecting a triangle. The intersection of two medians is the C.G. or centroid.

Medieval. Architecture of the Middle Ages, fifth to fifteenth centuries, Romanesque and Gothic. The term is usually applied to Norman and Early Gothic only.

Medina Cement. A similar cement to Roman, but lighter in colour.

Medulla. The pith of a tree.

Medullary Rays. The radiating bands of cells in timber that produce silver grain. See *Timber.*

Medusa. A proprietary waterproofing compound for cement.

Meeting Rails. The rails of sliding sashes that meet near the middle of the frame. See *Sash and Frame.*

Meeting Stiles. Middle stiles of folding doors. Shutting stiles.

Megohm. One million ohms.

Melamine. Formaldehyde plastics.

Mellowes. A registered metal-casement window, etc.

Mellowing. See *Souring.*

Melotone. A proprietary acoustical down.

Member. Any important piece in structural framework or timber framing, or an individual part of an "order" or moulding.

Membrane System. An elastic sheet of waterproof material placed on the face, or between layers, of concrete.

Memel Timber. *Pinus sylvestris*, or yellow deal. See *Redwood.*

Mempening. Malayan oak. There are numerous species of which some provide excellent wood similar to *Quercus rubra*. See *Oak*.

Mendip. A registered roofing tile, similar to Roman tiles.

Meranti. *Red Seraya (q.v.)*.

Merchantable. A term applied to specific grades of American lumber. The term implies it is suitable for constructional work. *Common merchantable* is a lower grade and *Selected M.* a higher grade.

Merchant Bar. Inferior wrought iron. The lowest quality used for smith's work. It is only fit for rough work and is hard and brittle.

Meridian Stresses. Vertical stresses in a dome.

Merlons. The projections between the embrasures in a battlemented parapet.

Meros. See *Femur*.

Merry-go-round. A machine for running the wires round a tank for prestressed concrete.

Mersida. Veneers with tough paper backing for wall hangings.

Merulius. See *Dry Rot*.

Mesh. 1. The many types of metal lathing used as a reinforcement for concrete. 2. A small space or interstice, as in a sieve or net, etc.

Mesowax. A proprietary waterproofer for canvas, etc.

Mesquite. See *Ironwood*.

Messmate. A species of Eucalyptus, *q.v.*

Messuage. A dwelling-house and the adjacent land and buildings belonging to it.

Metaform. Metal forms, or shuttering, for concrete. The standard unit is 2 ft. square and consists of 16 gauge sheet metal in a frame of 1 in. × 1 in. angle iron in which the dowel pins and clamps are placed for fixing the units.

Metal, or Metalling. Broken stone for macadam.

Metal Casements. Iron, steel, or bronze sashes with frames. The whole may be enclosed in a wood frame or secured direct to lugs in the brick or stonework, or merely built in. There are many patent types, Crittall's, Hopes, Williams, etc., and the sashes may be pivoted or hinged.

Metal Cramps. The various types of bent bars for securing two blocks of stone together or for fixing marble and thin facings to brick and concrete walls. There is considerable variation in size and shape, and

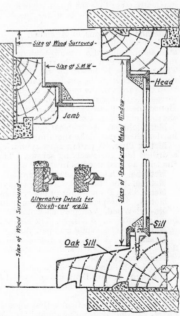

METAL CASEMENTS

they should be of non-corroding metal, or tarred and sanded iron. They are sunk in the surface and grouted. See *Masonry Fixings*.

Metal Faced Joinery. Woodwork with metal facings. Bronze and stainless steel, but other alloys may be obtained. See *Bronze sheathed*.

Metal Leaf. Very thin sheets of metal alloy used in decorating. They vary in colour from silver to copper, and are protected after application by cellulose lacquer or shellac solution to prevent oxidation.

Metallic Liquid. A proprietary liquid waterproofer.

Metal Lathing. Expanded metal for reinforcing concrete, and for plaster work.

Metal Trims. Thin metal skirtings, picture rails, cornices, door frames and architraves, etc., used as substitutes for wood in fire-resisting construction.

Metamorphic. Rocks that have been changed or made crystalline by great heat, pressure, or chemical action, below the earth's surface.

Metco. A proprietary wood-fibre wall-board.

Meter. An apparatus for automatically recording the quantity of a fluid passing through, as water, gas, and electric meters.

Method of Sections, or Moments. A method of calculating the stress in a single member of a truss by taking moments about a point in the truss. The principle is based on the fact that if a structure be cut by a plane,

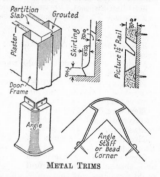

METAL TRIMS

then the forces in the members cut by the plane, together with the external forces on either part of the structure, form a system of forces in equilibrium. Hence the moment of the forces in the cut bars, about a point, is equal to the moments of the external forces acting on either side of the section plane.

Methylolurea. A chemical compound used to impregnate softwoods to harden them and make them moisture and fire-resistant. See *Urea*.

Metlex Plug. A patent split cylindrical tube of soft metal inserted into a drilled hole for receiving screws. The insertion of the screw opens out the tube so that it grips in the hole.

Metoche. The space between dentils.

Metope. The slabs between the triglyphs, in the Doric frieze, *q.v.*

Metre. The French unit of linear measurement (39·37 in.).

Mettal. A proprietary aluminium paint.

Me-Tyl-Wood. Veneered sheet metal. The metal is made porous for the bonding material to adhere. The sheets are waterproof.

Mews. Stables round an open yard.

Mezzanine. A low story introduced between the levels of two main floors. A floor beneath a theatre stage.

Mezzo-relievo. See *Relief*.

M.G. Abbreviation for *make good*.

M'gongo. A very light and soft wood from Nigeria. Used for insulating, cores for flush doors, and as a substitute for balsa. Weight about 12 lb. per c. ft.

Mica. A group of minerals that readily split into extremely thin, flexible flakes, capable of standing great heat. Muscovite, or potash mica, or talc, is transparent and used for lamp chimneys, small panels in stoves, etc.

Mica Flap A hinged plate of mica used in ventilators to allow currents to flow in one direction only. See *Fresh-air Inlet*.

Micarta Board. A decorative veneer. It consists of paper or cloth moulded at high temperature and pressure, and is permanently coloured and patterned to imitate wood, marble, etc. It may have different backings, or a steel core.

Micrometer. An instrument for registering measurements within very fine limits.

Micron. A unit of measurement in microscopic research. It is one millionth part of a metre.

Midden. A pit for the reception of faecal matter. An ash-pit.

Middle Third Rule. "The thrust on a rectangular section of a non-tensile material, as brick and stonework, must lie within the middle third, for stability."

Midfeather. An alternative name for the *withes* to flues, and for the *parting slips* to boxed sash frames, and for the *checker plate* under the oven of a kitchen range.

Midhurst Whites. Sand lime, or calcium silicate, bricks. Chemically combined sand and lime under superheated steam, which forms an insoluble hydrated silicate of lime. Strong, fire-resisting, and white throughout. About 8½ lb. weight.

Midloc. A registered casement fastener designed on the principle of the mortise lock.

Mihrab. The hall of prayer to a mosque.

Mild Clays. Loamy or sandy clays, containing free silica.

Mild Steel. Steels are malleable combinations of iron and carbon; usually produced by Bessemer, Siemens Martin, or cementation processes. In the Bessemer process, pig-iron is heated in a converter and a cold blast passed through to remove the carbon. The required amount of carbon, ·2 per cent, is added (speigeleisen), and it is then run into ingots. In the Siemens process, gas is used as fuel and an open hearth is used as in puddling. Scrap iron and steel are added to the ore and then it is tested for percentage of carbon. The *basic open hearth* process is the Siemens process with a specially lined converter of dolomite. Mild steel is used for structural steelwork. Tool steel has more carbon and may be tempered.

Milk of Lime. The milky liquid from stirred lime and water added to the mix for making silica bricks.

Mill. 1. A machine for grinding substances. 2. A building containing machinery, as a saw-mill.

Milla. An olive-grey hardwood with wide streaks from India. It is hard, heavy, strong, and durable. Fairly difficult to work. Polishes well. Used for superior joinery, flooring, etc. Weight about 48 lb. per c. ft.

Millboard. Stout pasteboard.

Mille. Often implies 1,200 in marketing, or ten "long hundreds."

Milled Lead. Lead rolled into sheets by machinery. The sheets are up to 35 ft. long and up to 7 ft. 6 in. wide. See *Lead.*

Mill-stone. See *Grit-stone.*

Mimbar. A form of pulpit in a mosque.

Minar. A turret, or lighthouse.

Minaret. A slender, lofty turret, with projecting balconies, from which people in Mohammedan countries are summoned to prayers by the priest.

Mineral Black. See *Pigments.*

Mineral Green. Carbonate of copper. A green pigment.

Minion. A local term for hoggin.

Minium. Red lead.

Minolith. One of the Wolman salts for preservation of wood. It protects against insects and fungi, and makes the wood fire-resisting.

Minster. A cathedral. The church to a monastery.

Minstrel's Gallery. A gallery to a large hall of a medieval mansion, from which the guests were entertained by music.

Minute. One sixtieth part of a degree, or of the lower diameter of a column.

Miraculum. A proprietary red-lead filler for anti-corrosive paints.

Miro. A brownish softwood from New Zealand. It is fairly soft and light, and strong but not durable. Resembles Rimu. Used for joinery, structural and constructional work. Weight about 36 lb. per c. ft.

Miscelia. A patent flooring tile with mosaic surface.

Miserere. See *Subsellia*.

Misericorde. A folding seat. See *Subsellia*.

Misering. Making trial borings with a specially shaped auger, or miser, to determine the strata of the earth.

Mitre. The intersection of two pieces meeting at an angle. The line of the mitre bisects the angle between the two pieces so that the corresponding members of a moulding meet on the mitre.

Mitre Arch. One consisting of two inclined stones. A triangular arch.

Mitre Block or **Box.** A rebated block or a long three-sided box, in which mouldings are placed to be cut for a mitred joint. The cuts are 45° for an angle of 90°.

Mitred Cap. A circular cap to a newel into which the handrail is mitred.

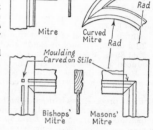

MITRE BLOCK

MITRES

Mitred and Cut String. A string cut to the outline of the steps and mitred with the risers.

Mitred Hip or **Valley.** One in which the slates or tiles are cut to meet each other on the line of intersection of the two roof surfaces.

Mitred Knee. See *Knee*.

Mitre Dovetail. See *Secret Dovetail*.

Mitred Tenons. Used where two tenons meet in a corner post.

Mitre Shoot. A special frame for steadying a moulding whilst shooting, or planing, a mitre.

Mitre Square. A bevel with the blade fixed at 45°.

Mitre Templet. A rebated appliance for guiding the chisel when forming small mitres, preparatory to making a scribed joint.

Mits. Abbreviation for *mitres*.

Mix. See *Batch*.

Mixer. A machine for the mechanical mixing of concrete. See *Concrete Mixer*.

Mixing Drum. See *Concrete Mixer*.

Mixing Platform. A prepared level surface of strong boards on which concrete is mixed by hand.

Mo. Abbreviation for *moulded*.

Moat. A deep trench, especially when filled with water as a protection to a building.

Moboron. Nigerian Cedar, or Agba, *q.v.*

Model By-laws. The codes, or sets, of by-laws issued by the Ministry of Health to guide local authorities in the control of building operations.

Modeller. One who prepares the clay model preparatory to the sculptor copying it in stone.

Modeller's Clay. Specially prepared clay for making models. Sometimes used by the plasterer for taking a squeeze, the clay being backed with plaster.

Modelling Tools. Curved boxwood tools of various sizes and shapes for shaping plaster mouldings and ornaments.

Models. 1. See *Fibrous Plaster*. 2. A small scale representation of a prospective building.

Modern System. A method of constructing tall chimneys. Special red shale tiles are moulded to the curve of the shaft to form a centering for a concrete filling.

Modillions. The enriched brackets under the corona of the Corinthian and Composite entablatures. Also see *Cornice*.

Modinature. The arrangement and distribution of the mouldings of an *order* or of a building.

Modular Ratio. The ratio of two elastic moduli. The ratio between the modulus of elasticity for steel to that for concrete $(E_s/E_c) = 30$ million $\div$ 2 million $= 15$, for a standard mix.

Module. The radius of a column at the bottom of the shaft. A unit of measurement in classic architecture.

Modulus of Elasticity. E. A measure of the elasticity of a material given by $\dfrac{\text{stress}}{\text{strain}}$. By Hooke's Law, stress $\propto$ strain $\therefore$ stress $= E \times$ strain, where E is a constant for the particular material. A good mnemonic is $E = \dfrac{f}{a} \div \dfrac{e}{l} = \dfrac{fl}{ea}$. An imaginary stress necessary to stretch the material to twice its original length, or compress to half.

Modulus of Rupture. A constant used in the calculations for rectangular beams and obtained by experimentally breaking test pieces of different materials. B.M. at rupture $=$ modulus of rupture $\times Z = f_r Z$. Also see *f*.

Modulus of Section. Z. The value, due to the size and shape of the section of a structural member to resist stress. See page 366.

Moisture Content. The percentage of moisture present in a material. The following values are important for wood. Limit for fungoidal growth 20 per cent, air dried 16 per cent, roof timbers 14 per cent, exterior work 16 per cent, interior joinery (ordinary heating) 12–14 per cent, (central heating) 8–10 per cent. When m.c. is less than 12 per cent the wood is *Kiln-dry*. The stiffness and strength vary almost inversely with change of m.c. A dry wood is nearly twice as strong as the same wood saturated. Wood is hygroscopic and the volume varies with the m.c. once the cell spaces are free of moisture, but some woods are very stable and change of m.c. has little effect when once seasoned. The percentage m.c. is given by the formula—

$$\text{m.c.} = \frac{\text{original wt.} - \text{oven dry wt.}}{\text{oven dry wt.}} \times 100$$

See *Shrinkage*.

Molave. A yellowish-brown hardwood sometimes called Philippine "Teak." It is not teak but has many of its qualities. Used for superior joinery, good-class construction, flooring, exterior work. Weight about 48 lb. per c. ft

Mole. A breakwater or stone pier.

Moler. Diatomaceous earth used for insulation bricks.

Moment of a Force. The turning effect of a force about an axis. It is expressed in compound units. A force of 10 lb. turning about a point 12 in. distant (perp. distance) has a moment of 10 lb. × 12 in. = 120 lb.-in.

Moment of Inertia. *I.* See *Inertia Areas* and page 366.

Moment of Resistance. *R.* The moments of the internal forces acting in the resistance areas of a section of a structural member.

Moment Scale. Any selected scale used to denote moments graphically.

Monarch. A patent dust and draught excluder consisting of a copper spring strip.

Monastery. A house for monks, for religious retirement. An abbey or priory.

Monastral Blue. A pigment of brilliant hue that can be used in either oil or water mediums, producing satisfactory greens and purples.

Moncrieff's Formula. Used in the design of mild steel columns.

Monel Metal. A white nickle-copper alloy used for sheathing wood and

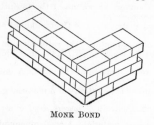

MONK BOND

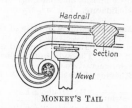

MONKEY'S TAIL

concrete as a protection against marine borers. It is also used for laundry and chemical equipment.

Monial. Same as *Mullion.*

Monk. A patent roof glazing.

Monk Bond. A modified Flemish bond with two stretchers and one header repeated in each course.

Monkey. The slip hook for releasing the ram of a pile driver. The term is often applied to the weight, or ram, also.

Monkey Dung. A mix of one part cow dung, one part lime, two parts sand, with ox hair. The mix is spread on canvas and tied round pipes for insulation.

Monkey's Tail. A vertical scroll to a handrail.

Monkey Tail Bolt. A bolt with long handle for the top of a tall door, to be operated from the floor.

Monkey Winch. An appliance for grubbing up trees when clearing sites.

Monkey Wrench. A spanner with adjustable jaws.

Monk's Park. One of the best Bath stones. Light cream colour, compact, and fine-grained. S.G. 2·2.

Mon'lithcrete. A fire-resisting floor reinforced by secondary beams with perforated webs, through which twisted steel flats are passed to give transverse bond to the concrete.

Monnoyer System. A method of constructing tall chimneys. Reinforced concrete segments are moulded and erected as masonry to form an octagonal shaft. A projecting hook engages with the adjacent block to form a continuous longitudinal rib and to engage with vertical reinforcing rods.

Monolithic. Formed of a single stone. A homogeneous mass.

Monoliths. Hollow pillars, with steel shoes, built up as they sink into the ground due to excavating the earth from inside. The plan may be square, rectangular, or octagonal; and they may be of concrete, brickwork, or masonry, with any number of wells according to size, but usually four. They are used where conditions do not allow for timbered trenches due to the presence of water. The wells are finally filled with concrete to form a solid pier.

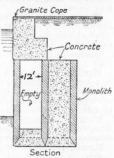

Section

MONOLITH

Monopavets. Colorpavets, but of natural colour.

Monopteral. A temple consisting of an arrangement of columns supporting a dome or cupola.

Monostyle. A single column or pier.

Monotower Crane. A crane with counterbalanced jib so that only one tower is required.

Monotriglyph. Intercolumniation with only one triglyph between those over contiguous columns.

Montant. A stile, *q.v.*

Monument. A structure to commemorate an event or person.

Moore's Bridge. A patent steel bridge screwed to the trimming joist and carrying a hearth for an upper floor.

Moorish. See *Saracenic Architecture.*

Moorish. or **Horseshoe, Arch.** Characteristic of Saracenic Architecture. The soffit consists of two segments struck from centres some distance above the springing. When one centre only is used the soffit is a little more than a semicircle and is called a horseshoe arch. See *Arches.*

Mopin System. A patent system of steel and concrete construction that allows for mass production methods. The characteristics are lightness, speed of erection, homogeneity, and economy.

Moppon. A proprietary asbestic-bitumen plastic compound, in red or black.

Mopstick. A handrail of circular section, except for a small flat on the underside. A return bead in brickwork.

Mora. A yellowish to reddish-brown variegated hardwood from B. Guiana. It is fire-resisting and very hard, heavy, durable, and strong. A substitute for greenheart. Used for structural work, flooring, joinery. Weight about 62 lb. per c. ft.

Mordant. Any material applied to a surface to make paint adhere.

Moresque. See *Moorish.*

Morin's Tables. Experimental values for the angle of repose and coefficient of friction of different materials.

Morocco. A figured rolled glass with small irregular pattern.

Mortar. A plastic mix of lime and/or cement with sand and water, for jointing and bedding bricks, stones, etc. It helps to distribute the load, makes the joints watertight, and, in most cases, unites the blocks. The addition of sand, crushed clinker, etc., assists crystallization in lime mortars and counteracts excessive shrinkage. Sand is added to cement mortar to increase the bulk and so lessen cost. The proportions vary, but usually two to three parts of sand to one of lime or cement, and sufficient water to make the mix plastic, are satisfactory. A cement-lime-sand mix in the proportions of 1 : 3 : 9 is used for re-pointing. See *Lime*, and *Sand.*

Mortar Mill. A large revolving pan containing two revolving rollers for crushing and mixing the materials for mortar.

Mortgagee. One who has had property conveyed to him as security for a debt. The liabilities of a leaseholder are transferred to a mortgagee in possession, unless arranged otherwise.

Mortise. A recess formed in one member to receive a projection, or tenon, on another member. See *Joints,* and *Tenon.*

Mortise Gauge. A joiner's tool for setting out parallel lines for mortises and tenon

Mortise Joints. The various types are: barefaced, box, chase, closed, double, dovetail, forked, foxtail, franked, halved, haunched, housed, loose, notched, oblique, open, pair of, shouldered, single, slot, stub, stump, sunk, tease, triple, tusk, twin.

Mortise Lock. A lock sunk in the edge of a door stile so that it is not visible on the face.

Mortising Machine. A machine for cutting mortises in timber. It may be operated by hand or by power. The cutting is done by chisel, hollow chisel with twist bit, or by chain cutter. Mult-mortisers will make a number of mortises in one operation. Gang-drilling machines can bore eighty holes in different directions in one operation.

Moruro. A purplish-brown hardwood, with light markings from C. America. Similar to mahogany and used for similar purposes, also used for structural work. Weight about 54 lb. per c. ft.

Mosaic. Surfaces formed by small pieces of material disposed in various designs on a cement bed. The material may be marble, glass (vitreous), or

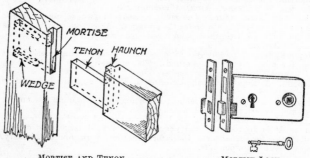

MORTISE AND TENON   MORTISE LOCK

pottery (ceramic). Roman, Italian, or cube, mosaic is built up of tesserae about ⅝ in. square. It is usually prepared in a studio by building up and attaching, face side down, to stout paper. The complete design is bedded and the paper removed when the cement is set. Venetian mosaic, or Terrazzo, consists of cement, sand, and chippings of the required material, laid in a plastic state, and rubbed down and polished when set.

Mosque. A Mohammedan place of worship.

Moss Bar. A patent reinforcement for concrete, consisting of a rolled steel joist with unequal flanges and perforated web, through which stirrups are passed to resist shear.

Motif. The principal theme, or subject, prevailing in a design.

Mottled. Variegated. Applied to wood that looks uneven, though smooth, due to reflected light on the uneven arrangement of the fibres, due to wrinkling of the growth rings. There are a number of variations known as fiddle, peacock, beeswing, plum, ram, and stop mottle.

Mouchettes. Narrow flame-like shapes between the *soufflets* and the containing arch in Flamboyant tracery.

Mould. 1. A thin templet, or pattern, used to mark the outline of the material before being wrought. 2. A receptacle, matrix, or form, in which plastic material is cast to the required shape. 3. A fungus.

Moulded. 1. A term applied to a piece of material on which a moulding has been wrought; or to panelled framing in which mouldings have been mitred round the panels. 2. Materials prepared by casting in a mould.

Moulded Work. A term used in masonry for labour entailed in working

MOULDS FOR PEDIMENT SPRINGER

mouldings. It is measured by girth × length, i.e. the actual surface area of the moulding.

Moulding. 1. The process of shaping material with machines or tools, or in a mould; or applying mouldings, as "moulding a panel." 2. Completing the setting process in resin-bonded plywood, structures, etc., by applying heat and pressure in an autoclave. See *Hot Press* and *Plastic Resins*.

Moulding Machines. The various machines used in woodworking, masonry, etc., for shaping the material, and which are driven by motive power.

Moulding Planes. The various planes (hollows, rounds, ovolos, beads, etc.) used by the joiner for shaping mouldings.

Mouldings. Long pieces of wood, or other material, of which the rectangular sections have been shaped into varied contours for ornamentation. 2. *Classic mouldings*: abacus, apopheges, astragal, cavetto, colarino, corona, cyma recta and reversa, listel, ovolo, torus, trochilus. The Grecian outlines were parts of the conic sections: ellipse, parabola, and hyperbola; the Roman outlines were circular arcs. Other common mouldings are: beads (quirk, sunk, return), chamfers, flutes, hollows, lamb's tongue, nosing, ogee (cyma recta), reeds, rounds, scotia. Different styles of architecture have their characteristic mouldings such as: billet, bird's beak, bowtell, cable, casement, chevron, dog-tooth, egg and dart, keel, scroll, wave, zigzag, etc. Mouldings are also named after the designer, or according to the period of common use. See page 212.

Mould Lines. The lines on a mould templet that coincide with the lines on the material when applying the mould.

Mould Oil. Used to coat formwork to prevent concrete adhering. Too liberal an application leaves a greasy surface after striking.

Mould Plate. The zinc contour of the mouldings, on a horsed mould, for plasterwork.

Mouldrite. See *Urea*. A synthetic resin that can be moulded into any required form.

Mould Stone. A moulded jamb stone.

Mound Breakwater. Rubble stone deposited by hopper barges to form a heap, or mound, until it appears above water level. It is the most primitive form of breakwater and only used where quarry waste and cheap labour are available. See *Breakwater*.

Mount. To plant, or fix, on to the face of anything.

Mountain Ash. *Eucalyptus spp.* Also called Tasmanian oak. The wood is light brown, fairly hard, heavy and durable, strong, resilient and elastic.

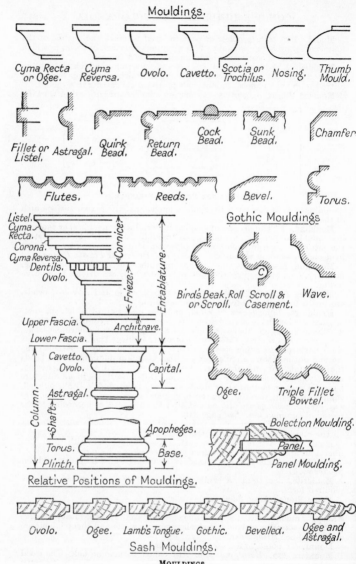

Mouldings.

Cyma Recta or Ogee. Cyma Reversa. Ovolo. Cavetto. Scotia or Trochilus. Nosing. Thumb Mould.

Fillet or Listel. Astragal. Quirk Bead. Return Bead. Cock Bead. Sunk Bead. Chamfer.

Flutes. Reeds. Bevel. Torus.

Listel. Cyma Recta. Corona. Cyma Reversa. Dentils. Ovolo. — Cornice.

Frieze. — Entablature.

Upper Fascia. Lower Fascia. — Architrave.

Cavetto. Ovolo. Astragal. — Capital.

Column. Shaft. Torus. Plinth. Apopheges. — Base.

Relative Positions of Mouldings.

Gothic Mouldings

Bird's Beak, Roll or Scroll. Scroll & Casement. Wave.

Ogee. Triple Fillet Bowtel.

Bolection Moulding. Panel. Panel Moulding.

Ovolo. Ogee. Lamb's Tongue. Gothic. Bevelled. Ogee and Astragal.

Sash Mouldings.

MOULDINGS

Figured wood used for superior joinery and fittings, and veneers. The wood of the British species, *Pyrus aucuparia*, also has many uses but it is seldom used for building purposes.

Mounting Steps. A block consisting of two or three steps for convenience when mounting a horse.

Mount Sorrel. A pinkish-brown granite from Leicestershire. Polishes well, but not good for building purposes. Weight 163 lb. per c. ft. Crushing strength 2080 tons per sq. ft.

Mouth. The entrance to the flue of a fireplace opening.

Movingue. Nigerian Satinwood.

Moving Walls. Patent partition walls operated by hydraulic-electric control.

M-Roof. A series of connected double-pitch roofs. See *Roofs*.

m.s. Abbreviation for *mild steel*.

Muckite. A proprietary bitumastic compound used for counter-tops, fancy articles, etc. The articles are moulded in a power press and stoved in an oven.

Mucuri. A reddish-brown hardwood from C. America. A hard, heavy, durable, and strong wood used chiefly for structural work. Weight about 54 lb. per c. ft.

Mudsill. A sole piece. Any foundation timber in or on the ground.

Muffity or Muffet. See *Stone Slates*.

Muffle. A thin coat of plaster placed along the contour or profile of a mould for running the preliminary coats of coarse stuff for a cornice or moulding. The muffle is then chipped off for running the finishing coat.

Muffled Sheet. A blown sheet glass with ripple pattern, used in leaded lights.

Muffle Kiln. One in which the articles are not in contact with flame or gases.

Mulgrave's Cement. See *Cement*.

Muller. A stone cylinder with rounded end, for grinding paint.

Mulleting. Gauging the edges of panels to the thickness of the plough groove.

Mullions, or Munnions. The vertical divisions in window openings and frames. See *Venetian Window*.

Multicolour Bricks. Hand-made, rough or smooth, sand-faced facing bricks. The faces are not uniform in colour due to the special care taken in mixing and burning to produce the varied effect. They are made in several thicknesses and are sometimes called *texture* bricks. See *Rustic Bricks*.

Multifoil. Tracery work having more than five cusps.

Multiply. Plywood having more than three plies, or veneers.

Muninga. A dark brown variegated hardwood from S. Africa. It is very durable, and hard, heavy, and strong. Used for superior joinery, flooring, etc. Weight about 45 lb. per c. ft.

Munnion. 1. See *Mullions*. 2. Sometimes applied to vertical sash bars.

Muntins. The vertical divisions in timber framing, between the rails.

Muntz Metal. A malleable brass used as a sheathing for under-water work. See *Elio*.

Mural. Belonging to or attached to a wall.

Muranese. Figured rolled glass in three patterns. They have high light diffusion with obscurity.

Murex. A London firm specializing in welding and welding equipment.

Muriatic Acid. Used to expose the aggregate, by removing the outer coat of cement, in concrete work.

Murite. A proprietary gypsum plaster.

Muscovite. A silicate of alumina and potash. See *Mica*.

Mushroom Construction. Reinforced concrete construction consisting of columns and floor slabs only.

Musiga. East African Greenheart.

Mutule. A modillion rectangular in section. A square block carrying the guttae in the Doric cornice.

Mvule. Iroko, *q.v.*

Mycelium. The spawn of fungi. See *Dry Rot.*

Myrtle. A Tasmanian hardwood resembling beech. Californian laurel. Both woods are highly decorative and used for superior joinery, flooring, and veneers.

N

N.A. Symbol for *Neutral Axis.*

Nailhead Moulding. Carved with a series of pyramidal or diamond-shaped projections like large wrought nail heads. It is a characteristic of Norman architecture.

Nails. Classified as wire, cut, or wrought, according to manufacture. Extreme sizes are called spikes (large) and sprigs, tacks or pins (small). The sizes rise by $\frac{1}{4}$ in. from 1 in. to $3\frac{1}{2}$ in., and then by $\frac{1}{2}$ in. to 6 in. They were originally designated by *penny,* a corruption of pound, because they weighed so many pounds per 1,000 nails, i.e. a 3 in. was a 10d. (tenpenny) nail because 1,000 weighed 10 lb. Another explanation is that it referred to the cost per 100 nails. There is a great variety of shapes, metals, and finishes, and they are often named according to the particular purpose for which they are used, as flooring, felt, slating, glazier's, etc.

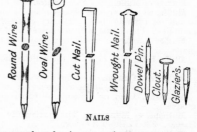

NAILS

Nail Set. See *Punches.*

Naked. Applied to the structural timbers of a building before the finishings are applied.

Naked Wall. One with a plane surface, having no projections or recesses.

Naos. The body of an ancient temple. Sometimes applied to the principal chamber of Greek temples. See *Temple.*

Naphthalene. Used to obtain the porosity in bricks of very light weight.

Naphthenates. Salts used in the treatment of wood. They may be copper or zinc and are excellent preservatives, and can be painted.

Naples Yellow. A yellow pigment. Antimonic anhydride heated with oxide of lead.

Nargusta. An olive-brown variegated hardwood from the W. Indies. Beautiful figure and used for decorative work, veneers, etc. Weight about 38 lb. per c. ft.

Narig. A yellowish to dark-brown variegated hardwood from the Philippines. It is very hard, heavy, durable, and strong, and used for nearly all purposes. Weight about 62 lb. per c. ft.

Narra. Philippine Padauk, *q.v.*

Narrow Ladies. Slates 14 in. by 7 in.

Narthex. Part of an early Christian church for penitents, divided from the main body by a screen; sometimes took the form of a vestibule or porch.

Nattes. An interlaced decoration used in early architecture.

Natural Bed. The surface of a stone that coincides with the laminae in stratified rocks, i.e. the bed on which the stone rested in the quarry.

Natural Cements. Similar to hydraulic limes, and requiring little preparation of the natural earth from which they are obtained. See *Cement*.

Natural Foundations. Earth requiring no preparation to sustain a structure.

Natural Seasoning. Cut timber stacked to allow the sap to harden naturally and the moisture to evaporate. The timber is skidded so that air is free to circulate round every piece. This method does not impair the strength and colour, but it is slow and costly compared with kiln seasoning.

Natural Slope. See *Angle of Repose*.

Natural Stone. Quarried stone used for building, as distinguished from pre-cast stone.

Nave. 1. The main body of a church. 2. The middle block, or hub, of a wheel through which the axle passes, and that carries the spokes.

Navel. A knob, or a boss on a shield.

Navvy. See *Steam Navvy*.

Naylorite. A silicic, liquid, cement hardener and waterproofer. It may be applied to surfaces when set.

n.e. Abbreviation for *not exceeding*.

Neat. A term applied to cement mortar without sand.

Neat House. A shippon.

Neat Size. Net size. Exact size after preparation.

Neat Work. The brickwork set out on top of the footings.

Nebule Moulding. A wavy moulding.

Neck or Necking. The connecting moulding between the capital and the shaft of a column, or the plain part between the mouldings and the shaft.

Necked Bolt. See *Cranked Bolt*. A necked latch is a Norfolk latch.

Needle Leaf. The name applied to the spine-like leaves of coniferous trees.

Needle Points. Needles without eyes, used for repairing small fractures in wood. They are easily broken to the required length.

Needles. 1. Horizontal timbers or steel sections, passed through a wall, and supported by dead shores, to support the wall above, during structural alterations. See *Dead Shores*. 2. The short piece passing through the wall plate into the brickwork and receiving the thrust from a raking shore, *q.v.*

Needle Scaffold. One supported by cantilevers thrown out from the face of a building.

Needling. Dead shoring.

Neon Tubes. Glass tubes nearly exhausted of air and containing neon gas, and used for electric lighting.

Nervures. The side ribs of a vaulted roof, to distinguish from diagonal ribs.

Neutral Axis. The longitudinal section of a beam or structural member where the stresses are neutral, i.e. where the change takes place between tension and compression. An imaginary line through the centres of gravity of the continuous cross-sections of a structural member.

Neutralizing. Preparing cement work for painting. A fluosilicate solution is satisfactory. Various acids and sulphates are used, and neutralize the lime effectively but they often prove harmful to the concrete.

Nevada. Patent glass lenses used in Lenscrete windows.

Nevastone. A registered name for kitchen equipment, sinks, etc.

Nevellator. A patent level for suspending from a stretched line.

New B.S. Sections. See *British Standard Sections*.

Newel or Newel Post. 1. The post carrying the handrail to a flight of stairs. See *Knee*. 2. The central column carrying the inner ends of the steps in a winding, or "solid newel," stairs. (*q.v.*).

Newel Cap. An ornamental top planted on a newel. Also see *Mitred Cap*.

Newel Joints. The joints between newel and string or handrail.

Newel Stairs. See *Dog-legged Stairs*.

Newel Wall. A division wall from ground level between flights of dog-leg stairs.

Newman's Tables. A table of safe resistances of different earths.

Newtonite. A proprietary waterproof sheeting for walls.

Newry. See *Granites*.

New Zealand. The woods exported include the following hardwoods: Beech, Black Maire, Honeysuckle, Puriri, Rata, Tawa. The softwoods include: Kaikawaka, Kauri, Matai, Miro, Podo, Rimu, Silver Pine, and Totara.

N-Girder. See *Roofs*.

Nib. A small projection on a casting, for fixing purposes, as on a tile. See *Eaves*.

Nibbed Tile. A tile with small right-angled projections at the top for hanging on to the batten.

Nib Rule. A rule, against which the nib on the horsed mould runs, when running cornices *in situ*. The rule is bedded on the wall with weak cement and sand and held by bricks bedded on top at intervals.

Niche. A small recess in a wall, usually with a "coved" head, spherical, ellipsoidal, or polygonal.

Nickel Steel. A strong tough steel of white colour and fibrous structure, with a higher elastic limit than carbon steel due to the small percentage of nickel. It does not corrode readily in sea-water.

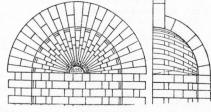

NICHE (BRICK)

Nicker. A wide mason's chisel used for forming a groove preparatory to splitting stone.

Nicoloid. A proprietary emulsion paint for interior or exterior wall surfaces.

Nidging. A particular method of preparing the face of granite with a hammer.

Nigerian. Applied to the West African woods from Nigeria: Ekki, Iroko, Mahogany, Makore, Movingue, Mansonia, Obeche, Opepe, Pearwood, Sapele, Satinwood, Sycamore, Walnut, Whitewood, etc.

Nigerian Mahogany. *Khaya spp.* Nigeria provides the bulk of the mahogany used in this country. There are several species and they are named after the port of shipment. Most of it is excellent wood, but there is considerable variation in quality. The better qualities are similar to Honduras and used for superior joinery.

Night Latch. A lock for an external door. A spring latch is operated by a sliding knob on the inside and by a key on the outside.

Nippers. See *Stone Tongs*.

Nipple. 1. A short piece of pipe with a male screw for connecting W.I. pipes. 2. The projections on the sections of radiators to ensure a water-tight connection.

Nissen Hut. A semi-cylindrical hut.

Nitaline. A proprietary hardened glass for wall decorations, etc.

Nitrification. Stone with the qualities of nitre due to the action of alkaline silicates, which are applied as preservatives but often prove harmful.

Node. The point where two branches of a curve cross each other to form a loop. A double or triple point. The intersection of several members in structural frames.

Noden-Bretteneau. A process for the electrical treatment of timber as a preservative.

Noel. A registered wood-block flooring.

No-fines Concrete. A term applied to concrete without fine sand. The aggregate is graded from $\frac{3}{8}$ in. to $\frac{3}{4}$ in. The large voids provide insulation and prevent capillarity but the concrete must be non-bearing.

Nog. A wood brick built in the wall for fixing purposes.

Nogging Pieces. Horizontal timbers placed between the studs to stiffen a partition. See *Bricknogged.*

Nogging Strips. Thin nogging pieces used to bond the bricks, in addition to stiffening the studs, in a bricknogged partition. They are bedded between courses of bricks, or pre-cast slabs.

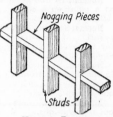

NOGGING PIECES

Nom. Abbreviation for *nominal*.

Nonagon. A plane figure, or polygon, with nine sides. For a regular figure, area = square of side × 6·182.

Non-bearing Wall. One that supports no load other than its own weight.

Non-conductors. Poor conductors of heat and electricity, such as wood, slag wool, etc.

Non-ferrous. Applied to alloys whose chief properties are due to other metals than iron. In some cases the alloy may contain up to 50 per cent iron. They include brass, bronze, copper-nickel, nickel-silver, lead or lead-base, tin-base, copper manganese, magnesium-base, aluminium-base, numerous cutting-tool and heat resisting alloys. They all resist corrosion.

Non-inflammable. Timber made practically incombustible by chemical treatment, such as : Burnettizing, injecting tungstate of soda or ammonium phosphate, etc. See *Fire-resisting Timbers.*

Non-slip. Applied to cement floors in which iron filings, or carborundum powder, have been trowelled into the surface. Slag pumice or fine lead cubes, mixed in the finishing coat, are also effective. Another method is to indent the surface, whilst it is plastic, with a special roller.

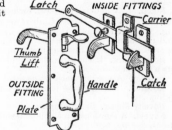

NON-SLIP FLOOR ROLLER   NORFOLK LATCH

Noral. A proprietary aluminium paste for anti-corrosion paint, also for an aluminium casting alloy.

Norfolk Latch. A fastening for a ledged door, operated by the thumb.

It consists of a lever, or *fall bar*, operated by a latch lift, or *sneck*, and controlled by a *keeper*, or carrier. The fall-bar, or latch, engages with a notched stop, or catch, on the frame. It is also called a thumb latch, or Suffolk latch.

Nori. A registered name for Accrington engineering bricks.

Normal. A line at right angles to a tangent at a point on a curve. A perpendicular. See *Ellipse* and *Circle*.

Normal Section. 1. A method used in the preparation of hand-rail wreaths, more suitable for those of fairly large radius. 2. A section along a normal.

Norman. A style of architecture prevailing in this country from A.D. 1066 to 1189. Low, massive columns, richly decorated semi-circular openings, "orders," zig-zag and chevron mouldings, were characteristic of the style.

NORMAN ENRICHMENT

Norman Slabs. Glass is blown into a square mould and then the sides are cut to form slabs.

Northern Pine. *Pinus sylvestris.* Redwood, Scotch fir, yellow deal or fir, red deal, Norwegian or Baltic pine, are some of the names given to this timber, which is one of the most useful timbers for constructional work and outside joinery. It is resinous, strong, stiff, and durable. Yellow colour with reddish markings. Weight about 41 lb. per c. ft.

North Light Truss. A roof with unequal pitches, having the steeper side glazed and facing north because of the more even light. See *Roofs*.

Norton Floor. A proprietary alundum tile for floor and stair treads.

Norusto. A preservative paint for iron.

Nose. 1. The projecting moulding on a rib for vaulting. 2. A drip or downward projection of a cornice. 3. The spout of a bib-cock. 4. The bottom of the shutting stile of a door or casement.

Nose Cock. 1. A cock for the burner of a gas bracket. 2. A bib-cock.

Nosing. The rounded projecting edge to a flat surface, as to a step, lead flat, window bottom, etc. See *Mouldings*. The outer vertical surface of a cornice

Nosing Line. An imaginary line touching the edges of the nosings in a flight of stairs.

Notch. 1. A trench, or groove, formed in the side of one timber to insert the side of another. If both timbers are notched it is called a double notching. The notches may be dovetailed. Notching is sometimes called *gaining*, and the term is loosely applied to trenching and housing. 2. Sometimes applied to the recess in stone jambs to door or window openings.

Notch Board. 1. The string to a flight of stairs. 2. An appliance for gauging a stream of water.

Notcher. A machine for cutting steel.

Novalux. Patent glass lenses used for reinforced concrete pavement lights, etc.

Novelty Siding. See *Sidings*.

Novoid. A powder with a colloidal silica basis, for making cement waterproof, oil- and acid-proof.

Nozzle. 1. An outlet to a C.I. eaves gutter. 2. The open end of a pipe. 3. A projecting spout.

N.R.M.E. Abbreviation for *notched, returned, and mitred ends.*

N-truss. A Linville truss. See *Roofs*.

Nu-bar. A patent collapsible gate.

Nudging. Applying the pressure to the clot of clay when moulding bricks.

Nulling. A carved or turned quadrant-shaped ornamentation in Jacobean oak furniture, etc.

Nursery. A place where plants, trees, etc., are raised and propagated. A play-room where young children are nursed and trained.

Nuseal. A proprietary bituminous waterproofer, applied with brush.

Nust. A proprietary paint for ironwork, to prevent corrosion.

Nut. See *Bolts* and *Lock Nuts.*

Nut Oil. Used in paints, for cheapness. Inferior to linseed and poppy oil.

N.W. Abbreviation for *narrow widths.*

O

②. Abbreviation for *two coats in oil.*

O. Abbreviation for *over.*

Oak. *Quercus.* Most useful hardwood for general purposes. Over 300 varieties. English oak is hard, tough, durable, but difficult to work. Austrian and Japanese oaks and American red oak are softer and have straighter grain, but not so durable. American white oak is similar to English. Wainscot oak is selected oak cut radially to show medullary rays, or silver grain. Rich light brown in colour, fumes and polishes well, or may be limed, corrodes iron. S.G. ·7 to ·9. See *Tasmanian* and *Silky Oak.*

Oast. A kiln for hop drying.

Obeche. Nigerian whitewood. Creamy white. Light, soft, easily wrought, and polishes well. Used for joinery, matchboarding, plywood, etc.

Obelisk. A slender, rectangular, monolithic tapering column, crowned by a small pyramid. They are of Egyptian origin, and examples exist of single stones up to 180 ft. high.

Oblate. See *Spheroid.*

Oblique Arch. See *Skew Arch.*

Oblique Cone, or **Cylinder.** One in which the axis is not at right angles to the base.

Oblique Grain. See *Spiral Grain.*

Oblique Joint. An angular joint not forming a right angle.

Oblique Planes. Auxiliary planes, used in geometry, inclined to both planes of projection, *q.v.*

Oblique Projection. A pictorial projection in which the elevation, section, or plan, is drawn, and then parallel projectors represent the other sides of the solid. The projectors may be at any angle, but usually 45° or 30°.

O.B.M. Abbreviation for *ordnance bench marks.*

Obscured Glass. Glass with ground or embossed surface to remove the transparency. The rough surface is prepared by a blast of fine sand, chemicals, by grinding, or by patterning in manufacture.

Obsidian. Dark vitreous volcanic rock.

Obtuse. Blunt. An angle between 90° and 180°.

Ochre. A pigment used in paints and in dips for glazing bricks. It is a fine clay, varying in shades of yellow, according to the hydrated ferrous oxide present. See *Pigments.*

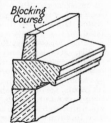

OBLIQUE PROJECTION
OF CORNICE

Octagon. A plane figure with eight sides. Area = square of side × 4·828 for a regular figure.

Octahedron. A regular solid contained by eight equal equilateral triangles. Area of surface = square of linear side × 3·464. Volume = cube of linear side × ·4714.

Octastyle. A temple or building with eight columns in front. See *Temple* and *Columns*.

Oculus. A round window.

Odeion, Odeon, or Odeum. A concert hall in early Grecian architecture.

Odoko. Nigerian Sycamore.

Odum. Iroko, *q.v.*

O.E. Abbreviation for *one edge.*

Ofe. A rich cedar-brown hardwood from Nigeria, also called Guarea and Pearwood. Hard, heavy, with glistening deposit. Not difficult to work, but liable to pick-up, and polishes well. Used for the same purposes as mahogany. Weight about 45 lb. per c. ft.

Offsets. 1. Horizontal breaks in walls to diminish the thickness. 2. A perpendicular to the chain line, used in surveying, to obtain the area of irregular projections. 3. A bend in a pipe.

OCTAGONAL BRICK
PIER

O.G. Abbreviation for *ogee.*

Ogee. A moulding with contour of contra-flexure, consisting of concave and convex arcs, resembling the letter S. Variously described as cyma recta and cyma reversa. See *Mouldings* and *Roofs*.

Ogee, or Ogival Arch. A pointed arch with the sides formed of two ogees. See *Arches*.

Ogee Gutter. A wood eaves gutter with cyma-recta contour on the face.

Ogmensteel. A patent method of reinforcing the concrete casings for steel-work.

Ohm. Unit of electrical resistance.

Ohm's Law. Gives the relation $E = I \times R$, where R = electrical resistance in ohms, E = pressure in volts, and I = current in amperes.

Oil Dag. A graphite lubricant with oil as the vehicle.

Oillet. See *Eyelet.*

Oil Mastic. See *Mastic.*

Oil Paints. Those in which the vehicle is mainly linseed oil.

Oil Proofing. Preparing cement work to resist acids. Proprietary silicates of soda and potash are mixed with the concrete. Hard waxes dissolved in methylated spirits are also used, for surface coatings, for ordinary floors.

Oils. May be of animal, vegetable, or mineral origin. Lubricating oils are usually of petroleum origin (from petrol to vaseline). Drying oils, used as a vehicle for paints, are of vegetable origin (linseed, poppy, walnut, hemp, sunflower seeds, etc.), although some of mineral origin are used in varnishes and special paints.

Oil Stain. Used for external work instead of paint where it is required to show the grain of the timber. It is usually varnished when dry.

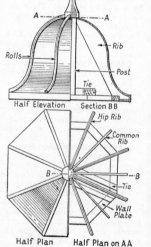

OGEE TURRET

Oil Stones. Used by woodworkers for sharpening tools. There are many varieties, both natural and artificial. Washita, Arkansas, Charnley Forest, and Turkey are the favourite natural stones; whilst carborundum and Indian (corundum and alundum) are popular artificial stones that may be obtained in various degrees of coarseness.

Okan. A reddish, seasoning to chocolate-brown hardwood from Nigeria. Very hard, heavy, durable, and tough. Difficult to work. Used for structural work, superior flooring, etc. Weight about 62 lb. per c. ft.

O'Kaycrete. A proprietary rapid hardening cement.

Okwein. A light to dark-brown hardwood from Nigeria. It is hard, heavy and spiral grain makes it difficult to work. Attractive appearance for superior joinery, etc. Weight about 48 lb. per c. ft.

Old Woman's Tooth. See *Router*.

Olon. Avodire, *q.v.*

Ombreé Finish. A variation in shade of a painted surface, usually varying from darker at bottom of wall to lighter at the top, by controlling the spray.

One-pipe System. A plumbing installation in which one pipe serves for soil and waste. The sanitary fittings have specially deep seals.

Onozote. A rubber compound expanded by gas into cellular form, for insulating.

Onyx. Marble, of stalactitic or stalagmitic formation. Various colours and markings.

Oolitic. Limestone composed of small rounded grains, like fish roe, Jurassic rocks, as Portland stone.

Opaline. A semi-translucent white glass.

Opalite. A vitreous compound made in various colours to imitate glazed bricks. It is bedded on the wall with special plaster, supplied by the makers of opalite.

Opaque Glass. Obscured. Not transmitting or reflecting light. A fluoride is added to the ingredients (fluorspar, sodium silicate fluoride, cryolite, etc.) and this forms minute white crystals which make the glass opaque.

Open Arca. See *Area*.

Open-cell Method. A process of pressure impregnation of wood. The antiseptic is retained in the cell walls only.

Open Channel. Semicircular drain pipes used in manholes; or surface channels used for conveying rain-water.

Open Eaves. A roof in which the overhanging rafters or sprockets are exposed to view.

Open Floor. One with exposed joists; not covered by a ceiling.

Opening. An aperture in a wall, etc., usually to receive a door or window.

Open Newel. Applied to a staircase with a rectangular well and newels.

Open Roof. One in which the principals are on view. No ceiling.

Open, or Spaced, Slating. Slates laid with spaces between, for economy. The maximum size of space should be *width of slate* −8 *in.*, to allow for a 4 in. side lap.

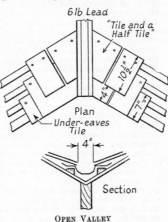

OPEN VALLEY

Open String. See *Cut String.*

Open Valley. One in which the lead work is exposed. See *Valley.*

Open Well. A stairs with rectangular well.

Opepe. A Nigerian rich golden-brown hardwood. Hard, heavy, strong, and durable. Used for superior joinery, structural work, strip flooring, etc.

Opg. Abbreviation for *opening.*

Ophite. Serpentine marble.

Opisometer. An instrument for measuring curved lines.

Opisthodomus. An enclosed space at the rear of a cell in a Greek temple.

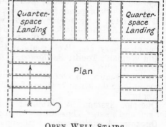

OPEN WELL STAIRS

Optical Square. A surveying instrument consisting of mirrors for reflecting the images of distant objects.

Oratory. A small private chapel, or a small chapel attached to a church.

Orb. A plain circular boss ; or stone panels for blank windows.

Orchestra. A space for the musicians in front of a theatre stage.

Order Form. A special form used by the builder or his foreman for ordering the delivery of goods.

Orders. 1. See *Architectural Orders* and *Columns.* 2. See *Arch with Orders.*

Ordinance. An arrangement of parts, in architecture.

Ordnance Datum. Particulars of heights and distances obtained from an Ordnance survey map.

Oregon Pine. See *Douglas Fir.*

Ores. Metals mixed with other substances, as obtained from the mine.

Organ Loft. A gallery containing an organ.

Oriel. A window projecting from the face of the wall and supported by corbels or brackets. The recess in the room formed by the oriel window.

Orientated. 1. A church with its chancel pointing east. Having an easterly direction, or having characteristics of Eastern countries. 2. A building arranged with due regard to the lighting.

Orifice. An opening or vent. The mouth of a pipe.

Orlet. A prefabricated concrete house.

Ormolu. A fine brass consisting of 75 per cent copper and 25 per cent zinc. Also gilt bronze.

Oro. A proprietary heat-resisting paint.

Orodo. A reddish-brown hardwood, with lighter streaks, from Nigeria. It is hard, heavy, strong, and hard wearing and used for flooring, etc. Weight about 48 lb. per c. ft.

Oroide. An alloy of copper and zinc.

Orpiment. King's yellow, *q.v.* Sulphide of arsenic used as a pigment.

Orthoclase. Potash felspar. $KAlSi_3O_8$. See *Granite.*

Orthographic Projection. Representing objects by plans and elevations. The various points are projected at right angles to the planes of projection, on the assumption that the eye is at an infinite distance from the object. See *Planes of Projection.*

Orthostyle. An arrangement of columns in a straight line.

O.S. Abbreviation for *one side.*

O.S. & W. Abbreviation for *oak, sunk, and weathered.*

Oubliette. A cellar with trap-door entrance.

Out and In Bond. Ashlar quoins and jamb stones arranged as alternate headers and stretchers.

Out Band. A jamb stone arranged as a stretcher, and recessed for a frame.

Outbuilding. See *Outhouse*.

Outcrop. The appearance at the earth's surface of an inclined strata of rock.

Outer Lining. The outside lining of a boxed frame for sliding sashes.

Outer String. The string farthest from the wall, in a flight of stairs.

Outfall. The place of discharge of a sewer.

Outhouse. A small building detached from, but belonging to, a dwelling-house. It may be attached but without direct communication from the house.

Outlet. A vent; an opening for discharging anything, as the nozzle to a gutter for discharging the water into the down-pipe.

Out of Wind. Not warped or twisted. A plane surface.

Outrigger. A cantilever projecting over the face of a building to carry a suspended scaffold or a large projecting cornice. The sleepers for a derrick crane.

Out to Out. Measurements taken to the outside edges of a piece of framing. See *Centre to Centre*.

Ova. An egg-shaped ornament carved on mouldings.

Oval. 1. Having the outline of an egg. 2. A false ellipse, *q.v.*

Overburden. See *Callow*.

Overcloak. The outer layer of lead worked over a roll to a gutter or flat.

Overdoor. A pediment. An ornamental finish to a door head.

Overflow. A pipe placed at the top of a sink or cistern to allow the water to escape when the discharge pipe or ball valve is out of order.

Overgraining. A thin, almost transparent, coat of raw umber mixed with beer, applied to a grained surface to give a greater variation of light and shade. A term applied to any additional work to a basic grained effect.

Overhand Work. Erecting brick walls from the inside. The method is to work from each floor and on trestle scaffolds, as the building rises.

Overhanging Eaves. Common spars continued over the wall plate to project over the face of the wall. Also see *Sprocket Eaves*.

Overhead Charges. The various charges, such as rent, office expenses, salaries, etc., added to the wages and cost of material, and other primary costs, when preparing estimates.

Overhead Screed. One supported above the lower concrete so that the top layer of concrete will be unbroken. See *Flying Screed*.

Overmantel. The top part of the ornamental front to a fireplace. See *Mantel*.

Oversailing. Applied to courses of bricks or stones that project over the face of the wall.

Oversailing Floor. One cantilevered over the support, common at first-floor level in half-timbered buildings

Over-story. See *Clerestory*.

Over-tiles. The top tiles, or imbrices, covering the joint of two under-tiles, or tegulae, of the Italian and Spanish types. See *Spanish Tiles*.

Ovolo. A convex moulding in the form of a quarter circle or ellipse. See *Mouldings*.

Oxide of Chromium. See *Chromes* and *Pigments*.

Oxide of Iron. Rust, or ferric oxide. A compound of iron and oxygen which produces red colour in bricks and tiles. An oxide is a compound of oxygen with some other element. See *Pigments*.

Oxide Paints. Those in which the colour is due to inorganic iron compounds. They are used chiefly for external work. The pigments vary in colour from dark red to purple and black.

Oxidizing. The oxygen of the atmosphere combining with metals, causing the corrosion of iron and the discoloration of "non-corroding" metals. See *Corrosion*.

Oxter Piece. A piece of ashlering. See *Ashlering*. A tie in a cottage roof between rafter and ceiling joist.

Oxy-acetylene Welding. The fusing of the joints of steelwork, instead of riveting, by means of the intense flame of acetylene burning in oxygen.

Oxy-chloride Cement. *Sorel cement.* A chemical combination of oxide and chloride of magnesia. It is used extensively for jointless floors as it sets extremely hard and will bind itself with almost any filler and aggregate. A basic hydrated magnesium oxychloride.

Oxylene. A process of chemically impregnating wood to make it fire-resisting.

Oylets. See *Eyelets*.

Ozalid. See *Blue Prints*.

P

P. A symbol for *total pressure*, and *buckling load* on a column.
p. Abbreviation for *prime, pitch*, and for *pressure per unit area*.

P. and C. Abbreviation for *parge and core*.

Pace. A portion of a floor raised above the general level.

Package Kitchen. A compact unit for small kitchens containing sink, refrigerator, water heater, cooker, and storage shelves. It is made of steel, aluminium, or wood.

Packing. 1. Any material used for filling, levelling, or making solid. 2. A non-conducting covering for pipes.

Pack Up. To raise and support anything on packings, or solid blocks.

Pad. 1. A handle. 2. See *Tool Pad*. 3. Soft material used as a cushion or seating. 4. A stone template.

Padauk. A hard, heavy, strong, durable, and stable hardwood from India, Burma, Africa, and Andaman Islands. It is deep crimson but seasons to nearly purple, and is variegated. It is rather difficult to work. Polishes well with careful filling. Esteemed for superior joinery, flooring, etc. Weight about 46 lb. per c. ft.

Padlock. A movable lock used with a hasp and staple for securing doors and gates from one side only.

Pad Saw. A saw with narrow blade that passes through the handle when not in use. It is used for small circular work.

PAD SAW

Pad Stone. See *Template*.

Page. A beam with parallel iron rods, along which the pallets slide when brickmaking. A folding wedge.

Pagoda. An Eastern temple for housing an idol.

Pail Closet. A privy with movable receptacle. See *Closets*.

Paint. A decorative preservative for wood, iron, etc. Oil paint usually consists of *base* (white and red lead, etc.), *vehicle* (oil), *colouring pigments*, *solvent* (turps), and *driers*. See *Inert Fillers*, and *Water Paint*.

Paint Brushes. There is a large variety, including: dusting, fitches, flat, distemper, knot, pencils, sash tools, turk's head.

Paintcrete. A proprietary hard-setting paint for concrete, stone, etc.

Paint Strip. Any material that will remove old paint, instead of burning off. There are several proprietary materials: Quickstrip, Nitromers, Nonflam, Varmover, etc.

Paired. Two of the same things matched on opposite hands.

Palaestra. A gymnasium.

Palafitte. A hut on piles and over water.

Pali. A light to reddish-brown hardwood from India. Fairly hard durable, and strong. Used for joinery, matchboarding, flooring, etc. Weight about 34 lb. per c. ft.

Palings, or **Pale Fence.** Vertical boards or stakes forming a fence or enclosure.

Palisade. A fence of pointed stakes or palings. Iron railings formed of pointed and ornamental vertical bars.

Palisander. Rosewood, *q.v.*

Palladian. A term applied to the Renaissance style of architecture, introduced by Inigo Jones about A.D. 1605.

Pallet. 1. A thin slip of wood built in a brickwork joint to serve as a fixing. See *Wood Slips.* 2. Thin boards on which bricks are placed to convey them to the hacks, after moulding.

Pallet Brick. A brick having one edge rebated to receive a fixing strip.

Palmettes. Carved ornamentation incorporating the Egyptian palm leaf. Sometimes applied to imitations of the honeysuckle flower.

Palosapis. A light-yellow hardwood from the Philippines. Fairly hard, heavy, strong, but not durable. Polishes well. Used for interior joinery, veneers, construction, etc. Weight about 36 lb. per c. ft.

Palu. A red to purplish-brown hardwood from India. Very hard, heavy, durable, and strong. Attractive grain figure, but too hard for ordinary purposes. Used for structural work. Weight about 64 lb. per c. ft.

Pampres. An ornament of vine leaves and grapes.

Pan. 1. A panel in half-timbered work. See *Post and Pane.* 2. A pedestal water-closet.

Panache. A pendentive, or spherical triangle, formed by a dome rising from a square base.

P. and S. Abbreviation for *planking and strutting.*

Pane. 1. A panel. 2. A sheet, or square, of glass cut to the required size.

Panel. A thin wide piece fitted between the members of thicker framing. Any flat surface sunk below the level of, or distinct from, the surroundings.

Panel Heating. Panels formed of coils of jointless piping bedded in walls and ceilings, through which hot water circulates for the invisible heating of rooms.

Panelled Framing. Frames and doors consisting of stiles, rails, and muntins mortised and tenoned together, with the spaces filled with panels.

Panelled Linings. Wide linings, round door and window openings, formed of panelled framing. See *Finishings.*

Panelling. Panelled surfaces.

Panel Mould. A mould for running panels in plasterwork.

Panel Mouldings. Small ornamental mouldings for mitring round panels. See *Mouldings.*

Panel Planer. A thicknesser, *q.v.*

Panel Saw. One shaped like a hand-saw, or crosscut, but with finer teeth like a tenon saw.

Panel Walls. Non-bearing external walls between pillars, and supported by beams. The clothing of skeleton-framed structures.

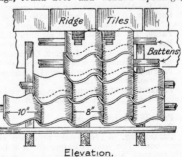

Elevation.

PANTILING

Pan-head. A cylindrical head to a rivet or bolt.

Panic Bolts. Patent bolts fitted to the exits of public buildings, and operated by pressure at the middle of the door.

Panisaj. Same as Hollock, *q.v.* from Assam.

Pannier. A corbel.

Pantheon. A temple dedicated to all the gods.

Pantile. A roofing tile shaped like a flat *S*, or ogee. Usual size, $13\frac{1}{2}$ in. by $9\frac{1}{2}$ in. A *square* requires about 180 tiles to a 10 in. gauge, and weighs about 8 cwt.

Pantograph. See *Eidograph.*

Pantry. A small adjunct to a kitchen, in which provisions, etc., are stored.

Pap. The outlet to an eaves gutter. A nozzle.

Paperhanging. The covering of walls with wall-paper.

Parabola. A section of a cone taken parallel to the side of the cone. See *Conic Sections.* Area = $\frac{2}{3}$ circumscribing rectangle. See *Arches.*

Paraboloid. A solid generated by a parabola revolving about its axis.

Paradise, or **Parvise.** The open space round a church. The cloisters or burial-place of a monastery.

Paragon. A patent roof glazing.

Paral. A yellowish-brown hardwood, with darker streaks, from India and Burma. It is fairly hard, heavy, durable, and strong, especially in shear. A crystal deposit dulls tools, when seasoned. Used for furniture, flooring, plywood, etc. Weight about 39 lb. per c. ft.

Parallax. The apparent change of position of one object relative to other objects when viewed from different positions.

Parallel Coping. One of parallel thickness, not weathered.

Parallel Gutter. See *Box Gutter.*

Parallelogram of Forces. The graphical method of finding the resultant of two forces, or the components of a force, by means of a parallelogram. If two forces act at a point and a parallelogram be formed, of which the two forces form adjacent sides, then the diagonal of the parallelogram is the resultant of the two forces.

Parament. The furniture, ornaments, etc., of a state room or apartment.

Paramount. A fireproof gypsum wall-board.

Parapet. 1. A protecting wall, breast high, to bridges, quays, etc. 2. The portion of the wall to a building carried above the foot of the roof.

Parapet Gutter. A gutter formed behind a parapet wall to a building.

Parastas. A pilaster.

Par. C. Abbreviation for *Parian cement.*

Parclose. A church screen with open tracery panels so that it only partly screens the enclosure. A parapet round a gallery.

Parge. The materials used for pargeting.

Pargeting. 1. Lining chimney flues with cement mortar, or with coarse mortar mixed with cow-dung. 2. Ornamental plasterwork formed by special tools or by stamping a repeat decoration before the plaster has set.

Parging. Same as *Pargeting* (1). The term is also applied to the facing of old walls with plaster.

Parian Cement. Powdered gypsum and dried borax and cream of tartar calcined and ground to a fine powder. Used in the same way as Keene's cement.

Paris White. See *Pigments.*

Parker. A registered concrete mixer.

Parkerizing. A proprietary process o coating ferrous metals to prevent corrosion. The iron is immersed in a boiling solution of an acid-metal phosphate. The process includes immersion in black dye followed by linseed oil, or shellac lacquer.

Parker-Kalon. Patent wood screws with more holding power than ordinary screws. The name is also applied to other screw fastenings.

Park Spring. A light-brown sandstone from Yorkshire. Used for general building, steps, etc. Weight 151 ib. per c. ft. Crushing strength 487 tons per sq. ft.

Parliament Hinges. Hinges shaped like the letter *H.* The knuckle projects considerably over the face of the door when closed. They are used to allow

PARQUET BORDER

PARLIAMENT HINGE

the door to open flat to the wall and to clear projections, such as an architrave.

Parlour. A sitting-room, or drawing-room, to a small dwelling-house. A private apartment for receiving friends in a convent, etc.

Paropa. A proprietary roofing consisting of a coating of bitumen and jute or wool felt, then a coat of hydrofuged mortar in panel form, with asbestized bitumen in the joints down to the bituminous coating.

Parpend. A through bond-stone faced on both ends.

Parpoints. Coursed rubble of gradually decreasing thickness.

Parquetry. Geometrical patterns formed of differently coloured woods; chiefly applied to flooring.

Parrell. The decorative parts of a chimney piece.

Parsonage. A manse. The official dwelling-house of a parson.

Part. One-thirtieth of a module, *q.v.*

Parting Bead. A thin slip, or bead, used to separate sliding sashes. See *Sash and Frame.*

Parting Sand. Dry sand placed between the two portions of damp sand to prevent them from adhering when placed together. Used when moulding metals.

Parting Slip, or Midfeather. A thin slip separating the weights for vertically sliding sashes. See *Sash and Frame.*

Parting Tool. A V-shaped gouge.

Partitions. Interior division walls. They may be of bricks, lath and plaster, timber, patent blocks, concrete, steel, or glazed framing; and they may be fixed, folding, sliding, or collapsible. There are many patent types of partition blocks: Frazzi, Grip, King, Cranham,

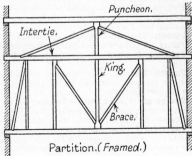

Partition. (*Framed.*)

Fosalsil, Shepwood, Trussit, Winget, etc. In steel-framed buildings they are non-bearing.

Partition Wall. Any internal wall not being a party, structural, nor a division wall. A non-bearing wall.

Party Colours. A term applied to surfaces painted in more than one colour.

Party Fence or **Wall.** One separating two adjacent properties or one standing on lands of different owners.

Pascal's Law. "When pressure is applied to any part of the surface of a fluid, a uniform and equal pressure is transmitted over the whole fluid."

Passage. A corridor. An alley.

Passings. The amount that one sheet of lead overlaps another in flashings, gutters, etc.

Pass Key. See *Master Key*.

Passover Offset. A bend in a pipe to allow it to cross another pipe in the same plane, so that both pipes may fit closely to a wall.

Paste. An adhesive for wallpapers made from household flour. It is prepared by mixing, say, 2 lb. of flour and one pint of cool water into a smooth paste. After standing for a short time half a gallon of boiling water is added while vigorously stirring, and allowed to cool. Paste ready prepared is obtainable.

Pat. A small thin cake of cement for testing purposes.

Patand. A plate, or sill, resting on the ground and carrying vertical timbers.

Patch. A repair to plywood.

Patent Boards. A term applied to the many proprietary building, or wall, boards, *q.v.*

Patent Floors. Fire-resisting floors in which some particular feature is patented, such as : Armoured Tubular, Bison, Broadmead, Corite, Cullum, Dawney, Diespeker, Fawcett, Fram, Frazzi, Hollocast, Homan and Rodgers, King, Kleine, Laing, Placet, Siegwart, etc.

Patent Glazing. See *Glazing*.

Patent Plasters. The many types of plasters having a gypsum base but with some chemical in combination that allows the compound to be patented. Most of them are substitutes for Keene's but some have special features. Well-known types are : Adamant, Astroplax, Barium, Bassett's, Battle Axe, Billingham, Carlisle, Gothic, Marbolike, Murite, Napco, Parian, Pioneer, Robinsons, Sirapite, Statite, Surfex, Thistle, Victorite.

Patent Plate. Sheet glass polished both sides.

Paterae. Flat circular ornamentation in soffits, friezes, transoms, etc., usually in the form of a cup or dish.

Patin. See *Patand*.

Patten. The base of a pillar or column.

Patter. A thick wooden float for tamping and levelling cement surfaces.

Pattern. A wood or plaster model for a casting.

Pattern Staining. Discoloration of ceilings due to unequal conductivity of plaster and the supporting timbers.

PATTER

Pavement. A hard surface forming a street, road, or yard, usually for pedestrians. It may be formed by bricks, stones, tiles, concrete, asphalt, wood blocks, etc.

Pavement Lights. Metal or reinforced concrete frames containing a number of small glass prisms for lighting basements, and forming part of the pavement. The prisms are scientifically designed to refract the light into the basement.

Pavilion. A one-story ornamental building for entertainments, or for sports fields.

Pavilion Roof. A roof with a polygonal plan. See *Polygonal Roof*.

Paving. 1. A pavement. 2. The act of forming a pavement or road.

Paving Flags. See *Flagstones*.

Paviors. 1. Specially hard-burned bricks, or stones (setts), for forming pavements. 2. One who lays pavements and roads of bricks or stones.

Pavit. A proprietary cold-process bitumen for paving.

Pawl. A cog that falls between the teeth of a ratchet to stop motion.

Paxboard and Paxfelt. Proprietary insulating materials.

P.C. Abbreviation for *prime cost* and *Portland cement*.

P.C.B.S. Abbreviation for *Portland cement British Standard Specification*.

Peak Joint. The joint at the apex of a truss.

Pearson's Formula. Used to find the wind pressure on inclined surfaces.

Pearwood. See *Scented Guarea* and *Oje*.

Pebble-dash. See *Rough cast*.

Peck. A slater's hammer.

Peckings. See *Place Bricks*.

Pecky. A term applied to timber with symptoms of decay.

Pedestal. 1. A base for a column, statue, or other ornament. 2. The framing supporting a desk or counter. 3. A bearing for a shaft, in machinery.

Pedestal Closet. The ordinary type of wash-down water closet, consisting of glazed earthenware pan and supporting pedestal.

Pediment. A triangular, or segmental, ornamental head to an opening, or over the entablature at the ends of buildings and porticos in classic architecture. It may be of stone, wood, brick, plaster, or concrete.

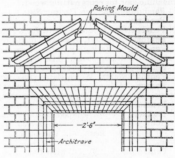

PEDIMENT (BRICK)

Peeling. The most economical method of preparing veneers. Also called rotary cutting. The log is prepared, and revolves against a long knife which peels the wood in a continuous length, like unrolling paper. The thickness varies from $\frac{1}{80}$ in. to $\frac{2}{3}$ in. according to requirements. Also see *Flaking*.

Peen. See *Piend*.

Peet's Valve. A patent stopgate valve for shutting off the cold supply to a domestic hot-water system. It is full bore and does not obstruct the flow.

Pegged Out. Applied to a building site that has been set out with pegs.

Peggies. Small slates from 10 in. to 14 in. long.

Pegs. 1. Pieces of wood driven into the ground where required when setting out a building site. 2. Wood pins or hooks for hats and coats. 3. Pins for securing wood joints.

Peg Stay. A casement stay fixed in position by pin and hole.

Pegui. A yellowish-brown hardwood from Brazil. Used as a substitute for greenheart. Weight about 62 lb. per c. ft.

Pele, or **Peel.** A fortified tower. A small donjon.

Pellet. A *little joiner* in the form of a tapered cylindrical plug, to cover a sunk screw, etc.

Pellet Moulding. One consisting of a series of small hemispherical projections.

Pelmet. An ornamental head to the inside of a window to hide the curtain and blind fittings.

Pen. A small enclosure or birds; a fold for animals.

Pencils. Very small paint brushes for lettering, etc.

Pendant. An ornamental finish at the bottom of a suspended post. Ornamentation suspended from a ceiling. A hanging ornament.

Pendant Post. The lower post at the foot of a hammer-beam roof truss.

Pendentive Dome. A dome over a rectangular chamber in which the corners are filled with pendentives to provide a circular seating for the dome.

Pendentives. The spherical triangles formed by the intersection of two vaults or by a dome springing from a rectangular base.

Pendulum Saw. A swinging machine-saw for cross cutting wood.

PENDENTIVE DOME

Pendulum Slip. See *Parting Slip.*

Penetrex. A proprietary colourless waterproofing liquid.

Pen-Farrell. A patent guard for a spindle machine.

Penfold. A pound, or cattle enclosure.

Pennant. A dark grey or bluish sandstone from the Bristol area and an excellent building stone. S.G. 2·7.

Pennycock. A patent roof glazing.

Penrhyn Slates. Welsh slates. See *Slates.*

Penryn. A light-grey granite from Cornwall. Used for constructional work. Weight 165 lb. per c. ft. Crushing strength 1250 tons per sq. ft.

Penstock. A reservoir valve. A sluice or floodgate.

Pentacle. A star-shaped figure formed by producing the sides of a regular pentagon. The name is also applied to a similar figure produced from a regular hexagon. The latter is more commonly used in diaper work.

Pentagon. A polygon with five sides. Area = sq. of side × 1·72, for a regular figure.

Pentane. A hydrocarbon occurring in petroleum.

Pentastyle. A series of five columns. See *Columns.*

Pent-house, or Pentee. 1. A projecting sloping hood to a window or door. canopy. 2. A habitable room, or rooms, on a flat roof.

Pentrometer. A needle for scientifically testing the penetration of asphalt.

Pent-roof. See *Lean-to.*

Perch. 1. A bracket or corbel. 2. A rack in which pieces of wood are arranged vertically for natural seasoning. A steer. 3. A measure of length of 5½ yards.

Percolex. A proprietary waterproofer for concrete.

Perfections. Applied to shingles 18 in. long.

Perflation. The rapid and complete changing of air in a building.

Pergola. An arbour or balcony; usually associated with rustic and trellis work, or espalier.

Peribolas. A wall round an ancient temple enclosing the sacred ground.

Perimeter. The bounding line of a figure. The distance round, or the sum of the sides of a plane figure.

Period. Applied to buildings, furniture, etc, characteristic of a past period: Jacobean, Queen Anne, Tudor, etc. The English periods are approximately: Tudor, 1485–1603; Jacobean, 1603–1649; Cromwellian, 1649–1660; Charles II, 1660–1685; William and Mary, 1689–1702; Queen Anne, 1702–1714; Early Georgian, 1714–1727; Adam, 1760–1792; the French periods are Renaissance, 1515–1600; Louis XIII to Louis XVI, 1610–1774; Directoire, 1795; Empire, 1805; Restoration, 1815.

Peripteral. A temple having a single row of columns all round. See *Columns* and *Temple*.

Peristyle. A range, or row, of columns round a building or courtyard. Also the space enclosed.

Perkin's Jig. A patent appliance for housing strings on a spindle machine.

Permacem. A proprietary waterproof cement paint. Also see *Cheecol*.

Permali. A registered name for an improved wood, *q.v.*

Permanent Set. The permanent distortion of a structural member due to over-stressing. See *Elastic Limit*.

Permanite. A proprietary damp-proof course.

Permason. A proprietary preservative for wood.

Permavent. A patent skylight providing ventilation with closed light. The lights are fixed direct to rafters and made in iron, steel, or copper.

Permoglaze. A proprietary liquid glaze for walls requiring a hard non-absorbent surface.

Permolite. A proprietary wood preservative from coal tar.

Permutit. A patented method of softening water by chemical reagents.

Perpendicular Style. (Fifteenth century.) The latest of the Gothic styles of architecture; also called *Late Pointed*. The special features were large window areas and the perpendicular lines in the tracery work.

Perpends. The vertical, or cross, joints on the face of brickwork and masonry. See *Keeping the Perps*. Also the parts of a brick wall erected at the ends to serve as a guide for the intermediate portion. Perpendicular marks on a roof to guide the slater, when fixing slates, tiles, etc.

PERPENDICULAR
TRACERY

Perpent Stone. A bonder, or through stone, in ashlar work.

Perron. A staircase outside a building. The outside steps leading to the entrance of a building.

Perrycot. Portland stone from a stratum lower than Whitbed.

Persienne. An exterior shutter with movable slats, for a window.

Perspective, or Radial Projection. The art of drawing, or representing, an object on a plane surface so as to appeal to the eye in the same way as the actual object, i.e. the relative positions and proportions appear the same, as in a photograph.

Perspex. A registered thermo-plastic made in large corrugated sheets for roof lights, etc.

Pervibration. The vibration of concrete in the interior of the mass, instead of on the surface.

Petaling. A light-brown hardwood from India. Very heavy, hard, durable, and strong. Used for constructional and structural work, furniture, etc. Weight about 55 lb. per c. ft.

Peterhead. A decorative red biotite granite from Aberdeen. Excellent for all types of constructional work. Polishes well. Weight 158 lb. per c. ft. Crushing strength 1,470 tons per sq. ft.

Peterlineum. See *Carbolineum*.

Petersburg or Petrograd Standard. A measure of timber of 165 cub. ft. Equal to 1,980 ft. board measure.

Petrails. Strong beams in timber-framed buildings. See *Post and Petrail*.

Petrifoid. A proprietary transparent waterproofing liquid.

Petrol Trap. Special drainage trap for garage yards, etc., to separate the oil before entering the drains.

Petropine. A composition for jointless floors.

Petrumite. A proprietary paint to imitate stone.

Petthan. A yellowish-red hardwood from Burma. Qualities and properties of greenheart. Weight about 48 lb. per c. ft.

Pew Hinge. A type of parliament hinge but with an ornamental *egg-shaped* knuckle.

Pews. Enclosed seats in churches.

Pewter. An alloy of tin and lead.

P.F. Abbreviation for *plain face*.

P.F.I. & R. Abbreviation for *part fill in and ram*.

Philbrick. A proprietary waterproofing solution.

Philplug Adapta. A patent wall plug, and other fixing devices.

Pholas. A marine insect very destructive to timber and concrete.

Phon. The unit of loudness of sound.

Phosphor Bronze. A copper-tin-phosphorus alloy. The phosphorus hardens the metal which may be obtained as sheet, tube, wire, or strip.

Phosphorus. A chemical element, P. It hardens cast-iron, whilst 1 per cent in wrought iron increases the welding properties. The smallest quantity in steel reduces the quality.

Photographic Prints. Reproduction of drawings by the camera. See *Blue Prints*.

Photometer. An instrument for measuring illumination, usually by comparison with a standard lamp of known candle power.

Photometry. The science that deals with the measurement and comparison of the illuminating properties of sources of light.

Photomicrograph. A photograph of a small object as viewed under a microscope.

Photostat. A patent duplicating machine.

Piazza. An open square, or space, surrounded by buildings. A covered path.

Pick. 1. A pointed hammer for dressing hard stone. 2. A heavy sharp-pointed tool with long handle for breaking-up earth when excavating. 3. A slater's hammer.

Picket. A pointed stake or peg driven into the ground. Narrow strips of wood used for open fencing.

Picking Up. Applied to wood when the grain tears away, or plucks up, due to cross or interlocked grain.

Pickling. 1. Treating old roof timbers against insects and decay. 2. Soda and hot water used for washing paintwork preparatory to re-painting. 3. Removing scale, etc., from a metal surface by immersing in sulphuric or hydrochloric acid. 4. Often applied to the Bethel process of preservation of wood.

Pictorial Projection. Any form of drawing (isometric, oblique, perspective, axiometric, planometric, etc.) that shows the three views of orthographic projection in one view.

Picture Plane. An imaginary plane on which the rays from the eye to any part of an object are intercepted to form a perspective drawing.

Picture Rail. An ornamental moulding round a room, from which pictures are suspended.

Piece. See *Wallpaper*.

Piece Mould Process. A method of casting concrete objects of an intricate or undercut character; the plaster mould is made in several pieces that can be withdrawn without destroying the undercut parts.

Piece-work. The term used when workmen are paid according to output.

Piedroit. A pilaster without base and capital.

Piend Check. A bird's-mouth rebate as in stone steps.

Piend Rafter. Hip rafters.

Pier. 1. A support, or pillar, for an arch, bridge, or beam. 2. A rectangular narrow projection on a wall to give additional support to a beam or other load. 3. Narrow solid work between openings. 4. A landing place projecting into the sea.

Pier Bonding. Arranging the bricks or stones to secure a pier to a wall.

Pier Glass. A long, narrow mirror fixed on a wall between two openings.

Pierre Perdue. The formation of under-water foundations by lowering blocks at random until the mass rises to the surface, as described for *Mound Breakwater.*

Pier Template. A stone cover on a brick pier to distribute the load over the whole section.

Pigeon Hole. A small space in a desk or other fitment for storing papers.

Pig Iron. The form in which cast iron is run from the furnace. The molten metal runs along a channel (sow) from which branch out a number of short channels. The latter are about 2 ft. long and form *pigs*, which are broken away from the sow. See *Cast Iron.*

Pig-lug. See *Dog-eared Fold.*

Pigments. The colouring matters used in paints, etc. They may be of mineral, animal, or vegetable origin, and natural or artificially produced. The following pigments are in common use. Natural earths: Paris or English white (whiting), red oxide of iron, graphite and mineral blacks, ochre, sienna, umber, and vandyke brown. Chemical compounds artificially produced include (whites) zinc and antimony oxides, white lead, lithopone, titanium white and oxide; (reds) iron oxides, Venetian red Indian red, purple oxide, red lead, burnt sienna, lead chromate, vermilion; (greens) Brunswick greens, lime green, oxide of chromium; (yellows) lead chromes, zinc chromes, synthetic ochres; (blues) ultramarine blue, lime blue, prussian blue.

Pigs. 1. The blocks of lead or cast iron as prepared from the ores. See *Pig Iron.* 2. Pieces of iron about 15 in. × 6 in. × 2 in., between which are driven wedges to break away blocks of stone along the natural joints, when quarrying Portland and similar stone.

Pilaster. A thin, rectangular pier projecting from the face of a wall.

Pilaster Capping. An ornamental projecting top to a pilaster.

Pilcher's Stop Rot. A proprietary preservative for timber.

Pile. A long pointed support of wood, concrete, or steel, driven into poor soil to provide foundations for structures, or to form a protected enclosure for under-water work, etc. Concrete piles may be prepared and driven into position, or they may be poured *in situ* by first driving a cylindrical shell, which is withdrawn when the concrete is poured. There are many patent types. See illustration on page 234. See *Bearing, Friction,* and *Sheet Piles.*

Pile Driving. The act of driving piles to the required depth, usually by means of a machine called a pile driver, which is provided with an automatically falling weight, or ram, which drops on a helmet, or dolly; or with an automatic steam or air hammer. The safe load on piles may be determined from $L = \dfrac{Wh}{8d}$, where L = safe load in cwts., W = ram in cwts., h = fall of ram in in., and d = penetration of pile at last blow. The operation of driving is often assisted by the water-jet method, *q.v.*

Piling . 1. Driving piles. 2. A system of piles in position. 3. Stacking wood for seasoning.

Pillar. A detached support, or column, rectangular or polygonal in section. All steel columns, stanchions and struts. (*London Build. Act*, 1932.)

Pillaring. Planes of weakness, in addition to the planes of cleavage, that give *grain* in slates, which causes the slates to break easily.

Pillow. 1. See *Bolster*. 2. The baluster side of the Ionic capital.

Pilot Nails. Nails used temporarily when erecting formwork.

Pilot Nut. A temporary tapered nut to protect the thread of a bolt whilst being placed in position.

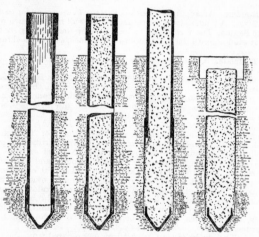

VARIOUS STAGES OF *In-situ* CONCRETE PILING

Pilot Piece. A projection at the foot of a post to protect it from vehicles.

Pimco. A patent clip for fixing building boards.

Pin. 1. A cylindrical metal fastening for the joints of structural steelwork. 2. A fine wire nail with very small head. 3. The steel wire through the knuckle of a hinge. See *Butt Hinge*. 4. A cylindrical wooden peg driven through a mortise and tenon joint to give additional security, as at the corners of sashes. 5. See *Dovetail*. 6. See *Pinning Up*. 7. To wedge the end of a timber in a wall. 8. See *Pin Knot*.

Pinch Bar. See *Crow Bar*.

Pines. *Pinus spp.* There is a large number of important softwoods called pine, with a distinguishing prefix, as white, red, pitch pine, etc. The sources of supply are widespread, about 25 species occurring in North America alone. The woods are usually fairly light, soft, stable, resinous, and durable, but there is considerable variation in the species. They are used for all kinds of joinery and carpentry. The best known are : Baltic redwood, chil, hoop, jack, kauri, Korean, loblolly, long leaf, Oregon, Parana, pitch, ponderosa, Quebec, red, Scots, shortleaf, Siberian, silver, sugar, Weymouth or white or yellow, etc. Also see *Matai, Podo, Rimu*.

Piney. A dark-red hardwood, with darker markings, from India. Very hard and strong, fairly heavy and durable. Not difficult to work and polishes well. Suitable for superior joinery, constructional work, veneers etc. Weight about 37 lb. per c. ft.

Pinhead Morocco. Similar to Morocco glass, but smaller pattern.

Pin Hinges. Similar to butt hinges but with a loose pin that can be easily removed so that the door can be taken away without removing the screws.

Pin-holing. Minute holes in varnished and enamelled surfaces, caused by dampness or steam, or by diluting proprietary brands with turps or white spirit.

Pinion. 1. A small toothed wheel used in conjunction with a larger one for driving shafting. Also called geared or spur wheels. See *Winch.* 2. A patented pantile.

Pin Joint. A joint secured by a single pin.

Pin Jointed. Applied to a pillar having its ends adequately restrained in position only. See *Fixed Ends.*

Pin Knot. A small knot not more than ¼ in. diameter. In some gradings it may be up to ½ in.

Pinnacle. An ornamental terminal to a roof or tower. A pointed turret.

Pinning. Securing the ends of steps, etc., into a wall. Fixing securely. See *Pin* (4).

Pinning Up. *Making good* the horizontal joints between old work and new work, by inserting pieces of slate or tile and running in cement. See *Under-pinning.*

Pinoleum Blind. An outside blind that protects from the sun's rays but allows for the admission of light and air.

Pin Rail. A wooden rail with pegs for hats, coats, etc.

Pins and Line. See *Line Pins.*

Pintle. A bolt or pin on which some part turns.

Pinus sylvestris. See *Northern Pine.*

Pioneer. 1. Proprietary cellular anhydrite blocks for partitions, etc. Easily cut and faced to receive special lining paper to avoid plastering. 2. A patent hard-faced plaster.

Pipe Clip. A fastening for a pipe. They are distinguished as saddle, hook, and holderbat.

Pipe Duct. A channel or passage prepared to receive pipes.

Pipe Hook. See *Wall Hook.*

Pipe Nail. See *Fastenings.*

Pipes. Pipes are used for conveyance of air, gas, water, drainage, etc., and may be of iron, lead, concrete, asbestos cement, stoneware, etc., and they may be lined with various materials to suit the particular requirements. They are named according to the purpose for which they are used : air, antisiphonage, flue, gas, hot and cold water, overflow, rainwater, service, soil, ventilation, waste, etc.

Pipe Stopper. An expanding appliance for stopping the end of a drain when drain-testing.

Pipli. A greyish-brown hardwood from India. Like Satin Walnut. Not difficult to work and polishes well. Used for joinery, fittings, and constructional work. Weight about 37 lb. per c. ft

Piscina. 1. A niche containing a sink, in a church, or the sink only. 2. A fish pond.

Pisé de terre. Cob, or dried clayey earth, sometimes used for the walls of country cottages. Usually rammed between shuttering whilst plastic.

Pisolite. A coarse-grained oolitic limestone.

Pistol Brick. A dogleg brick.

Pitanco. A proprietary plastic insulating material.

Pitch. 1. The distance apart of the rivets in a riveted joint. 2. The amount of advance in one revolution of a helix or screw thread. 3. The ratio of the rise to the span in a roof. See *Roofs*. 4. The inclination of roofs, stairs, etc. 5. A thick, dark resinous substance obtained from distilling tar.

Pitch Block. A temporary piece cut to the pitch of stairs and used under wreaths, etc., whilst cutting to shape.

Pitch Board. A triangular templet used for setting-out strings for stairs. The sides of the triangle represent the rise, going, and pitch of the steps.

Pitched Roof. One with an inclination of more than 20 degrees.

Pitcher. A rough granite paving-stone.

Pitcher Tee. A three-way pipe connection with curved junction.

Pitch Faced. Hammer dressed, or rock faced, coursed rubble. The perimeter, or edges, of the face is brought into one plane by knocking off

Minimum Pitch	Materials
$\frac{1}{2}$	Thatch, Shingles
$\frac{1}{2 \cdot 6}$	Plain Tile
$\frac{1}{3}$	Small Slates
$\frac{1}{3 \cdot 5}$	Asbestos Tiles
$\frac{1}{4}$	Ordinary Slates, Pan Tiles
$\frac{1}{5}$	Large Slates, Roman Tiles
$\frac{1}{8}$	Corrugated Sheets, Felt
$\frac{1}{80}$	Lead, Zinc, Copper

PITCHES OF ROOFS

the superfluous stone, in the form of rough chamfers, with a pitching tool.

Pitching. 1. Facing the inner slope of a reservoir with dry stone, about 1 ft. thick, to above water level. 2. Arranged at an inclination. Not vertical.

Pitching Block. A planted piece on a post to simplify the cuts and to provide a bearing for the thrust from a strut or rafter.

Pitching Piece. 1. A rough bearer receiving the carriages for a flight of stairs. The ends are fixed in the wall and newel. In a geometrical stair the wall end only is fixed so that it acts as a cantilever. 2. The timber receiving the tops of the rafters in a lean-to roof.

Pitching Tool. A mason's chisel, about 9 in. long, with a bevelled non-cutting edge about $\frac{1}{4}$ in. thick for breaking away superfluous stone.

Pitch Mastic. A composition of calcareous aggregate bound with coal-tar pitch. Sometimes an admixture of silicious aggregate is added and a proportion of chalk. It is used for jointless floors and may be coloured and wood filled.

Pitch Paper. Damp-resisting underlining for wall-paper.

Pitch Pine. *Pinus spp.* A golden yellow, resinous timber with deep red markings. Very strong, stiff, and durable, and obtained in large sizes. Some specimens are very curly. Shrinks slowly and greatly. From the Southern States of U.S.A. S.G. ·8. There are several species marketed as pitch pine. Longleaf is the best. Other species are: Cuban, Florida loblolly, shortleaf, and Southern yellow pine.

Pith, or **Heart.** The soft spongy substance at the centre of exogenous trees.

Pit-hill. Baked clay produced by spontaneous combustion of the rubbish heaps at pit-heads.

Pit Sand. See *Sand.*

Pitting. Blemishes in a plastered surface.

Pittsburgh. A process for making drawn sheet glass. Similar to Foucault process but without fireclay die.

Pitt's Patent. A method of fixing door knobs to the spindle.

Pivot. A pin on which something rotates.

Pivoted Sashes. Sashes that rotate on centres, or pivots, fixed just above the middle of the height.

Pix. Same as pyx, *q.v.*

PIVOT AND SOCKET

Pl. Abbreviation for *plugged.*

Place Bricks. Defective stocks, generally only used for temporary work. See *Grizzles.*

Placing Plant. A mechanical contrivance by which concrete is hoisted

INSLEY CONCRETE PLACING PLANT

up a mast in a bucket and then discharged down metal chutes to any required position on the job.

Plafond. The soffit of the larmier to an entablature.

Plagioclase. A lime-soda triclinic felspar in granite and other rocks.

Plain Tiles. The ordinary flat tiles, having just sufficient camber for bedding and usually with two nibs for hanging on the battens. The sizes are 10½ in. × 6½ in. × ¾ to ½ in., and 600 tiles to a 4 in. gauge are required per square, weighing about 15 cwt. See *Eaves*.

Plan. The top view or horizontal section of an object as projected in orthographic projection. Anything drawn or represented on a horizontal plane, as a map or the horizontal section of a building.

Planceer. The soffit of an opening, cornice, or stairs. The soffit of the corona to a Gothic cornice.

Planceer Pieces. Horizontal timbers to carry the soffit boards or plaster for overhanging eaves.

Plane. 1. A stock, or body carrying a cutting iron, for shaping the surface of timber. Bench planes include jack, smoothing, and try-plane. Most of the joiners' planes can be obtained in wood and metal. Other kinds are: badger, bead, Bismarck, block, bullnose, chamfer, chariot, compass, fillester, hollow, jointer, matching, moulding, ogee, old woman's tooth, ovolo, panel, plough, rebate or rabbit, reed, router, rounds, scraper, scrub, scurfing, shave, shoulder, side rebate, spokeshave, toothing, trenching, universal. Machines have superceded moulding planes, of which there is a big variety. Patent types include Stanley, Sargent, Bailey, Norris, and Preston. 2. A flat surface. 3. The Scotch name for sycamore.

Plane Concrete. Not reinforced.

Planed. Applied to a wood surface that has been wrought smooth by hand or machine. Surfaced.

Plane Geometry. Geometry dealing with problems in one plane only, i.e. two-dimensional, or without thickness.

Plane of Cleavage. The planes along which laminated rocks may be easily split. The best example is slate, which is easily split into thin layers to form roofing slates.

PLANES OF PROJECTION

Plane of Rupture. An imaginary plane forming a triangular prism of earth that tends to overturn an earth retaining wall. The plane bisects the angle between the natural slope of the earth and the wall. See *Angle of Repose*.

Plane of Saturation. The natural level of the water in underground formations.

Planes of Projection. The planes of reference used in orthographic projection: vertical (V.P.), horizontal (H.P.), and end vertical planes, on which the elevation, plan, and end view are shown respectively.

Plane Tiles. See *Plain Tiles*.

Plank. 1. Sawn softwoods over 11 in. wide and between 2 in. and 6 in. thick. A hardwood plank is from 1½ in. by 9 in., over 8 ft.l ong. 2. See *Wreath*. 3. Large stone slabs used for fences.

Planking. Temporary timbering for excavations. Also loosely applied to thick flooring boards, scaffold boards, temporary floors of planks or deals.

Planning. The art of designing buildings and their approaches.

Planometer. An instrument for measuring areas.

Planometric. A pictorial representation of an object in which the plan is first drawn, and then oblique projectors, drawn at any convenient angle, are used to show the front and side of the object.

Plant. The builder's materials used in constructional work : scaffolding, cranes, mixing machinery, ladders, etc.

Planted. A term applied to mouldings and other ornamental features that are attached and not formed in the solid.

Plantex. A registered type of tracing cloth.

Plaque. A circular or elliptical ornamental plate for a wall.

Plasbestos. A proprietary bituminous emulsion waterproofer applied cold.

Plaskon. Urea-formaldehyde thermo-setting resin.

Plastaleke. A proprietary bituminous waterproof compound.

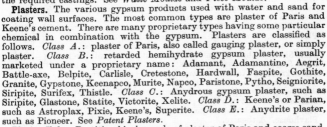

PLANOMETRIC PROJECTION

Plaster. A general term for plastic material for finishing wall surfaces ; usually lime and sand. See *Plasters.*

Plaster Board. Wall-board used as a substitute for plastered surfaces. They are gypsum products but vary considerably with regard to composition and surface texture according to manufacture. The advantages of this type of board are : little expansion and contraction, cheapness, damp-resisting, and permanency.

Plasterer's Laths. See *Laths.*

Plasterer's Putty. See *Lime Putty.*

Plastergon. A proprietary wood-fibre wall-board.

Plaster Mould. Moulds for casting cement-objects. They are made of plaster of Paris, fibre, laths, and battens, and are formed from a model of the required castings. See *Waste Moulds* and *Piece Moulds.*

Plasters. The various gypsum products used with water and sand for coating wall surfaces. The most common types are plaster of Paris and Keene's cement. There are many proprietary types having some particular chemical in combination with the gypsum. Plasters are classified as follows. *Class A.:* plaster of Paris, also called gauging plaster, or simply plaster. *Class B.:* retarded hemihydrate gypsum plaster, usually marketed under a proprietary name : Adamant, Adamantine, Aegrit, Battle-axe, Belpite, Carlisle, Cretestone, Hardwall, Faspite, Gothite, Granite, Gypstone, Keenapco, Murite, Napco, Paristone, Pytho, Seigniorite, Siripite, Surifex, Thistle. *Class C.:* Anydrous gypsum plaster, such as Siripite, Glastone, Statite, Victorite, Xelite. *Class D.:* Keene's or Parian, such as Astroplax, Pixie, Keene's, Superite. *Class E.:* Anydrite plaster, such as Pioneer. See *Patent Plasters.*

Plaster Slabs. Partition blocks made of plaster of Paris and coarse sand. They are used for lightness and are usually perforated, vertically and horizontally. Strong reeds are sometimes used for reinforcement.

Plastic. A term applied to semi-liquid material that can be moulded to any required shape.

Plastic Glues. The large range of resin-bonding materials used for ply-wood and mass production generally. There are several groups. (*a*) Thermo-setting resins : phenol-formaldehyde, urea-formaldehyde, soya-formaldehyde, melanine, etc. (*b*) Thermo-plastics, including acryl-polymers, cellulose derivatives, vinyl polymers. (*c*) Casein plastics. (*d*) Natural resin glues. They are extensively used for all classes of woodworking, but they require expensive equipment and expert control. Thermo-setting resin (urea-formaldehyde) is applied in liquid form as *hot-* or *cold*-press resin. The former requires cooking, or moulding, in an autoclave. The resin allows for a time interval of up to three days between spreading and cooking. Cold-press resin does not require heat, but the time interval is only about 15 minutes. The time interval depends upon the hardener, which is added just before use. Changes and improvements are continually taking place as a result of intensive research. Electric currents are now used for curing and new methods are being evolved both for wood and metal.

Plasticimeter. An instrument for measuring the plasticity of materials.

Plastic Paint. Proprietary preparations for wall surfaces, in various colours, that can be worked in any form of low relief decoration or textural finish.

Plastics. The popular name applied to a wide range of materials like Bakelite, Urea, Erinoid, etc., *q.v.* They can all be moulded or extruded and retain their shape. See *Plastic Glue*.

Plastic Strain. The condition when material is strained beyond the elastic limit.

Plastic Wood. A paste for filling holes in wood. Also see *Improved Wood*.

Plastoglage. A proprietary synthetic paint applied with spray gun.

Plastoleum. A proprietary floor covering.

Plastumen. A proprietary bituminous asbestos compound for roof repairs, etc.

Plat. A landing to a staircase.

Plat Band. 1. A flat fascia or string course. 2. A door lintel.

Plate or **Platt.** 1. A horizontal timber on a wall to distribute the load from other timbers. See *Wall Plate*. 2. See *Tracery*.

Plate Dowel. A fastening for the foot of a post on stone or concrete. See *Fastenings*.

Plate Girder. Built-up girders formed by riveting together steel plates and angle irons or other standard sections.

Plate Glass. The usual type for large panes, as for shop fronts. Polished plate is free from blemishes and may be obtained in sheets of over 100 sq. ft. area and from $\frac{3}{16}$ in. upwards in thickness. See *Glass*.

Plate Lock. A stock lock, *q.v.*

Plate Rack. A rack near a sink in a scullery, in which the crockery is placed after washing. See *Kitchen Fitments*.

Plate Shelf. A shelf around a room, about picture rail level.

Plate Staple. See *Striking Plate*.

Plate Tracery. An elementary form of tracery work in window openings, in which a thin screen wall inside the main arch is pierced by holes of geometrical design. See *Tracery*.

PLATE TRACERY

Platform. A part of a floor raised above the level of its surroundings.

Play. Same as clearance, *q.v.*

Playhouse. A theatre.

Plenum System. A method of heating large buildings. Air is cleansed in an air-washer, drawn over a battery of steam or hot-water pipes, and passed through distributing flues to different rooms by means of an electrically driven centrifugal fan.

Plethora. A disease in trees due to the uneven distribution of the sap.

Plimber or **Plimberite.** A proprietary building board of pulverised softwood chips bonded with synthetic resin, and fibrous reinforcement, from $\frac{3}{8}$ in. to $1\frac{1}{4}$ in. thick, and up to 8 ft. 6 in. by 4 ft.

Plinth. 1. A projecting surface at the bottom of a wall, column, etc. 2. A plain, thin, planted piece of wood at the bottom of a wall or fitment. 3. A plain, flat, square block forming a support for a column or pedestal. See *Mouldings*.

Plinth Block. A block at the foot of an architrave, against which the plinth or skirting abuts. See *Finishings*.

Plinth Course. A course of chamfered bricks forming the top course to a plinth in brick walls.

Plot. A piece, or patch, of ground.

Plotting. 1. The act of planning or laying-out. 2. A term used when setting-out a series of connected points forming a graph. See *Graphs*. The illustration shows the plotting of a point *a*, when $x = 12$ and $y = 14$.

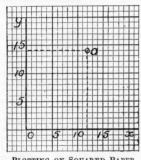

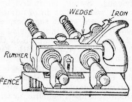

PLOTTING ON SQUARED PAPER   PLOUGH

Plough. An adjustable plane for forming grooves in timber.

Ploughed. Grooved: usually to receive a tongue or panel.

Plug. 1. See *Pipe Stopper*. 2. Wedge-shaped pieces of wood or lead driven into brickwork or masonry joints to serve as fixings. 3. To *stop* a hole with any solid substance. 4. See *Feathers and Plug*.

Plug Centre Bit. One in which the "centre point" is in the form of a cylindrical plug to insert into a previously bored hole, so that it serves as a guide whilst boring a larger hole round it.

Plugging. The removal of mortar from brickwork joints with a plugging chisel and the driving in of wood plugs, for fixing joinery, etc. See *Drills*.

Plug-it. A patent wall plug.

Plug Point. A power point for electrical appliances.

Plug Tap. A water tap with a conical spindle, or plug, having a hole in one direction only, so that a half turn of the spindle releases or stops the flow of water. It has been super-seded by the screw-down tap for water supply.

Plug Tenon. A stub tenon at the foot or head of a post to prevent lateral movement. It serves the same purpose as a dowel.

PLUG TENON

Plumb. Vertical.

Plumbago. Graphite, *q.v.*

Plumber's Solder. An alloy of lead and tin (70 per cent and 30 per cent) that fuses at a lower temperature than lead. It is used with a flux (resin and spirits of salts) to join lead pipes.

Plumbing. 1. The art of working in lead. 2. Arranging vertically.

Plumbing Iron. An elliptical-shaped iron bulb with a crook handle. It is heated and used by the plumber when wiping joints.

Plumbing Unit. A prefabricated unit for domestic dwellings in which all the essentials for hot and cold water supply are concentrated.

Plumb Line. 1. The string, or line, carrying a plumb-bob. 2. A line perpendicular to the plane of the horizon.

Plumb Rule and Bob. A parallel straight-edge of pine, about 5 ft. × 4 in. × ¾ in., from the top of which is suspended a heavy metal bob that swings in a hole at the bottom. The instrument is used for determining perpendiculars.

Plummet. A plumb-line or plumb-bob.

Plunger. A solid piston, as to a force pump.

Pluvex. A bituminous damp-proof course, with fibre base and lead lining. Also a plastic compound.

Plydek. A proprietary plastic flooring compound.

Plymax. A registered metal-faced plywood.

Plysyl. A proprietary permanent plywood formwork for concrete walls and floors. Also a dry floor unit designed on the stressed skin principle.

Plywood. Strong laminated boards consisting of from 3 to 36 layers of veneer with the grain of adjacent layers at right angles to each other. The layers are glued together, with waterproof glue in good qualities. Three-ply is from 3 mm. thick to 6 mm. thick, four- or five-ply is 6 mm. to 12 mm. Above 9 mm. thick there may be any number of plies. The qualities are AA, A, B, and BB. First qualities are free from defects on both sides, second qualities on one side, and inferior plywood has defects on both sides. The timber may be birch, alder, Oregon, gum, Norwegian pine, maple, Okoumé, oak, walnut, mahogany. Australian plywood is chiefly of hoop pine, maple, silky oak, and walnut. Plywood may be faced with any ornamental timber, bakelite, or with metal. Standard sizes are up to 10 ft. long and 5 ft. wide, but larger special sizes may be obtained. See *Slicing*, *Peeling*, and *Laminwood*.

P.M. Abbreviation for *purpose made*.

P.M.C. Abbreviation for *plaster moulded cornice*.

Pneumatic. Applied to any machine or contrivance actuated by compressed air.

Pneumatic Caisson. Caissons for under-water work in which the lower or working chamber is filled with compressed air to keep the foundation free from water.

Pneumatic Paint Brush. Shaped like an ordinary paint brush but the paint is forced among the bristles as a fine jet under pressure. It has an advantage over spraying as the paint is brushed into crevices.

Pneumatic Tubes. Tubes for conveying messages, worked by an electrically driven pump that controls the required pressure and vacuum.

P.O. Abbreviation for *planted on*.

Pocket. 1. The hole in the pulley stile of a *sash and frame* window for the insertion of the weights into the box. 2. A space in a chimney flue where cold air can gather and so prevent the smoke from rising. 3. A space filled with resin, as found in many resinous trees. 4. A space formed in a wall to receive the end of a beam, etc.

Pocket Chisel. See *Sash Chisel*.

Pocket Piece. The loose piece in a pulley stile to provide access to the weight box. See *Pocket*.

Pock-marking. Same as pin-holing.

Podium. A continuous low wall, or pedestal, supporting a number of columns.

Podo. The most useful softwood of the S. Hemisphere. They are excellent woods, moderately light, soft, strong, and durable. Straight, fine, even grain, and easily wrought. Stain and polish well. Used in joinery, flooring, etc. The African species are called yellowwood. Australian and New Zealand species are Brown pine, Kahikatea, Matai, Totara, etc. American species are not exported.

Poilite. Cement and fibrous asbestos, or mineral flax, made into tiles and sheets for roofing and wall-boards.

Point. A mason's chisel with very narrow cutting edge.

Pointed Arch. See *Arches* and *Vaulting*.

Pointed Architecture. Gothic Architecture.

Pointing. Finishing brickwork or masonry joints with richer mortar than that used for bedding the bricks or stones.

Pointing Machine. A frame for transferring the positions from the model to the actual object, in sculpture.

Points. Saw teeth, *q.v.*

Poisson's Ratio. *Transverse strain* ÷ *longitudinal strain* = n, where $n = \frac{1}{3}$ to $\frac{1}{4}$ for most materials.

Flat Joint

Poitrail. Petrail.

Polar Diagram. The vectors to the load line, or force diagram, from which the funicular, or link, polygon is obtained, in graphical problems, by drawing lines, or links, parallel to the vectors.

Flat Joint Jointed.

Weather Struck Joint

Pole. 1. A selected point to which the vectors radiate from the various points on the load line, in graphical problems. 2. A long, slender, round piece of timber. A market form of tapering round timber from 24 ft. to 35 ft. long, and up to 6 in. diameter. Used for scaffolding, flag poles etc. 3. A measure of length of 5½ yd.

Wrong Method

Keyed Joint

Pole Plate. A horizontal timber usually supported by the ends of the tie beams of roof trusses, and carrying the feet of the common rafters. See *Box Gutter*.

Recessed Joint

Poling Boards. Short vertical boards used for the timbering of excavations. See *Timbering*.

Bastard Tuck

Polished Plate. See *Plate Glass*.

Polishing. See *French Polishing*.

Poll. To split, especially a flint.

Tuck Joint

Polled Face. Flint walling with split, or knapped, flints showing on face.

POINTING

Pollapas. See *Urea*.

Pollard. Applied to the timber from trees that have been frequently lopped; as pollard oak.

Polygon. A plane rectilineal figure having more than four sides pentagon, hexagon, heptagon, octagon, nonagon, decagon, etc.

Polygonal Roof. See *Roofs*.

Polygonal Rubble. Kentish rag, rubble-walling. The faces of the stones are polygonal in shape.

Polygon of Forces. See *Vector Polygon*.

Polyhedra. Solids having many sides. There are five regular polyhedra : tetrahedron, cube, octahedron, dodecahedron, and icosahedron.

Polystyle. A building with many columns. See *Columns*.

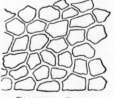

POLYGONAL RUBBLE

Polyvino. A proprietary plastic tile flooring. It is supplied as tiles 12 in. x 12 in. by $\frac{1}{8}$ in. or in 12 yd. rolls 3 ft. wide. The material is non-slip, hard wearing, and resistant to fire, water, oils, and acids.

Pommel. A spherical ornament, as a ball finial to a turret roof.

Pontoon. A bridge formed across a series of floating structures, such as boats or air-tight cylinders.

Pony Girder. A secondary girder supported by cantilevers only, for carrying the front portion of a large balcony.

Poon. A reddish-brown hardwood, with darker streaks, from India, Burma, Andaman I. It is moderately hard, heavy, durable, and strong. Interlocked grain but not difficult to work, and polishes well. Used for construction, joinery, etc. Weight about 38 lb. per c. ft. Another species, called Poonspar, is a better wood and also used for structural work, also called Malabar red pine.

Poor Limes. Those containing more than 15 per cent of impurities insoluble in acids.

Poplar. A white or pale brown, soft, close-grained hardwood. Durable if kept dry. Many species, but little used in building. Weight about 32 lb. per c. ft. See *Canary Whitewood*.

Poppet. 1. The top frame supporting the pulleys for a hoist. 2. A short piece of wood. 3. Applied to numerous small mechanical parts.

Popping. Defects in plastered surfaces due to using lime not thoroughly slaked.

Poppy Head. An elaborate carved ornament on the raised ends of pews in old churches.

Porcelain Ware. Sanitary fittings made of white clays mixed with pottery waste, and enamelled.

Porcella. A proprietary fire-resisting paint.

Porch. A roofed approach, or shelter, to a doorway.

Porous. Applied to materials that readily absorb water.

Porphyry. Ornamental granites.

Portable Machines. There is a big variety of electrically driven machines for use on the site; saws, planes, sanders, drills, etc.

Portail. The principal entrance at the front of a church.

Portal. 1. The smaller of two gates at the entrance to a building. 2. An arch to a gateway. 3. An entrance passage. 4. A prefabricated house.

Portal Bracing. A type of gusset bracing for tall steel-framed buildings, in which an arch is formed, of plates and angles, between the columns and under the floor girders.

Portal Strut. A modification of portal bracing in which a lattice girder replaces the plate bracing. It is less expensive but gives a level soffit and is less adaptable to architectural treatment.

Portasaw. A registered power-driven portable saw for tree-felling, and for ripping or crosscutting.

Portcullis. A strong grating of wood or iron sliding vertically in coulisses to close a gateway.

Portico. A porch with the roof supported by columns. They are distinguished according to the number of columns, as tetrastyle (4), hexastyle (6), octastyle (8), decastyle (10).

Porticus. See *Peristyle*.

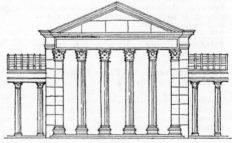

PORTICO

Portland Cement. A mixture of lime, silica, and alumina, in approximately the proportions of 60, 20, and 10 per cent respectively, with oxides, alkalis, etc. The mixture is formed into slurry and tested, then placed on drying floors, after drawing off the water, and burnt into clinkers at about 1,400° C. It is then ground to the required degree of fineness and cooled. Modern plants are speeded up by means of a battery of wash mills and perforated conveyers so that the cement is prepared in about six hours. It is used extensively in building for foundations, walls, floors, and roofs, as concrete or mortar. It is also used in combination with bitumen, rubber, wood, etc., for floor urfaces. See *Cement*.

Portland Stone. An oolitic limestone. See *Limestones, Whitbed* and *Basebed*.

Post. 1. A vertical timber acting as a support, or in constructional framework. 2. A term used for rocky and other unsuitable materials in clay used for fire-bricks.

Post and Pan, or Pane. Half-timbered buildings with *panes*, or panels, of brickwork or lath and plaster.

Post and Petrail. The same as *Post and Pan*.

Postern. A small door or gate serving as a private entrance, usually at the rear of a building.

Posticum. The rear part of a temple. See *Temple*.

Pot Colour. Applied to glass in which the colour is uniform throughout and not flashed.

Potential, Difference of. Voltage in electricity.

Pots. Crucibles, holding about 2 tons, used for melting the materials for plate glass.

Pottsco. Molten slag specially prepared for light-weight concrete aggregate.

Pounce. Fine pumice powder.

Pounded. See *Tamping*.

Powell-Wood. A process of preserving timber by impregnating with a saccharine compound.

Power Station. A building for generating electricity.

Poyntell. Paving in lozenge form.

P.P.C. Abbreviation for *plain plaster cornice*.

Pram. A wheelbarrow with the body supported on trunnions, for concrete.

Pratt Truss. The American name for the Linville, or N-truss. See *Roofs*.

Preamble. The introduction to bills of quantities stating the quality of materials and workmanship for each particular trade.

Precast Blocks and Lintels. Concrete blocks cast before fixing, as distinct from *in situ* work. See *Armoured Tubular Floor*.

Precast Stone. Concrete faced to imitate natural stone. See *Cast Stone*.

Precelly. South Wales slates. They are sold as randoms and rustics.

Precipitation. The chemical treatment of sewage.

Precipitation Tank. One in which chemicals are used, as distinct from a *settling* tank, *q.v.*

Predella. A platform or step on which a church altar rests, or a raised shelf at the back of an altar. A kneeling or foot stool.

Preece Test. A test for zinc coatings on iron

Prefabricated. Applied to buildings in which the construction is designed in factory-made, or mass-produced large units, so that the minimum of assembly is required on the site.

Preliminary Operations. A term applied to the preparation of a site before actual building operations commence: levelling, excavating, erection of temporary huts and sheds, hoarding, water supply, access to site, etc.

Presbytery. The east end of a church, near the altar, reserved for the officiating clergy. (R.C. Church.) The house of a parish priest.

Presdwood. A proprietary, strong, water-resisting wall-board.

Preservation. The treatment of building materials to arrest decay.

Surface treatment is usual. Paint, tar, bituminous compounds, cement, galvanizing, and several patented methods (Bower's, Barff's, Angus Smith's) are used for metal. Paint, creosote, charring, etc., are preservatives for timber; but impregnation with salts or tar-oil, under pressure, is much more satisfactory, by such methods as Bethel's, Blythe's, Boucherie, Burnettizing, Kyanizing, Noden-Bretteuneau, Powell-Wood, René, Celcurizing, Wolman, etc. See *Wood Preservation*. There are a large number of patent preparations for surface treatment of brickwork; and a number for

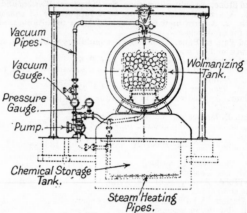

PRESERVATION OF TIMBER

stone, such as Fluate, Hemingway's, Browning's, Ransome's, Church's, and Zerelmy's solutions, etc. See *Cementone, Colopakes, Knapen, Roxor*. Also see *Silicon-ester, Ethyl-silicate*, and *Steam Cleaning*.

Presomet. A proprietary bituminous paint for iron.

Presotim. A proprietary, decorative, wood preservative.

Press. 1. A detached cupboard. 2. A set of bookshelves in a library. 3. Any appliance for enclosing or squeezing. 4. A machine used in the manufacture of plywood, etc. 5. A wardrobe for linen, etc.

Pressed Bricks. Those moulded under considerable pressure in metal moulds before burning. Better qualities are called Engineering bricks. They have smooth faces, frogs, and sharp regular arrises.

Pressed-fibre Boards. Synthetic wood sheets from vegetable products, or fibrous materials, finely shredded in digesters, and pressed into sheets of different thicknesses. They are more stable than wood, and insulating. See *Wall-boards*.

Pressed Glass. Units that are moulded and pressed, as pavement lights, tiles, etc.

Pressed Steel. Steel objects shaped by pressure of the cold metal.

Prestcore. A registered type of piling. The piles are built-up *in situ* of precast units, to avoid vibration and where headroom is limited.

Prestex Trap. A patent lavatory trap with removable cap and a deep seal.

Pre-stressed Concrete. Reinforced concrete in which a large number of high tensile pre-stressed wires take the place of steel rods. High-tensile steel rods are also used and pre-stressed by means of jacks.

Presweld. A patent steel-lattice framework for roofs.

Pricking Up. Scratching the first coat of plaster on laths, to form a key for the second coat. The pricking-up coat is the first coat of plaster on laths.

Prick Post. 1. The same as *Queen Post*. 2. A light post between the main posts of a wood fence.

Prie-dieu. A low desk for the priest during prayers.

Primary Clays. Those found at considerable depth at the place of formation by the decomposition of rocks. They are usually *lean* clays.

Primary Flow and Return. The circulation pipes between boiler and cylinder in a low pressure hot-water system, *q.v.* They should be of slightly larger diameter than the secondary circulation pipes because of greater incrustation.

Prime. Applied to the best grades of wood.

Prime Cost. The actual cost of producing a particular piece of work excluding profits.

Prime-cost Sums. Amounts to be paid by a builder to nominated sub-contractors or suppliers of specified materials.

Priming. The first coat of paint on woodwork, plaster, etc. It is a thin paint of red and white lead to penetrate all holes and crevices so that the stopping will adhere and to provide a hard surface. It fills the pores and is a preservative. Priming for ironwork has red lead only.

Princess. 1. Roofing slates, 24 in. × 14 in., requiring about 100 per square to a 3 in. lap. 2. Posts or vertical bolts, in a roof truss, of less importance than, and in addition to, King and Queen posts. See *Roofs*.

Principal Posts. The door posts in a timber-framed partition.

Principals. The trussed frames carrying a roof. The *principals*, or *principal rafters*, are also the inclined members of the truss, carrying the purlins. See *Roofs*.

Principles of Drainage. The important features of a good drainage system. See *Drains*.

Principles of Line. A term used in carving, and art generally, in which *light and shade* is subservient to outline.

Print. 1. See *Blue Print*. 2. A plaster cast of a flat ornament.

Printing. The mechanical reproduction of drawings. See *Blue Prints*.

Prior Glass. A special glass used in Heyward's leaded lights.

Priory. A monastery governed by a prior.

Prising. Moving by leverage.

Prism. 1. A solid with uniform section throughout its length. 2. A triangular block of glass for refracting light.

Prismatic Compass. A surveying instrument used for rough surveys.

Prismatic Glass. Glass with saw-tooth section on one side to refract the light.

Prism Lights. See *Pavement Lights*.

Prismoid. A solid that approaches to the form of a prism but with unequal parallel ends.

Prismoidal Formula. A mensuration formula for solids with parallel ends. Vol. = $\frac{1}{6}L (A_1 + 4A_2 + A_3)$; where L = length and A = sectional area at the ends and middle.

Privy. See *Closet*.

Prodorite. A proprietary compound for acid-proofing concrete. Also a refractory cement, and a bituminous dressing.

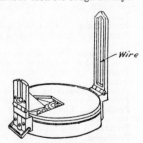

PRISMATIC COMPASS

Profiles. 1. Temporary wooden guides for the erection of brickwork. 2. The outline, or contour, of anything.

Progress Chart. A chart showing the progress of the different trades during the erection of large build-

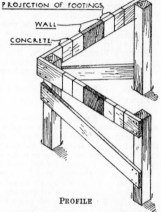

PROJECTION OF FOOTINGS

WALL

CONCRETE

PROFILE

ings. They are essential, under modern conditions, to avoid the holding up of any particular trade. Penalties are usually incurred by failing to conform to the progress chart.

Progressive Kiln. A kiln for seasoning timber. The timber, up to 2,500 c. ft., moves through the kiln on trolleys, under progressive variations of temperature and humidity. Kilns in which the timber is stationary are called Compartment kilns. See page 246.

Progress Report. A weekly report by the clerk or foreman stating the progress of the work, and any special observations he thinks necessary.

Projecting Quoins. The quoins of buildings projecting over the face of the wall, as an ornamental feature.

Projection. 1. See *Orthographic, Axiometric, Isometric, Oblique,* and *Planometric.* 2. The overhang of a moulding, cornice, etc. Applied to something jutting out or prominent with reference to another thing.

Projectors. Imaginary lines connecting corresponding points in the different views in Orthographic Projection. See *Planes of Projection.*

Promenade. A place set apart for walking for exercise, amusement, or show.

Pronaos. The space in front of an ancient temple. A portico. See *Temple.*

Prop. 1. A timber post. See *Centering.* 2. A rigid support not forming a structural part.

Property Room. A room near the stage of a theatre for storing scenery, etc.

Proprietary Materials. Materials registered under a trade name, that gives a firm the sole right of manufacture or selling.

Props. Adjuncts to a play on a theatre stage.

Propylaeum. A vestibule, court, or portico, at the gates of a building.

Propylon. A gateway before the main entrance (pylon) to an Egyptian temple.

Proscenium. The stage of a theatre between the scene drop and the orchestra. Also the frame to the curtain.

Prostyle. A range of columns in front of a building, forming a portico.

Protector Apparatus. A sanitary patent for cleansing hospital bed pans.

Protex. A proprietary emulsion for damp-proofing.

Prothyrum. The passage, or corridor, from the vestibulum to the atrium in ancient Roman architecture.

Protion. A proprietary red bituminous waterproofer.

Protractor. An instrument for setting out angles. The most common form is a semicircular piece of celluloid graduated in degrees.

Provision. A direction, in a bill of quantities, to include a fixed sum in the estimate for work not clearly defined.

Prussian Blue. Ferro-cyanide of iron. See *Blue Pigments* and *Pigments*.

Pseudo. A prefix signifying that a thing is modified, false, incorrect, or an imitation. It is often applied to architectural features that are not true to the classic styles from which they are derived.

Ptd.A. Abbreviation for *pointed arch*.

Pteroma. The space between the walls of the cell to a temple and the column of the peristyle.

Pteromata. Side walls of a Greek temple.

Ptg. Abbreviation for *pointing*.

P-trap. A trap in sanitary pipes shaped like the letter P.

Public Health Acts. See *Acts*.

Puddle. Clay mixed with water and sand into a plastic mass. It is used to prevent the admission or leakage of water, as in a puddled dam or round a manhole.

Puddled Dam. Two parallel rows of sheet piling with puddled clay between to prevent the access of water to *underwater work*.

Puddling. Producing wrought iron from cast iron in a reverberatory furnace. The process eliminates the carbon and other impurities, and is called pig boiling when pig iron is used direct. 2. Applying puddle.

Pudlinks. Putlogs, *q.v.*

Pudlo. A proprietary waterproofer for concrete, in powder form.

Puff Pipe. The short pipe fixed from the outgo of the trap, to a W.C., to the anti-siphonage pipe. An anti-siphonage pipe.

Pug. The plastic material for bricks.

Pugging. 1. Material with non-conducting properties used to prevent the transmission of sound through floors. Various materials are used : coarse plaster ; lime plaster with sawdust, brickdust, fine slag, or coke breeze ; slag wool ; Cabot's quilt ; etc. Also called deafening and deadening.

Pug Mill. A machine with revolving knives to convert clay, for brick-making, to a uniformly plastic state.

Pug Piles. Dovetailed piles.

Pull. 1. A cord, handle, or lever, for ringing a bell. 2. See *Drawer Pulls*. 3. An appliance providing a grip for the hand.

Pulley. A grooved wheel revolving on a pin in a block, and used for changing the direction of the pull on a rope, for raising loads.

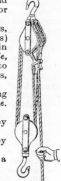

Pulley Blocks. Movable blocks containing several pulleys, or sheaves. They are used in pairs (top and bottom blocks) with a continuous rope passing over the pulleys alternately in the top and bottom blocks. The arrangement is called *tackle*, and is used for raising heavy loads. The ratio of weight to effort is $W = nE$, where n = number of pulleys in both blocks, neglecting friction. See *Hoisting*.

Pulley Stiles. The stiles of a frame, for vertical sliding sashes, carrying the pulleys and pocket. See *Sash and Frame*.

Pulley Tackle. See *Pulley Blocks*.

Pullman Balance. A patent spring in the form of a pulley to replace weights for balanced sashes.

Pulsometer Plant. A vacuum pump for raising water by steam pressure.

Pulpit. An elevated enclosed stage for the preacher in a church.

Pulvinated. Convex or cushion shaped. A swelling.

Pumecrete. A proprietary pumice stone for fire-resisting floors and partitions.

PULLEY
TACKLE

Pumice. A light and very porous stone of volcanic origin.

Pummel, or Pommel. A long wooden shaft with a heavy wide cast-iron base for ramming, or consolidating, concrete or earth. See *Ram.*

Pump. A mechanical contrivance for raising water or extracting air. There is a large variety for water : chain pump, siphon, reciprocating (bucket, plunger or ram, and piston), rotary, centrifugal, pulsometer, hydraulic, etc. The type generally used for keeping excavations free from water is the Contractor's diaphragm pump.

Punched Work. Masonry dressed with a punch.

Puncheon. 1. A short post acting as a king post in the top part of a trussed partition (see *Partition*), queen post roof truss, and in Gothic roofs; sometimes called a *quarter* or key post. 2. A block or short post placed on a pile when the head is lower than the fall of the ram. 3. Short stout timbers supporting the horizontal frames in a box coffer dam, or the walings in excavations.

Punches. 1. Steel tools for driving nails below the surface, preparatory to stopping or for making holes in materials by pressure instead of by rotation. 2. A *centre punch* has a conical point for marking metal preparatory to drilling holes. 3. A *handrail punch* is cranked and used for tightening up handrail bolts. 4. A mason's tool like a punch. 5. See *Punching.*

Punching. Forming the nail holes in slates, or rivet holes in metal. The holes are formed by pressure and not by rotation. See *Drilling.*

Punky. Applied to soft spongy heartwood, or to wood showing signs of decay.

Punning. Ramming, or consolidating, concrete, earth, etc.

Purbeck. Cream or grey limestones from Dorset. The former is a freestone and is hard and durable and used for paving and steps. The latter is called Purbeck marble. It is shelly, polishes well, and is used for decorative work.

Purfied. Applied to sculptured tracery work of a delicate character. Having a decorated border. Ornamented with crockets.

Purimachos. A proprietary plastic fire-cement.

Puriri. A greyish-brown hardwood, also called New Zealand Teak. It is not a teak but has many of its qualities, with attractive figure, but too hard and heavy for ordinary purposes. Used chiefly for structural work. Weight about 63 lb. per c. ft.

Purlin. A horizontal beam supported by the principal rafters of roof trusses and carrying the common rafters between the wall plate and ridge. See *Roofs* and *Centering.*

Purlin Roof. A cottage roof, where the purlins are carried by cross walls instead of by roof trusses.

Purpleheart. A dark brown to purplish hardwood from Br. Guiana. Also called amaranth, violet wood, palisander. It is very hard, heavy, strong, and durable, with fine close grain, wavy, rowey, and with distinct rays. Used for decorative work or where strength and durability are essential. Weight about 62 lb. per c. ft.

Purple Oxides. See *Pigments.*

Purpose-made Bricks. Bricks specially moulded for their particular purpose, instead of being cut and rubbed to shape. They are often purposely made for arches and circular work generally.

Push Tap. A water tap actuated by a spring knob or lever.

Pussur. A wine-red hardwood, with variegated streaks, from India. It is hard, heavy, strong, and very durable. Not difficult to work and polishes well for superior joinery and fittings. Weight about 46 lb. per c. ft.

Putlogs. Short bearers carried by the wall and the ledgers of a scaffold, and supporting the scaffold boards.

Putty. Plastic material, used for stopping, jointing, etc., that sets hard.

Glazier's and painter's putty is made from whiting and linseed oil, with a little turps for drying. See *Lime Putty* and *Mason's Putty*.

Putty Powder. Oxide of tin used for polishing glass.

Puzzolana. Of volcanic origin, and forming a mixture analogous to cement. It contains an excess of silicates and too little lime.

Pycastyle, or **Pycnostyle.** Columns spaced $1\frac{1}{2}$ diameters apart. See *Columns*.

Pyinkado. A reddish-brown to rich red hardwood from India. Very hard, heavy, strong, and durable, beautiful figure, and polishes well. Used for piling, structural work, and superior joinery. Also called ironwood. Weight about 62 lb. per c. ft.

Pyinma. A light-red to reddish-brown hardwood from India. It is moderately hard, heavy, and strong, and fairly durable. Not difficult to work and polishes well. Used for superior joinery. Weight about 42 lb. per c. ft. The wood from the Andaman Islands is similar but superior.

Pylon. 1. The gateway to an Egyptian temple. 2. A slender lattice tower. 3. A leaf of a double door or gate.

Pyramid. A solid having a rectilinear figure for its base and its sides triangles, so that it tapers to a point at the apex. The base may be triangular, square, or polygonal. Vol. $= \frac{1}{3}$ base $\times$ height. See *Frustum*.

Pyrelide. A patent fireproof door, shutters, etc.

Pyrene. A registered metal finish for kitchen equipment. The process chemically bonds paint to sheet metal. It is also applied to heating equipment.

Pyrex. A proprietary heat-resisting glass.

Pyrites, or **Marcasite.** Ironstone nodules, race, or kidneys, causing black fused spots in bricks. A mineral in which sulphur is combined with iron, copper, or other minerals.

Pyroc. A proprietary light-weight plaster that can be applied by spraying. It consists of vermiculite, lime, and cement, and is easily applied, fire-resistant, and insulating.

Pyropruf. Similar to Pyrok but with greater fire-resisting qualities.

Pyroxylin Sheeting. Placed over windows to exclude the sun's glare.

Pyruma. A proprietary fire-cement in the form of putty.

Pyx. An ornamental box for the consecrated Host in the Roman Church.

Q

Q.P. Abbreviation for *quartered partition*.

Q.S. Abbreviation for *quick sweep*.

Q.T. Abbreviation for *quarry tile*.

Q. T. Window. A patent mechanical method of controlling the opening of windows.

Quadra. 1. A square frame or border. 2. A plinth to a podium.

Quad, or **Quadrangle.** A rectangular space surrounded by buildings, as a cloister. Usually associated with a college or school.

Quadrant. 1. Quarter of a circle. An arc subtending 90°. 2. A surveying instrument for obtaining altitudes. 3. A casement, or fanlight, stay.

Quadriga. A decorative feature surmounting a monument in the form of a four-horsed chariot.

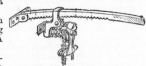

QUADRANT FOR FANLIGHT

Quadrilateral. A plane figure with four sides. See page 252.

Quadripartite. Divided into four parts. Applied to examples of Early English vaulting having diagonal and transverse ribs.

Quagginess. A defect in timber consisting of shakes at the centre.

Quantities. See *Bill of Quantities*.

Quantum. A patent flushing valve that releases the amount of water required, automatically and silently.

Quarrell. 1. A lozenge, or diamond-shaped pane of glass or piece of stone. 2. A glazier's diamond. 3. A small piercing in the tracery of a window. A quatrefoil.

Quarries. Common paving tiles; made in various colours and sizes.

Quarry Face. The natural face of stone, as it is quarried.

Quarry Pitched. Applied to stone that is roughly squared in the quarry.

Rectangle. Rhombus. Rhomboid. Trapezoid. Trapezium.

QUADRILATERALS

Quarry Sap, or Water. The natural moisture found in freshly quarried stone. As the sap evaporates it forms a protective coat on the stone.

Quarry Tiles. See *Quarries*.

Quarter. 1. See *Puncheon*. 2. A square panel. 3. See *Quarterings*.

Quarter Bond. The ordinary bond in brickwork formed by a 2¼ in. closer.

Quartered. Oak logs cut into quarters and then sawn radially to obtain *silver grain*. The term is applied to any log cut radially to obtain figure and stability.

Quartered Partition. A partition formed of quarterings.

Quarterings. 1. Small scantlings used as studs in a plastered partition. 2. A deal, or plank, sawn into four equal pieces. 3. Small stuff, square in section, suitable for first fixings.

Quarter Round. An ovolo, *q.v.*

Quarter Space, or Pace. Landings that are only as wide as one flight of the stairs, or half the width of the staircase. See *Open Well*.

Quarter Turn. A wreath turning through 90°.

Quartz. A crystalline silicon dioxide, SiO_2. An important constituent of granite, and sandstone, and practically indestructible. Often coloured by metallic oxides. Variations are known as flint, agate, amethyst, rock-crystal.

Quartzite. Decorative stone harder than granite. It is quarried like slate.

Quatrefoil. A tracery panel with four foils and cusps. See *Tracery*.

Quay. A wharf. A place for the loading and unloading of vessels.

Quebracho. A rose-red, variegated hardwood from C. America, also called Ironwood. It is very hard, heavy, durable, and strong, and polishes well, but very difficult to work and too hard for ordinary purposes. Used

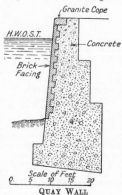

QUAY WALL

for structural work and cabinet work. Weight about 63 lb. per c. ft. There are similar woods also called Quebracho.

Queen Anne Arch. See *Venetian Arch* and *Arches.*

Queen Bolts. Bolts used instead of posts in a Queen-post truss.

Queen Closer. See *Closer.*

Queen Posts. The two vertical posts in a Queen-post roof truss, or in any framed truss with two principal posts. See *Roofs.*

Queens. Slates 36 in. × 24 in. Often irregular in size and sold by weight. Requires about 37 per square at 3 in. lap.

Queensland Walnut. A beautiful hardwood, like black walnut from Australia. It is difficult to work but used for all kinds of fittings, superior joinery, electrical appliances, etc. Weight about 45 lb. per c. ft.

QUETTA BOND

Quercus. The botanical name for the various species of oak.

Quetta. A reinforced brickwork bond. The arrangement of the bricks forms vertical cavities in which steel rods are placed and the spaces filled with concrete as the work proceeds, thus forming a series of reinforced vertical posts.

Quicklime. See *Lime.*

Quickstryp. A proprietary paint remover.

Quick Sweep. Circular work of comparatively small radius.

Quilt. See *Cabot's Quilt.*

Quirk. A narrow sinking, or groove, forming part of a moulding. A deep indentation.

Quirk Bead. A semicircular moulding with a quirk. Used to break the joint between two boards, as in matching, or to ornament a salient corner. When it is formed on both adjacent faces it is a return, or double-quirk, bead. See *Mouldings.*

Quirk Cutter. A special tool for forming quirks on circular work.

Quoin. The external angle of a wall.

Quoin Bonding. The arrangements of the bricks or stones for a return wall.

Quoin Header. The first brick at the quoin, showing a header on the face of the wall and a stretcher on the return wall.

Quoit Terminal. An inverted vertical scroll or monkey tail at the bottom-end of a handrail, to give warning when approaching the foot of the stairs.

R

R. Abbreviation for *render, reaction* and for *moment of resistance.* **r.** A symbol for *ratio of reinforcement* in concrete beams.

Rabatment. A term used in geometry when a surface is rotated, or revolved, about a trace into a plane of projection, so that its true shape is shown.

Rabbet. Same as *Rebate.*

Race. 1. Pieces of limestone in brick earth. Also see *Pyrites.* 2. A water-course to a water-wheel. A rapid current. 3. A fenced space for driving-belts or ropes. 4. A groove or channel in which something slides.

Rack. 1. A toothed rail to engage with a pinion. 2. An open frame to receive articles, as a place of storage. 3. A frame on which anything is stretched. 4. A sliding pulley for an endless cord to a window blind.

Racked. Temporary timbering braced to prevent distortion.

Racking. 1. Frames distorted through strain. 2. *Setting back* each course of brickwork, during erection, to be completed, or joined up to, later.

Rack Rent. Rent that is greater than two-thirds of the net annual value of the property. An excessive rent.

Rad and Dab. 1. Blocks of clay and chopped straw between split laths, as a substitute for a brick-nogged partition. 2. See *Wattle and Dab*.

Radial Box. Similar to reducing box, and used for securing bricks whilst being cut to special shapes.

Radial Bricks. Purpose-made bricks for tall chimneys, and other circular work. They are often perforated.

Radial Projection. Perspective drawing.

Radial Shakes. Radiating shakes at the heart of a log or balk. See *Timber*.

Radial Square. A special tool for drawing radial lines on circular work.

Radial Step. Same as winder, *q.v.*

Radian. An angle of 57·3°.

Radiating Surface. The amount of surface required for radiators, in steam-heated rooms. The required surface area is obtained from the formula $S = (30A + 8B + C) \div 270$; where A = glass surface, B = wall surface, C = cubic contents of room. All measurements in feet.

Radiation. The emission and diffusion of rays of light or heat from a body. See *Light* and *Heat*.

Radiator. A part of a heating apparatus that radiates heat. There is a great variety of hot water, steam, gas, oil, and electrical radiators, but the usual form for hot water consists of a number of cast-iron sections, or columns, assembled with screwed nipples. Another type consists of pipe coils. See *Low-pressure System* and *Panel Heating*.

Radius of Gyration. K. A value used in the design of columns, and denoting a relative measure of stiffness. It depends upon the sectional area, and the distribution of the material about an axis. $K = \sqrt{\dfrac{\text{Inertia}}{\text{Area}}}$.

Radius Rod. A rod equal in length to the radius of the required circle and used for describing or testing circular arcs.

Rads. Abbreviation for *radiator*.

Rafters. The timbers carrying the coverings of a roof. Also called common rafters and spars. See *Principal Rafters*. Also see *Centering*.

Raft Foundation. A site for a building, covered with reinforced concrete and usually supported on piles.

Rag and Stick Work. Same as fibrous plaster, *q.v.*

Rag Bolt. A bolt with a jumped-up, jagged, dovetailed end for fixing in stone or concrete with molten lead or cement. See *Masonry Fixings*.

Raglet. A narrow groove in masonry to receive the edge of an apron flashing.

Raglin, or Raglan. A local term for slender ceiling joists.

Rags. Slates 36 in. × 24 in., requiring about 37 per square at 3 in. lap

Rag Stone. Coarse grained sandstone.

Ragusa. A proprietary asphalt.

Rag Work. Rough rubble walling built of thin stones.

Rail. 1. A horizontal or inclined member of a frame or fence. A bar protecting an enclosure. 2. A steel bar forming a track for vehicles with flanged wheels.

Railings. An open iron protection, or fence, for an enclosure.

Railway Roof. An umbrella or cantilever roof, used on island platforms. See *Roofs*.

Rainhill. A red sandstone from Lancashire. It is excellent stone for general building. Weight 160 lb. per c. ft. Crushing strength 351 tons per sq. ft.

Rainwater Head. A flaired-out top to a stack of rain-water pipes, for the collection of water from roof gutters.

Rain-water Pipes. See *Down Pipes*.

Raised Girt. See *Girts*.

Raised Panel. A panel thicker at the middle than at the edges, for ornamentation and strength. See *Fielded Panel*.

Raising Plate. A pole plate, or a wall plate. A horizontal timber carrying the feet of other timbers.

Rake. 1. A slope, or inclination. 2. An implement with long teeth and a long handle used for mixing plaster.

Rake Board. An inclined board. A barge board.

Raker. 1. A tool for cleaning mortar joints preparatory to repointing. 2. See *Raking Shores*. 3. A wind brace. 4. An occasional tooth, in a crosscut saw, with deep gullet. See *Spearfast*. 5. The strong timbers between purlins to carry a dormer.

Raking Balusters. Applied to those with head and foot blocks shaped to the pitch of the stairs, as in Jacobean architecture. Also applied to metal balusters not vertical.

Raking Bond. A course with the bricks laid obliquely across the wall to give longitudinal bond. It is used about every fifth course, between stretchers, and in thick walls. See *Herring Bone* and *Diagonal Bond*.

Raking Copings. Copings on an incline, as a gable wall. See *Gable*.

Raking Cornice, or **Moulding.** One inclined to the horizontal and having horizontal returns. See *Pediment*.

Raking Course. See *Raking Bond*.

Raking Out. Cleaning out mortar joints preparatory to pointing.

Raking Pile. One that is not vertical.

Raking Risers. Applied to risers that slope inwards to give more foothold for narrow treads.

Raking Shores. The inclined stays, or struts, supporting a defective wall, or one undergoing structural alterations. See *Shoring*.

Raking Surface. A twisted surface, as the soffit of a winder in a geometrical stone stairs with flush soffits.

Ram. 1. The weight to a pile-driver. 2. To consolidate loose material with a rammer, or pommel. 3. See *Hydraulic Ram*.

Ramp. 1. A vertical curve in a moulding or handrail. See *Knee*. 2. A concave sweep from a lower to a higher level. 3. An inclined floor, runway, or road, usually curved in plan.

RAKING SHORES

Ramp and Twist. A term applied to masonry that involves *twist* in the working of the stone. Surfaces that both rise and curve.

Rampant. Applied to an arch, or vault, with its springings, or abutments, on different levels. See *Arches*.

Rampant Centre. A centre for a rampant arch.

Rampart. 1. A parapet wall. 2. An incline formed into wide steps for the use of horses. 3. The defence walls of a town.

Rance. A shore, or prop.

Rand. A margin, or border.

Random Rubble. Walling built of irregular unsquared stone not in courses. When it is levelled up, about every 12 in. in height, it is called *random rubble built to courses.*

Random Slates. Slates of varying widths, or ungraded, as used for diminishing courses.

Random Widths. Applied to bundles of shingles of varying widths.

RAMMER

RANDOM RUBBLE

Range. 1. A grate (with an oven) for a kitchen. It usually has a boiler for the hot-water supply. 2. An arrangement of buildings in lines or tiers.

Ranging Bond. Long horizontal strips of wood built into the joints of walls as fixing fillets.

Ranging Poles. Poles, 6 to 10 ft. long, with steel shoes and with differently coloured sections, used in surveying for sighting, or ranging, long straight or level lines.

Rankine's Formulae. The various empirical and mathematical formulae evolved by Professor Rankine for the design of columns, retaining walls, etc.

Rank Set. Applied to plane irons that are set coarsely for rough work. The reverse of fine set.

Ransome's Process. A proprietary preservative for stone, that precipitates calcium silicate.

Rapaloid. A proprietary cellulose enamel.

Rapid Hardening Cement. Portland cement with a greater than usual CaO content. See *Alumina Cement.*

Rapier. A registered concrete mixer.

Rap-rig. A patent type of scaffolding built of light timbers quickly assembled and adjusted. It is more suitable for interior work but may be used for any type of work.

Rasp. Like a coarse file, but with teeth instead of furrows. Used on wood and lead.

Rata. A dark-red decorative, hardwood from New Zealand. Also called Ironwood. Very strong and durable; hard and heavy. Difficult to work and used chiefly for structural work, etc. Weight about 46 lb. per c. ft.

Ratchet. A small projection to engage with a toothed wheel to check the movement. A mechanical contrivance for converting a to-and-fro movement into a circular movement, as on braces, screwdrivers, etc.

Rate of Feed. A term used in the machining of wood, stone, etc., stating the speed that the material advances in feet per minute through the machine.

Rat Tail. A round tapered file.

Rat-trap Bond. A 9 in. brick wall with the bricks on edge, leaving a 3 in. cavity between opposite stretches. The method is weak but gives a cheap and light wall. Also called silver-lock bond.

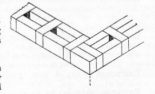

RAT-TRAP BOND

Raven. A proprietary gypsum wallboard encased with a specially processed fibre for one-coat plastering.

Ravenite. Compressed wool felt-fibre and bitumen for roof coverings.

Raw. Applied to linseed oil when not boiled, and to sienna and umber when not burnt.

Rawlplug. A patent plug in the form of a hollow cylindrical tube of compressed fibre. The screw, when inserted, opens out the plug so that it grips tightly round the drilled hole. There is a great variety of sizes. The firm also specializes in numerous patent fastenings, toggle bolts, and appliances for speedy erection of shuttering: Rawloops, Rawlties, Rawlhangers.

Raymond Pile. A concrete pile formed by driving a spirally-tapered shell and filling it with concrete. The shell is telescopic and is left in the ground.

Rayrad. A registered design of concealed radiator.

Rayvecto. A patent heating panel.

R.C. Abbreviation for *rough cutting*.

Rcw. Abbreviation for *rough cutting* and *waste*.

Rdm. Abbreviation for *random sizes*.

Reaction. The counteraction of a force, or a number of forces, to obtain equilibrium. Specially applied to the upward resistance of the wall against the downward pressure of the supported load. See *Equilibrium*.

Reading Drawings. The art of understanding or interpreting drawings, especially in orthographic projection.

Reagent. A substance, in chemistry, used to detect the presence of another substance by reaction.

Realgar. Bi-sulphide of arsenic. A red pigment.

Reamer. A bit, for a brace or machine, for enlarging holes or for making tapered, or conical, holes.

Reamy Glass. Glass with irregular surface and definite striations.

Rear. See *Rere*. A rear arch is a separate arch behind the face arch.

Rear Dorse. A fireplace without breasts in an external wall.

Rear Vault. The small vault, or arch, formed by placing the glass near to the outer face of thick walls. The vaulted space connecting an arched opening with the arch in the inner face of the wall.

Reb. Abbreviation for *rebated*.

Rebate. 1. A rectangular recess formed along the edge of a piece of material to receive a frame, door, sash, or another piece of material. See *Finishings*. 2. A kind of hard freestone used for pavements. 3. An implement to beat up mortar.

Receding Cornice. One that sets back from the face of the main structure, instead of projecting.

Reception Wall. A thin wall to retain vertical damp-proofing material. See *Damp-proofing* and *Retention Wall*.

Recess. 1. A small niche or cavity in a wall. 2. The set-back, or rebate, to an opening, or jamb, in a wall, to receive a door or window frame.

Recessed Arch. One arch within, and behind, another. See *Compound Arch*.

Reciprocal Diagram. See *Stress Diagram*.

Reconstructed Stone. See *Cast Stone*.

Rectilinear. A plane figure bounded by straight lines.

Redalon. A proprietary retarding agent, for concrete surfaces that are to be plastered.

Red Cedar. See *Western Red Cedar*.

Red Clay. From felspathic rocks containing iron oxides, which combined during conversion into plastic clays. Produces red bricks.

Red Deal. See *Northern Pine* and *Redwood*.

Red Eyne. Indian Redwood. A rich dark-brown hardwood with lighter streaks; hard, heavy, strong, and durable. Difficult to work, but polishes well and very decorative. Used for superior joinery, heavy constructional work, etc. Weight about 62 lb. per c. ft.

Red Gum. A variegated reddish-brown hardwood from U.S.A. It is fairly hard, heavy, and strong, but not durable. The heartwood is called Satin Walnut and used for superior joinery, etc. The sapwood is called Hazel Pine. Weight about 36 lb. per c. ft.

Red Hand. A registered trade mark for several roofing felts.

Red Lead. Pb_3O_4. An oxide of lead used chiefly in priming coats. It is produced by twice heating of lead in an open furnace, first to litharge and then to red lead. See *Pigments*.

Red Oxide. Earth pigments used in paints. The colour is due to the presence of iron. See *Pigments*.

Red Pine. A Canadian pine very similar to Baltic redwood.

Red Sanders. Indian Red Sandalwood or Coralwood. Orange red to almost black in colour, and very hard, heavy, and durable. Difficult to work, but polishes well. Used for decorative joinery, structural work, etc. Weight about 63 lb. per c. ft.

Red Seraya. See *Seraya*.

Red Stringybark. See *Eucalyptus* and *Stringybark*.

Red Tulip Oak. A beautiful pale-pink to reddish-brown hardwood from Queensland. Rather difficult to work and glue, but fumes and polishes well. Used for superior joinery, bentwork, electrical fittings, flooring, etc. Weight about 51 lb. per c. ft.

Reducing Box. A specially prepared frame, used when shaping bricks for a niche hood.

Reducing Pipes. Tapered pipes, for connecting lengths of different diameters.

Reduction in Bulk. The reduction that takes place when dry materials are mixed for concrete. The reduction is from a quarter to one-third of the original bulk.

Redundant Frames. A structural frame with more members than are necessary to keep it stable in equilibrium, when subjected to the external forces.

Redwood. The standard name for *Pinus sylvestris*, previously called red or yellow deal, and numerous other names. It is resinous, fairly light and soft, and pale-reddish brown with distinct rings. It is fairly strong and durable, and used extensively for carpentry and joinery, etc. Weight from 25 to 39 lb. per c. ft. There are numerous other timbers called redwood, but with a distinguishing prefix, usually denoting source of origin, and they are chiefly hardwoods. See also *Sequoia*.

Reeded Glass. A registered obscured glass with ribs $\frac{1}{2}$ in. wide with $\frac{1}{16}$ in. relief, producing strong horizontal or vertical lines. Cross-reeded glass has ribs on both sides giving an appearance of small squares. See *Obscured Glass*.

Reeds. A series of sunk beads on the face of the material. See *Mouldings*.

Re-entrant Angle. An internal angle. One pointing inwards.

Refectory. A room for refreshment or meals in a college or monastery.

Reflectalyte. A registered rolled glass specially designed for lighting fixtures.

Reflection. The throwing back of heat, light, and sound rays. Highly polished surfaces give regular reflection, irregular surfaces give spread reflection, and matt surfaces give diffused reflection. Any ray striking a polished surface at an angle to the normal to the surface (angle of incidence), is reflected at an equal angle on the other side of the normal (angle of reflection).

Reform Lanterns. A patent type of top lighting.

Refracting Glass. Prismatic glass for refracting light rays in the required direction.

Refraction. When a ray of light passes from one medium to another of different density, as from air to water, air to glass, etc., the direction of the ray is bent, or refracted.

Refractive Index. The ratio of the sine of the *angle of incidence* to the sine of the *angle of refraction*. Air to glass = $\frac{3}{2}$, glass to air = $\frac{2}{3}$, air to water = $\frac{4}{3}$, water to air = $\frac{3}{4}$.

Refractory. Applied to materials that are difficult of fusion, and that can resist very high temperatures.

Refractory Cement. Cement used for resisting very high temperatures.

Refrigerators. Specially constructed chambers with low temperatures for the preservation of food; also apparatus and plant for making ice.

Refuge. See *Safety Island.*

Refuse Destructors. See *Destructors.*

Regalite. A proprietary jointless magnesite flooring.

Register. A louvre ventilator for wall or ceiling.

Register Stove. Open grates for heating living rooms, consisting of a cast iron frame, furnace bars, and fireclay back.

Register, To. To indicate a particular point or part.

Reglet. A small rectangular moulding. A listel or fillet. Also a rectangular groove or raglet.

Regrating. Re-dressing hewn stone.

Regula. See *Reglet.* Specifically applied to the fillet under the triglyph in the Doric Order, *q.v.*

Regular Coursed Rubble. Masonry walling with the courses varying in height from 3 in. to 9 in.

Reinforced Brickwork. Metal reinforcement to strengthen longitudinal bond. The usual forms are hoop

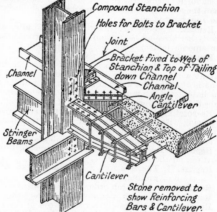

REINFORCED CONCRETE CORNICE

REINFORCED BRICKWORK

iron, wire-netting, expanded metal. The iron must be coated to prevent corrosion: tarred and sanded, galvanized, or specially treated in manufacture, or buried in cement concrete. Tie irons are also a reinforcement. See *Wall Ties* and *Quetta Bond.*

Reinforced Cames. Cames, for leaded lights, with a steel core.

Reinforced Concrete. The steel reinforcement in concrete is intended to strengthen the tensile resistance, hence the position in the concrete is important and must be carefully designed. The steel may be any form of bar, of which there are many patents, or expanded metal. See *Expanded Metal*.

Reinforced Glazing Bars. The many patent forms of roof-glazing bars usually consisting of a lead-lined steel core. See *Glazing*.

Reins of Vault. The sides, or walls, supporting a vault.

Releasing Key. A tapered member for easing shuttering on the completion of the work.

Relief. Prominent, carved, ornamentation is "in relief." There are three kinds: high, half, and low relief (or alto, mezzo, and basso relievo), according to the degree of prominence.

Relieving Arch. An arch over a lintel to carry the weight of the wall above. See *Rough Arch*.

Reliquary. An ornamental chest or casket for sacred relics.

Renaissance. A style of architecture, prevailing in the early part of the sixteenth century, that attempted to replace Gothic by the classic architecture of Rome and Greece.

Rendered Laths. Plasterer's laths riven, or split, and not sawn, to obtain the maximum strength. See *Laths*.

Render and Set. Two-coat plaster work on walls.

Render, Float and Set. Three-coat plaster work on walls. See *Float*, and *Setting*. Three-coat work on laths is called "lath, plaster, float, and set."

Rendering. 1. Applying the first coat of plaster to a wall. 2. Bedding of slates in hair mortar to provide a firm bed. Also used as an alternative term for *torching*. 3. Cementing wall surfaces to make them waterproof. See *Damp-proofing*.

Rengas. A variegated rich-red hardwood also called Borneo Rosewood. Fairly hard, rather heavy and difficult to work. Dust is an irritant. Used for superior joinery, etc. Weight about 43 lb. per c. ft.

Rentokil. An insecticide for wood.

Repairs. Restoring to a sound condition. It is difficult to define to what extent a tenant is liable under a loosely defined *repairing lease*. If the repairs alter the character of a building it is at the landlord's expense, but the liability of the tenant extends to the rebuilding of a wall, floor, or any subsidiary part, even to the extent of a main wall.

Re-pointing. Raking out the old mortar in brick joints, for about $\frac{1}{2}$ in., and replacing with new. Cement and sand is too strong for such work and it should be tempered with lime, depending upon the kind and quality of the brickwork. The suggested mix is 1 c, 3 *l*, and 9 sand for old brickwork.

Repoussé. Raised ornamentation on metal, formed by hammering, or beating, on the back.

Reprise. 1. The return of a moulding for an internal angle. 2. A stool for a jamb or mullion.

Requisition. A written order for materials.

Rere Arch. An arch with level soffit over splayed jambs.

Reredos. An elaborately carved screen in the chancel of a church. It often serves as a background to the altar. An altar piece.

Rere Vault. See *Rear Vault*.

Resac. A light to dark-brown hardwood from Malay. Durable and strong very hard and heavy. Not difficult to work and used for superior joinery, structural work, etc. Weight from 44 to 62 lb. per c. ft.

Reservoir. A place where water or other liquids are stored for use as required. An artificial lake used for communal water supply.

Reset. To correct the set of a saw.

Residency. The official house of a government officer, especially in India.

Resilience. The work done per unit volume of a material in producing

strain. The act of rebounding through being elastic. The property of a material to recover from strain, due to its elasticity.

Resilient Chair. Spring steel chairs supporting a secondary floor from the main floor for sound-proof studios.

Resin. Amber, gum animé, copal, etc. Fossilized vegetable products. The gums exude from trees, and the moisture, or oil, evaporates, leaving resin. See *Varnish*.

Resin Bonded. Applied to woodwork jointed by resin products, instead of animal glue. See *Plastic Glues*.

Resinous. A term applied to coniferous trees containing resin. The various pines and firs.

Resistance. The ability of a material to resist change of shape when loaded. See *Moments of Resistance*. The properties of a material that resist any test to which it may be subjected. See *Tests*.

Resistance Welding. A process in which the metals are placed in contact and allowed to take an electric current. The joint heats up rapidly, because of the high resistance at that point, and then the pieces of metal are forced together, and the current switched off. The four types are: flash or butt, seam, spot, and projection welding.

Resolution of Forces. The replacing of one force by any number of other forces, or components, that together have the same effect.

Respond. A pilaster matching, or pairing with, another pilaster or column, as when forming the springings for an arch. A half pillar.

Respond Newel. A half-newel on the wall to pair with the newel.

Restaurant. A place for the sale of refreshments, or meals.

Rest Bend. See *Duck Foot Bend*.

Rest House. A hostel. A hut remote from any other buildings to provide shelter for travellers.

Restlight. A patented *sun-light* glass.

Resultant. A force that replaces a number of other forces. The result of compounding two or more forces.

Resweld. A proprietary resin-bonded plywood. It is good for exterior work, formwork, etc.

Retaining Walls. Walls resisting the pressure of earth. $k =$ constant, and depends upon nature of ground. They may be of brick, stone, reinforced or mass concrete. See *Dam*, *Wedge Theory*, and *Angle of Repose*.

Retarders. Materials that hinder or prevent the setting of cement. They are applied to the shuttering so that the cement can be easily removed to expose the aggregate or to provide a rough surface for plastering. Sugars, peat water, carbons, oils and fats, clay, tannic acid, also many patent *dopes*, or *emulsions*, are used for the purpose.

Retardo. A patent retarder for cement.

Retention Wall. A thin wall built outside the wall of a building and between which a waterproofer (molten asphalt or Hygeian rock) is poured, for a vertical damp-proof course. The retention wall is usually below ground level and is $4\frac{1}{2}$ in. thick, with a $\frac{1}{2}$ in. to 1 in. cavity.

Reticulated. 1. Ashlar prepared with a smooth face and then worked with a number of irregular sinkings, pockets, or reticules. The margins between the sinkings are from

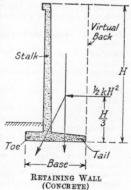

RETAINING WALL
(CONCRETE)

¾ in. upwards, and regular in width. 2. Masonry built of diamond-shaped or polygonal stones.

Retrochoir. The part of a cathedral behind the high altar.

Return. 1. A continuation of a member or moulding in another direction, usually at right angles. 2. The pipe that returns the water back to the boiler. See *Low-pressure System*.

Return Bead. See *Mouldings*.

Returned End. The end of a moulding shaped to the contour of the moulding.

Revalé. Applied to masonry mouldings, etc., that are completed *in situ*.

Reveals. The sides of a door or window opening between the frame and the face of the wall, and at right angles to the face of the wall.

RETICULATED

Reversal of Stress. Fluctuation from one kind of stress to an opposite kind in structural members, i.e. tension to compression. Materials *fatigue* quicker under these conditions.

Reverse. A templet shaped to the reverse of a required moulding, and used for testing the correctness of the moulding whilst working it to shape.

Reversible Hinge. A screen hinge that allows the wings to fold in either direction.

Revetment Walls. Retaining walls.

Revolving Doors. Doors for public buildings that revolve by slight pressure when one enters or leaves the building. They are usually divided into two or four compartments. The latter are draughtproof, and consist of four wings revolving round a central axis, in the form of a turnstile.

Revolving Shutters. Narrow strips of wood or metal strung on thin metal bands or hinged together to form a continuous surface, to protect a window or to close an opening. When not in use, the shutter, by means of springs, winds round a cylinder, and the whole is hidden in a special box framed above the opening.

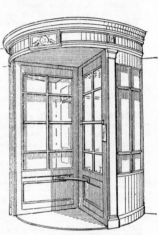

REVOLVING DOORS

Rex. A proprietary paint-brush cleaner.

Rexilite. A proprietary bituminous roofing felt.

Rexine. An imitation leather used for wall coverings, etc.

Rexoid. A proprietary bituminous roofing felt.

R.F.R. Abbreviation for *return fill and ram*.

R.F.S. Abbreviation for *render, float, and set*.

Rhenish Bricks. Very light bricks, called "floating bricks." Calcareous bricks with dolomitic lime as the binding agent.

Rhodesian Mahogany. A fairly hard and heavy hardwood from South Africa, used extensively for superior flooring. It is not a mahogany, but it has some of the characteristics of a hard mahogany and polishes well. Now called Copalwood. Weight about 45 lb. per c. ft.

Rhodesian Teak. A very hard, heavy, strong, and stable hardwood. It is not a teak but it has many of the qualities of teak and is used for similar purposes. It is excellent for superior block flooring. Also called **Mukusi** and **Zambesi Redwood.** Weight about 52 lb. per c. ft.

Rhomboid. A parallelogram with adjacent sides unequal and its angles not right angles. Area = base × vertical height. See *Quadrilaterals.*

Rhom Brick. A patent brick, rhomboida lin shape, but with the appearance of an ordinary brick on the face work.

Rhombus. As rhomboid, but all four sides equal.

Rhone. A local term for an eaves gutter.

Rib. 1. The curved framework of a centre. 2. A curved rafter. 3. The intersecting members of vaults. 4. Projecting bands ornamenting ceilings. 5. The beam portion of a reinforced concrete T-beam. 6. See *Ribbing*.

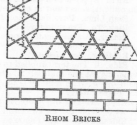

RHOM BRICKS

Rib and Panel Vault. Vaulting in which a framework of ribs support panels of thin stone, called cells, webs, or infilling. The ribs form a series of arches, and are usually moulded below the panels and rebated on the top edge to receive the panels.

Ribands. Timbers that run along the caps of the struts to steady the beam sides in formwork.

Ribbed Framing. Timber-framed buildings with curved struts and ornamental timbers, in addition to the main members.

Ribbing. 1. A wavy or corrugated surface of wood caused by variation of shrinkage of spring and summer wood. 2. Ribs in panel vaulting.

Ribbing Up. Building up laminated circular work.

Ribbon Board. A wide board

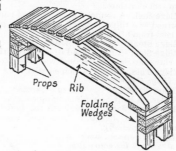

RIBBON BOARD

RIB CENTER

housed into the studs, in balloon framing, to carry the ends of the joists.

Ribbon Building. Houses stretching along a main, or arterial, road, with no development behind. Opposite to estate building.

Ribbonite. Soft lead woven into twisted hanks and used for caulking the joints of socketed pipes.

Ribbon Rail. A light metal core, or rail, tying the tops of the metal balusters together. It is usually covered by a wood handrail.

Rib Center. A center for a segmental arch in which the rib is shaped from one piece of wood.

Rib Mesh. Expanded metal.

Ribmet. Expanded metal lathing for floors, brickwork, plasterwork, etc., where a light reinforcement is required.

Rib-ply. A registered plywood with ribs formed during manufacture to make the sheets rigid.

Rich. Applied to a plastic mix having a high proportion of the essential ingredient, as cement in concrete.

Rich Lime. Pure lime. See *Lime*.

Rick, To. Framing forced out of shape. Distorted.

Ricker. A small scaffold pole.

Riddle. A coarse sieve.

Riddling. See *Sieving*.

Ride. 1. The term applied to a joint that is open at the ends and hard, or touching, at the middle. 2. Also applied to a door that touches the floor as it opens. The remedy is to *cock*, or use rising butts. 3. A directed way for horse riding.

Rider. 1. The top shore, in a system of raking shores, that rests on a short back shore. 2. The band of a *band and gudgeon*.

Rides. Pins, or gudgeons, to carry bands, or strap hinges.

Ridge. 1. The highest part, or apex, of a roof. See *Crested Ridge*. 2. The horizontal timber to which the tops of rafters are fixed. See *Roofs*. 3. The internal angle, or nook, of a vault.

Ridge and Valley Roof. An M-roof, *q.v.*

Ridge Board. A timber from 1 in. to 1½ in. thick against which the tops of the rafters are fixed at the ridge of a roof.

Ridge Course. The last course in a slated or tiled roof. The slates are cut to the required length as in the first course at the eaves.

Ridge Covering. The covering at the ridge to overlap the intersection of the slates or tiles. It may be of V-shaped or half-round tiles, zinc, lead, sawn stone or slate.

Ridge Roll. A rounded piece of wood laid on the ridge, over which lead is turned as a ridge covering. Tingles are placed under the roll to prevent the lead from lifting.

Ridge Stone. See *Apex Stone*.

Ridge Stop. A cap of lead dressed over the ridge and up the wall where the ridge intersects a vertical wall, to make the junction watertight.

Ridge Tiles. Specially shaped tiles used to make the ridge watertight. See *Ridge Covering*.

Ridley Sheeting. See *Dovetail Sheeting*.

Riffler. A bent rasp for shaping concave surfaces.

Rift Sawn. Sawn timber with the annual rings running at right angles to the face of the stuff. Quartered logs converted into smaller stuff. Timber cut radially.

Rig. A space over a theatre stage for controlling scenery.

Rig, To. Setting up anything temporarily, as a scaffold. To equip with the essentials for doing a job.

Rigger. 1. A scaffolder. 2. See *Outrigger*.

Right. The term applied in geometry to regular figures, as a *right cone*.

Crested Ridge.

RIDGE TILE

RIM LOCK

Right Angle. An angle of 90°. See *Three-four-five Rule*.

Right-hand. See *Hand*.

Rigid. Firm. Able to withstand change of shape.

Rigidal Mansard. A registered general purpose corregated aluminium cladding sheet up to 12 ft. by 2 ft. 8 in.

Rigidity. See *Shear Modulus*.

Rigifix. See *Column Guards*.

Rilievo. See *Relief*.

Rim. The outer ring of framework. A raised edge or border, especially if circular. The back and sides of a drawer.

Rimer. See *Reamer*.

Rim Latch. A fitting for keeping a door closed. The spring latch is in a metal case and is operated by a spindle and knobs. The case is screwed on the inside face of the door.

Rim Lock. A latch and lock combined in one metal case, requiring a pair of knobs and spindle for the latch, and a key for the lock.

Rimu. New Zealand red pine. Moderately light, soft, and durable. Strong and stable. Light brown, variegated, with straight, fine grain. Easily wrought, and polishes well. Used for superior joinery, etc. Weight about 36 lb. per c. ft.

Rind Galls. A defect in timber caused by torn-off branches being covered with later rings not uniform with the rest of the timber.

Ring Course. The outer course in an arch of several courses in depth.

Ring Fence. A narrow circular fence on a spindle machine, for circular work.

Ringing Engine. A simple form of pile-driver operated by manual labour.

Ring Latch. A fall bar, or latch, operated by a drop ring handle.

Rink. A building with a special floor for skating.

Ripper. 1. A slater's tool for cutting the nails when removing broken slates. It consists of a long thin steel blade, hooked at the end to grip the nails, and a cranked handle. 2. A rip saw, i.e. one for cutting timber with the grain.

Ripping. Sawing timber along the grain, usually with a *rip-saw* which has teeth specially shaped for the purpose.

Ripple. A beautiful figure in wood caused by compression in the growing tree buckling the fibres. It is common in Sycamore and also called fiddleback.

Rippled. A registered figured rolled glass.

Ripsnorter. A registered portable saw, and other electrically driven tools.

Rise. 1. The difference in height between the two ends of an inclined member. 2. The vertical distance between two consecutive stair treads. 3. The vertical distance from the springing of an arch to the middle of the soffit. See *Step* and *Arches*.

Risen Mouldings. Mouldings projecting above the face of the framing. "Bolection" mouldings without rebates.

Riser. 1. The vertical front of a built-up step. 2. The deep stone that builds up the unit of snecked rubble.

Rising Butts. Special hinges with helical knuckle joint that causes the door to rise as it opens, to clear floor coverings, etc.

Rising Main. The cold water supply from the main continued vertically, to supply each floor as required.

Ritolastic. A proprietary bituminous liquid water-proofer.

Ritzide. A proprietary plastic jointless flooring, of bitumen and leather pulp.

Ritz-plazzo. A cheaper type of Ritzide with wood filling instead of leather.

Rive. To tear or split.

RISING BUTT

Riveted Joints. These are named according to the **arrangement** of both plates and rivets. The rivets may be single, double, etc., according to the number of rows. If they are arranged triangularly they are *zigzag*, and if in straight lines, *chain*. The plates may lap, or butt with either one or two cover plates. Hence they are "double-riveted lap (or butt) joint with chain (or zigzag) riveting, etc." Formulae, for single riveted lap joint: $\frac{\pi}{4}d^2f_s$ (rivet shear) $= (p-d)tf_t$ (plate in tension) $= dtf_c$ (bearing of rivet) $= 3dtf_s$ (plate in shear), where p = pitch, t = thickness of plate. See f.

Rivets. Cylindrical wrought iron or steel pins, usually with spherical, or snap, heads, used for the joints of structural steelwork and boilerwork. The shank is prepared with a head, made red hot, inserted into the hole, and the free end is clenched, or hammered, into a head by means of a cup-shaped swage. It contracts as it cools and grips the joint securely. The heads may be snap, conical, pan, or countersunk. For constructional work rivets are $\frac{5}{8}$ in., $\frac{3}{4}$ in., or $\frac{7}{8}$ in. diam.; for plate work $d = 1\cdot2\sqrt{t}$, where t = thickness of plates. The pitch is a matter for design, but it should be between 3 in. and 6 in. in constructional work. When conditions and materials are suitable the rivets may be cold and driven and shaped by hydraulic pressure or by hammering.

Ro. Abbreviation for *rough*.

Roach. A strata of the Portland beds. It is limestone formed of shoals of shells, and merges into whit-bed.

Road Metal. See *Metal*.

Robertson. A registered corrugated, or veed, protected steel sheet, for roofing.

Robin Hood. A blue-grey sandstone from Yorkshire. Used for general building, dressings, steps, etc. Weight 144 lb. per c. ft. Crushing strength 574 tons per sq. ft.

Robinson Cement. A fire-resisting plaster, for interior work only, having a plaster of Paris base. It is made in three qualities and does not expand or shrink. It is slow setting but becomes very hard and strong after a few hours.

Roble. The name applied to numerous hardwoods from S. America. They are excellent woods, of various shades of brown, with many of the properties of oak, and used for all purposes.

Robot Fireman. A patent automatic stoker for furnaces.

Roche Abbey. A cream speckled limestone from Yorkshire. Used for dressings and selected stone for general building. Weight 139 lb. per c. ft Crushing strength 250 tons per sq. ft.

Rock Elm. A Canadian elm, providing the bulk of supplies in England. It is excellent wood and more suitable for joinery than common elm, but not so durable and strong.

Rocker. See *Rocking Frame*.

Rocker Hinge. A patent hinge used in Hope's metal casements.

Rocket Tester. A smoke rocket, sometimes used for testing drains. It is lit and inserted through a trap and emits a dense pungent smoke. See *Drain Testing*.

Rock Face. Masonry with the face of the stones prepared with a rough rock-like surface. The method is common for engineering masonry, and for plinths and quoins in ashlar work.

Rock-faced Bricks. A specially rough face to facing bricks to give an appearance of rough hewn stone

Rocking Frame. Specially prepared oscillating moulds for vibrating concrete. The rocking, or wave-like, motion causes quick setting and consolidation.

Rock Wool. A vibration absorber used for acoustical purposes. Slag wool, *q.v.*

Rococo. Ornamentation consisting of a meaningless assemblage of scrolls and shell work and confused detail; prevalent in the time of Louis XIV and XV. A contemptuous term applied to over-elaborate decoration.

Rod. 1. A board upon which the work is set out full size, and from which the material is set out preparatory to construction. 2. A measure of reduced brickwork. Rods = $\dfrac{\text{Area of wall} \times \text{No. of } \frac{1}{2} \text{ bricks in thickness.}}{272 \times 3}$

A rod = $306\frac{1}{4}$ c. ft. = $11\frac{1}{3}$ c. yd. and requires about $4\frac{1}{2}$ thousand bricks, $3\frac{1}{2}$ c. yd. sand, 1 c. yd. lime, 160 gal. water; or 19 cwt cement and

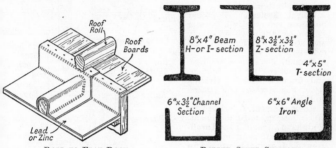

ROLL TO FLAT ROOF   ROLLED STEEL SECTIONS

57 bushels sand for a 1–3 mix of cement mortar. Ordinary brickwork weighs about $15\frac{1}{2}$ tons. 3. See *Cleaning Rods*.

Rodding. Cleaning drains of obstructions by means of rods coupled together, with the necessary appliance attached to the end of the rods.

Rodding Eye. An eye to a drain pipe to allow for the insertion of cleaning rods.

Roe. Ornamental short broken stripes in mahogany, and similar woods.

Roebling. A patented type of constructional work common in America. It embraces fire-resisting floors, partitions, etc.

Roe Stone. Oolitic stone.

Rog. A proprietary paint for new cement work.

Rok. A roofing felt consisting of fibrous sheets saturated with patent waterproof compound and coated with bitumen.

Rokalba. Rok, faced with asbestos on one side.

Roll. 1. See *Ridge Roll*. 2. A Gothic moulding consisting of a round, broken horizontally at the middle with the top half projecting. Often used as a drip. See *Mouldings*. 3. Ornamentation in the form of a cylinder. 4. The turned-back edge of sheet lead.

Roll and Weathered. A Gothic coping unequally weathered and with a roll along the top.

Roll Capped. Applied to ridge tiles having a roll along the apex.

Roll-cap System. The method of covering flat roofs with zinc, with practically no soldering and no nails through exposed surfaces. This allows for free expansion and requires a careful arrangement of welts, clips, saddle pieces, and cappings to wood rolls.

Rolled Glass. Glass in a plastic state pressed by heavy rollers into sheets. If left from the rollers it is called rough plate. See *Glass*.

Rolled Steel Sections. See *British Standard Sections*.

Roller. A heavy cylinder for consolidating loose material, or for propelling or crushing. A cylinder under a heavy object to assist movement.

Roller Bearings. Hardened steel rollers used to reduce friction. They are used in heavy bearings for machines, and under the free end of large steel strusses to allow for expansion and contraction.

Rolling Loads. Moving loads on structures, as vehicles on bridges.

Rolling Shutters. Same as revolving shutters, *q.v.*

Rolyat System. A patent domestic one-tank hot-water system. A tank serves for supply and storage, i.e. for tank and cylinder.

Roman. The style of architecture practised by the Romans and founded on the Grecian style. See *Architectural Orders*, and *Columns*.

Roman Cement. A dark brown, quick setting, natural cement. Nodules of ferruginous and argillaceous limestone are burnt and ground. It cannot be worked up a second time.

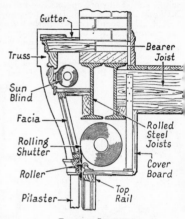

ROLLING SHUTTERS

Romanesque. A general term for the architecture derived from the Roman and prevailing in Europe before the introduction of Gothic.

Roman Mosaic. Also called *Cube Mosaic*. It is made up of tessarae about $\frac{1}{2}$ in. to $\frac{3}{4}$ in. square. See *Mosaic*.

Roman Mouldings. Mouldings derived from the Roman style of architecture. They consist of arcs of circles. See *Mouldings* (2).

Roman Tiles. *Single.* Bridgwater tiles made in various sizes and formed with a wide flat water-way and a bold roll. A common size is $13\frac{1}{2}$ in. × 10 in. *Double.* These are about $16\frac{1}{2}$ in. × $13\frac{1}{2}$ in., and have a second roll down the flat water-way, so that each tile has two rolls and two flat water-ways. If the lap is 3 in. and the gauge $13\frac{1}{2}$ in., it requires 85 tiles per square.

Rones. Same as gutters, *q.v.*

Rood. 1. A carved figure of Christ on the Cross. 2. A measurement of area. Four roods = 1 acre.

Rood Arch. An arch or beam between the choir and nave of a church.

Rood Loft. A small gallery over the chancel screen in a church.

Rood Screen. One separating the presbytery and choir from the nave of a church.

Rood Tower. A steeple placed at the intersection of transept and nave, to a cruciform church.

Roof Beam. A tie beam.

Roof Boarding. Close boarding, usually tongued and grooved, on rafters. It is covered with felt, or sarking, and battened for slates or tiles. See *Counter Battens*.

Roof Clippy. See *Cripple*.

Roofing. The act of covering with a roof, or the roof itself, or the materials forming the roof.

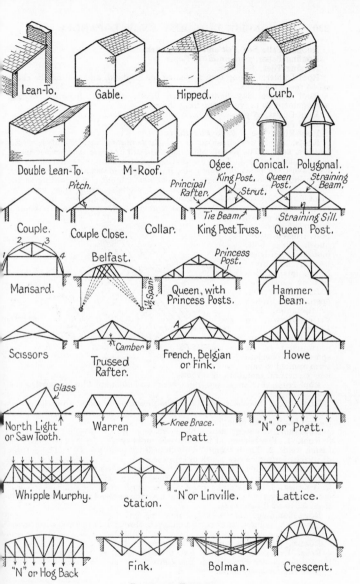

Lean-To. Gable. Hipped. Curb.

Double Lean-To. M-Roof. Ogee. Conical. Polygonal.

Pitch. Couple. Couple Close. Collar.

Principal Rafter. *King Post.* *Strut.* *Tie Beam.* King Post Truss.

Queen Post. *Straining Beam.* *Straining Sill.* Queen Post.

Mansard. Belfast. *Princess Post.* *½ Span.* Queen, with Princess Posts. Hammer Beam.

Scissors. *Camber.* Trussed Rafter. *A* French, Belgian or Fink. Howe

Glass North Light or Saw Tooth. Warren. *Knee Brace.* Pratt. "N" or Pratt.

Whipple Murphy. Station. "N" or Linville. Lattice.

"N" or Hog Back. Fink. Bolman. Crescent.

ROOFS AND TRUSSES

Roofing Felt. See *Felt* and *Sarking Felt.*

Roofing Paper. See *Building Paper.*

Roofing Tiles. See *Tiles.*

Roof Ladder. A long board, cleated for foothold, and laid on the slates when repairing roofs.

Roof Lights. Sky lights, lantern lights, dormers, lucarnes, or any arrangement to provide light from the roof.

Roof Outlet. A rain-water head with grid for a flat roof.

Roof Pitch. See *Pitch.*

Roof Rolls. See *Rolls to Flat Roofs.*

Roofs. Roofs are classified as: *single* (common rafters only); *double* (rafters, and purlins resting on walls); *trussed* or *triple-membered* (trusses, purlins, and rafters). They are distinguished according to the shape as: flat, lean-to or pent, gable or span, hip, curb or mansard, M-roof, V-roof, north-light or saw-tooth, ogee, domical, conical, etc.; or according to the materials of the covering, as: asphalt, lead, felt, slates, tiles, etc. The most important distinction is according to the arrangement of the constructional members (which may be timber, steel, or composite): couple, couple close, collar, purlin, king-post or bolt, trussed-rafter, queen-post or bolt, mansard, collar-beam, hammer-beam, lattice, Belfast or bowstring, north-light, arched, open, Fink, French or Belgium, Linville, Howe, de Lorme, etc.

ROOF TERMS

Roof Trees. Common rafters, or spars. An obsolete term applied to various structural members of a roof.

Roof Truss. A triangulated framework of wood or steel for supporting the roof coverings.

Roof Work. Specially applied to the plumbing required to make a roof watertight with flashings, aprons, etc., and the covering of gutters, flat roofs, etc., with zinc, copper, or lead.

Roo-Shor. A patent adjustable shore, or prop.

Root. 1. The bottom of the cross-sectional space prepared to contain a fusion weld. 2. The bottom of a tenon near the solid wood. 3. The bottom of the cut forming a saw-tooth or screw thread.

Roove. A washer. A metal plate for clenching the tail of a rivet, etc.

Ropala. A chocolate-brown hardwood of the silky oak family, from C. America. Darker streaks make it a handsome wood for superior joinery, etc. It is very strong and durable and used for exposed structures. Weight from 52 to 62 lb. per c. ft.

Rope Moulding. Same as cable moulding, *q.v.*

Ropy. A painted surface not smooth through faulty brushwork, or running of the paint.

Rosace. A rose-like ornament, or a rose window.

Rose. 1. A decorative plate for door furniture. 2. A perforated nozzle for a pipe. 3. See *Strainer.* 4. See *Rosette.*

Rose Bit. A countersink

Rose Nails. Wrought nails with projecting diamond, or rose-shaped, heads.

Rosendale Cement. An American natural cement containing about 30 per cent of clay.

Rosette. Paterae shaped or carved to imitate a rose for ornamentation.

Rose Window. A circular window with radiating bars or tracery. See *Marigold*.

Rosewood. Numerous fragrant woods are marketed under this trade name ; some of them are also called Blackwood, but there are many alternatives. The best known is from India, and this is also called Bombay Blackwood and Sissoo. The wood is dark purplish-brown, variegated, highly decorative, and costly. It is used for superior joinery, veneers, parquet floors, etc.

ROSE ORNAMENT

Rosin. The solid left after distilling turpentine. See *Resin*.

Rosser. A registered name for stormproof windows, and other mass-produced joinery.

Ross of Mull. Pink, red, and grey granites from Argyllshire. They polish well and are used for monumental work, and for bridges, etc. Weight 175 lb. per c. ft. Crushing strength 1100 tons per sq. ft.

Rostrum. A platform, in a hall or place of worship, to accommodate several people. A pulpit.

Rot. See *Dry Rot, Wet Rot, Decay*, and *Preservation*.

Rotahoist. A hoist specially adapted for hoisting bricks.

Rotary Cutting. See *Peeling*.

Rotdoom. A proprietary wood preservative against insects and fungi.

Rotinoff Pile. A system of piling, consisting of units of hollow shells that are driven and then filled with concrete.

Rotor. A force vector, in graphic statics.

Rotten Stone. Pumice stone.

Rotunda. A building circular in plan and usually with dome.

Rouge. A polishing powder, usually oxide of iron.

Rough Arch. Arches formed in half-brick rings of uncut bricks. They are generally covered by other work, and used to relieve a horizontal member of the superincumbent load.

Rough Axed. Arches formed of ordinary bricks roughly axed to shape. They are first cut with hammer and bolster, and then axed with a scutch.

Rough Brackets. Short pieces of board nailed on the sides of the carriage to support the middle of stair treads.

Rough Carriage. See *Carriage Piece*.

Rough Cast. Cement surfaces covered with small coarse material,

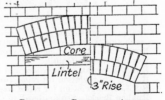

ROUGH AND RELIEVING ARCHES

such as marble or granite chippings, crushed gravel, Derbyshire spar, gravel, brick hogging, etc. The loose material is thrown on to the cement surface whilst it is in a plastic state.

Rough Cast Plate. Glass that has been cast and rolled but not ground and polished.

Rough Cutting. Applied to brickwork that is left from the trowel or bolster, as when fitting round steelwork, etc.

Rough Grounds. See *Grounds*.

Roughing Out. 1. Preparing the groundwork before running the finishing coat of mouldings in plasterwork. 2. Cutting away superfluous wood preparatory to forming any required shape of moulding.

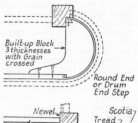

Rough Plate. Glass with irregular and dull translucent surface, as left from the rollers. See *Rolled Plate Glass*.

Rough String. A carriage to a flight of stairs.

Round. 1. As *Torus*. A nosing. 2. A joiner's plane for forming hollows. 3. A rung of a ladder.

Rounded Steps. Curtail, Bullnose, and Round End steps, *q.v.* Also see *Commode Step*.

Round Elbow. A connection, for W.I. pipes, in the form of a quarter circle.

Roundels. 1. See *Crown Glass*. 2. A small disc. 3. Same as *astragal*.

Round-end Step. A step with a semicircular end. A drum-head step.

Router. A joiner's tool for levelling the bottom of trenches or grooves.

ROUND-END STEPS

Rove or Roove. A metal plate for clinching the tail of a rivet, etc.

R.O.W. and P.F. Abbreviation for *rake out, wedge, and point flashings*.

Rowlock. A decorative brickwork panel or frieze, consisting of a circle

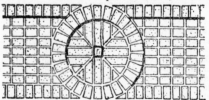

ROWLOCK AND CHECKER BOARD

of headers, enclosing a panel of bricks arranged radially, and continued on each side with checkerboard pattern of all headers.

Roxor. A method of cleaning stone by soft soap and steam.

Royal Blue. See *Granite*.

R.S. Abbreviation for *render* and *set*.

R.S.F. Abbreviation for *rough sunk face*.

R.S.J. Abbreviation for *rolled steel joist*. See *Cramp for Masonry*.

Ruabon Bricks. Strong, hard, impervious, facing bricks. They are very smooth and regular, with the appearance of red terra-cotta. They have lost favour due to the demand for sand-faced bricks.

Rubber Cone. Used for the joint between W.C. pan and the flushing pipe. When the joint is made the ends are wound with copper wire.

Rubber-faced Tiles. Poilite tiles with rubber vulcanized on the face, and used for floors.

Rubber Floorings. Various proprietary compositions, with a rubber base, used for jointless floors. The composition usually consists of rubber (latex), about 40 per cent, granular fillers and colouring matter, about 60 per cent.

Rubbers. Soft bricks from sandy loams, for cutting and carving. They are fine-grained, uniform, and obtainable in various colours.

Rubbing Bed. A masonry machine with a revolving table, up to 14 ft. diam., for stone and marble dressing.

Rubbing Stone. A flat circular piece of York stone used for rubbing plane surfaces on bricks.

Rubbish Shoot. 1. See *Shoot.* 2. See *Tips.*

Rubble. 1. Stones of irregular size and shape used in masonry walling. 2. Broken bricks, etc.

Rubble Ashlar. Compound walls with ashlar face and rubble backing.

Rubble Walls. All types of masonry walling inferior to ashlar. The various types include random (uncoursed, and built to courses), squared (uncoursed, built to courses, and regular coursed), snecked, polygonal, flint, and Lake district.

Rubercrete. Multiple layers of Ruberoid surfaced with bituminous macadam.

Ruberoid. A proprietary bituminous roof covering.

Rubislaw. A greyish-blue granite from Aberdeen. Fine grain and polishes well. Very durable and used for constructional work. Weight 163 lb. per c. ft. Crushing strength 1,288 tons per sq. ft.

Ruby Wood. A decorative hardwood from India, and of various shades of red in colour.

Rudenture. The same as *Cable Moulding.* A carved or plain staff, or cylinder, placed in the flutings of a column, or other hollow.

Rueping. An empty-cell method of wood preservation, with creosote.

Rufereen. A proprietary tough waterproof felt.

Rufflette. A patent curtain rail for metal windows.

Rule Joint. A knuckle joint *q.v.* A hinged joint similar to the joint of a two-fold rule.

Run. 1. A gangway, especially when used for a wheelbarrow. 2. Applied to a saw-cut not true to a required line. 3. Linear measurement. 4. A term applied to plastic materials that have been sieved. 5. See *Curtaining.* 6. The horizontal distance between the ends of an inclined member.

Rung. A stave of a ladder.

Runner Fittings. Rollers for sliding doors and casements.

Runners. The bearers along which anything slides. The supports for a drawer. 2. Vertical poling boards for very loose soil. They are splayed at the foot, and driven as the soil is excavated. 3. The bearers, between which the folding wedges are driven, for a centre for an arch. 4. See *Plough.* 5. Horizontal timbers supporting joists for shuttering. 6. Faulty slates with projections near the head that prevent good bedding.

Running Mouldings. Preparing plaster mouldings *in situ,* by running a specially prepared horsed mould over the plastic material.

Running-off. Applying the final coat of plaster (lime putty) to a moulding.

Running Planks. 1. A gang plank. 2. Strong timbers along which the top centering for a culvert is moved to a new position.

Running Rules. Strips of wood, fixed temporarily, used as guides for the horsed mould when running mouldings.

Running Screeds. Narrow bands of plaster, used instead of rules, on which the mould runs.

Running Shoes. Pieces of zinc fixed on the horse of a mould, to protect the wood and to make it run easily.

Running, or Siphon, Trap. An intercepting trap to a drain in which inlet and outlet are on the same level. It is a bad type as the flow is sluggish and allows for deposits in the bend.

Runtree. The head to a quarter partition.

Runway. Same as run (1).

Rupture. Same as upset, *q.v.*

Rust. Oxide of iron. The red coating, or scales, on iron due to exposure to moist air.

Rusticated. A term used in masonry for several ashlar finishings that project and are of a rough character. A projecting stone is a *rustic*.

Rustic Bricks. Rough-faced bricks. The roughness is due to sand blasting, impressing the clay with a pattern, or cutting with a wire to remove the surface of the clay. Careful mixing and burning also produces varied colour tones. See *Multicolour Bricks*.

RUSTICATED
WORK

Rustic Slates. Multicoloured slates. The variegated appearance is due to chemical changes along the cleavage planes. They are used chiefly for dimishing work.

Rust Joints. The joints of cast-iron pipes made water-tight by caulking with a composition of iron filings and sal-ammoniac.

Rustless Steel. See *Stainless Steel.* It resists corrosion and acids, and is used extensively for fittings, furniture, and for facing joinery work.

Rust Pocket. A loose cover to a gulley at the foot of a ventilating stack of pipes, to remove the rust deposited from the pipes.

Rutger's Process. Pressure preservation of wood in which a mixture of creosote and zinc chloride is generally used.

R.W.B. Abbreviation for *rebated weather boards*.

R.W.P. Abbreviation for *rain-water pipe*.

Rybat. A rebate, or a jamb stone recessed for a frame. See *Inband* and *Outband*.

S

S. Abbreviation for *stop*, and for *shear*. Also for *side, square,* and *surface*.

Saaflux. A patent electric lighting system. The units are easily detached for cleaning, etc.

Sabicu. A strong, durable, hardwood from Cuba, used for landing stages, decking, etc. It is also used as a substitute for mahogany because of its excellent qualities and appearance. Weight about 52 lb. per c. ft.

Sabinite. A patent acoustic plaster. Dr. Sabine was the pioneer of the science of acoustics.

Sabot. An iron skid or runner.

Sacrarium. 1. The part of a church within the altar rails. 2. A piscina.

Sacristy. A vestry to a church.

Saddle. 1. A block used as a seating for circular work, when moulding on the spindle. Also called a *cradle*. 2. See *Apex Stone* and *Gable*. 3. A large roll on a flat roof to divide the surface into bays. 4. A horizontal piece on top of a post to provide a seating for the end of a beam, etc. A bolster or head-tree. 5. A thin projection on the floor under a door to allow for clearing the floor coverings.

Saddle Back. 1. See *Saddle Joint*. 2. A tower roof with two opposite gables.

Saddle Back Board. A raised and rounded floor board under a door, so that the open door will clear a carpet.

Saddle Back Coping. A coping weathered both ways forming a triangular top.

Saddle Bars. Metal bars fixed between masonry jambs and mullions to strengthen leaded lights.

Saddle Bead. A specially shaped glazing bead for a curved bar. It serves for two adjacent panes of glass.

Saddle Boiler. A hot-water boiler to a kitchen range, shaped like an inverted U. See *Low-pressure System.*

Saddle Clip. A pipe clip for small diameter pipes.

Saddle Cramp. A cramp with long shoes for cramping up over curved or projecting surfaces.

Saddle Joint. A rounded "stop" to the joint for a weathered cornice, to throw the water away from the joint. Also called a *horsed* or *saddle-back* joint. A timber joint to prevent the lateral movement at the foot of a post and to provide a good bearing surface.

Isometric View

Section

SADDLE JOINT

Saddle Piece. A shaped piece of lead to cover the junction of two intersecting valleys.

Saddle, or **Saddleback, Roof.** A tower roof with gables.

Saddle Scaffold. A scaffold sitting astride a ridge and resting on the roof. It is used for chimneys. See *Straddle Poles.*

Saddle Stone. See *Apex Stone.*

Safe. 1. See *Lead Safe.* 2. A box or cupboard for preserving food. 3 A burglar-proof box or chamber.

Safe Edge. A plain edge to a file. It has no cuts, or teeth.

Safe Load, or Safe Stress. A proportion of the breaking, or ultimate, strength of the material, and depending upon the selected factor of safety.

Safety Arch. A relieving arch, *q.v.*

Safety Devices. The many appliances on machines to safeguard the machinist, to satisfy Home Office Regulations. They include: cages, riving knives, fences, guards, grippers, circular blocks, pushers, etc.

Safety Glass. Glass reinforced with wire; or 3-ply glass with the middle ply of transparent celluloid or mica. Also see *Glass.*

Safety Island. An isolated raised portion of a road to assist pedestrians to cross the road with safety.

Safety Lintel. A substitute for a relieving arch. A wood, concrete, or steel lintel placed behind a stone lintel or soldier arch to relieve it from the weight of the wall above.

Safety Valve. A device that automatically allows the steam to escape from a boiler when it exceeds a certain pressure. The valves may be *lever, dead weight,* or *spring* valves.

Sag, or **Sagging.** The deflection, or bending, of a member between the points of support. Bending downwards in the middle.

Sag Rod. A hanger.

Sailing Course. See *Oversailing.*

Sail-over. To project over. Oversailing.

St. Andrews × Bond. See *English Bond* (*Cross*).

St. Bees. A red sandstone from Cumberland. Used for genera building, dressings, etc. Weight 135 lb. per c. ft. Crushing strength 507 tons per sq. ft.

St. Petersburg or Petrograd Standard. A measure of timber, of 165 c. ft.

Sal. A hard, heavy, strong, durable wood from India. Excellent for piles, structural work, etc. Weight about 50 lb. per c. ft.

Salamander. Pulped asbestos made into boards and embossed. It is a fire-resisting decorative wall covering.

Sal-ferricite. A proprietary waterproofer.

Salient. Applied to an external angle, or to a projecting member.

Salle. A hall or large room.

Sally. 1. A bird's mouth. The end of a timber cut with an interior angle. 2. A timber with hole to control a rope for a bell. 3. The projecting tongue of a scarfed joint. 4. Same as Salient.

Salmon Gum. A dark-red species of Eucalyptus, *q.v.* It is excellent wood with some beautiful figure. Weight about 62 lb. per c. ft.

Salmon Wood. A variegated reddish-brown hardwood from India. It is very hard and strong, fairly durable, and polishes well. Used for superior joinery, etc. Weight about 47 lb. per c. ft.

Salon. A drawing- or reception-room.

Saloon. A large public room used for the reception of company, for dancing, or for the sale of liquor. 2. A large communal cabin on board ship. 3. A railway carriage for sleeping or dining.

Salt Cake. Sulphate of soda for glass manufacture.

Salt Dip, or **Body.** A thin coating on bricks for glazing. It consists chiefly of washed clay to assist the salt to give a bright uniform glaze.

Salt Glazed. Glaze on bricks, drain pipes, etc., caused by throwing salt into the fire-holes of a down-draught kiln.

Samar. Used as a substitute for teak.

Samel Bricks. See *Grizzles.* Imperfectly burnt bricks or tiles.

Sanatorium. A place or building for convalescents or consumptives.

Sanctuary. 1. A church; or the part of a church where the altar is placed. Same as Presbytery. The holy of holies in Jewish synagogues. 2. A person's private room.

Sand. Fine particles of stone, especially silicious stones, used in mortar, plaster, etc. It should be graded, clean, and sharp, and either pit or river sand. Sand is used to increase the bulk for cheapness, to counteract shrinkage, to increase cohesion, to reduce fattiness, and to allow the air to penetrate the mortar to increase the setting power of lime. Sea sand should not be used because the salt causes efflorescence. Leighton Buzzard white sand is the best.

Sandalwood. Green and yellow. Decorative hardwoods from E. Indies.

Sand Bin. A box for washing sand free from loam and other impurities.

Sand Blasting. Roughening surfaces by sand ejected under pressure, from a gun-like mechanical appliance. It is used for exposing the aggregate in concrete, for engraving glass, etc.

Sand Cushion. See *Helmet.*

Sanded Sheet. Blown sheets of glass slightly roughened by sprinkling sand on the flattening table.

Sanders. Machines for sandpapering wood. The machines are designed for all purposes from large framing and floors, to small cylindrical objects. The various types are: bobbin, disc, drum, belt (overhead or vertical), dowel, portable, etc. Drum sanders may sand both sides at once, and may have up to eight drums.

Sand Faced. Sand-moulded bricks, made by using sand in the mould instead of water (see *Slop-moulding*). The bricks may be machine or hand-made, and are extensively used for facings.

Sanding. Cleaning up the surface of timber by glasspaper. The term is more usually applied to machine papering.

Sand-lime Bricks. See *Lime-sand Bricks* and *Midhurst Whites.*

Sand Moulding. See *Sand Faced.*

Sandpaper. Includes all kinds of abrasive papers, except emery cloth, used by hand or machine. Glass, flint, garnet, or aluminous oxide, etc., is the abrasive, and is glued on paper or cloth, and in sheets or rolls to suit the machines. The abrasive is crushed and sifted to the required grade and affixed by glue to the cloth or paper. See *Glasspaper* and *Abrasives.*

Sandstone. Stones consisting of grains of sand (silica) cemented together by various binding materials : silicic acid, calcium carbonate, iron oxide, clay, etc. The colour varies considerably, from cream to red, and bluish-grey. There are many excellent stones for building purposes. Bramley Fall, Corsehill, Craigleith, Darley Dale, Forest of Dean, Hailes, Mansfield, Bristol Pennant, Rainhill, Wilderness, etc.

Sandwich Beam. A flitched beam.

Sandwich Construction. Applied to light wood roof trusses, etc., built up of single and double members with the single members sandwiched between the double members.

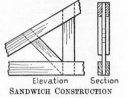

Elevation   Section

SANDWICH CONSTRUCTION

Sanitary Appliances. The various implements for cleaning and testing drains. The term is also used for the accessories used with drain pipes for an efficient drainage system : interceptors, grease and other traps, stoppers, special shapes of pipes, gulleys, etc.

Sanitary Fittings. Water closets, lavatory basins, slop sinks, bidets, baths, etc., and the accessories associated with them.

Sanitary Ware. Glazed earthenware used for sanitary fittings.

Sanitation. The many principles and rules applied to buildings and their surroundings for the preservation of health. These cover : light, water, cleanliness, ventilation, and especially the removal of waste matter.

San Juan. A light-brown, lustrous, hardwood from C. America. Light and soft but durable ; rather woolly but not difficult to work. Used for construction and joinery. Weight about 30 lb. per sq. ft.

Sanodar. A proprietary roofing felt and lining, inodorous and insulating.

Santa Maria. A red hardwood, with darker stripes, from C. America. It is like Poon, and known by many other names. It is fairly hard and heavy, very strong and tough. It polishes well and is a good substitute for good quality mahogany. Used for superior joinery and structural work. Weight about 45 lb. per c. ft.

Santobrite. A proprietary anti-fungi chemical solution for plaster, paste, distemper, glue, etc.

Sapele. A West African light-coloured mahogany. Numerous names ; sometimes called Gold Coast cedar. There are two main species, each of which has a *man* and *woman* tree. It is excellent wood with all the characteristics of mahogany, not difficult to work and polishes well. Esteemed for superior joinery, veneers, etc. Weight about 38 lb. per c. ft.

Sapodilla. A decorative hardwood from Honduras.

Sapote. The name applied to several similar hardwoods from C. America. They are hard, heavy, durable, and strong, and used for superior joinery and structural work. Weight about 54 lb. per c. ft.

Sapwood. The outer portion, or later annual rings, of timber trees. It is not so dense or strong as the heartwood and more readily attacked by decay and disease, but it is easily treated and so made durable. The difference in colour, especially in hardwood, usually prohibits its use in polished work. See *Timber*.

Saracenic Architecture. Moslem, or Mohammedan architecture prevailing in the East. It is associated with Mohammed, and dates from the seventh century. There is great local modification, but the chief characteristics are the bulbous dome and the minaret. The Moorish developed peculiarities of its own sufficiently to be a definite style. The horseshoe arch, and the wealth of intricate detail, consisting chiefly of gilded mosaic and elaborate tracery work, are characteristics.

Sarcophagus. A stone coffin.

Sargent. A patent steel plane with wood sole ; and a registered *steel square*

Sarking Boards. Close boarding to carry slates or tiles.

Sarking Felt. Bituminous felt placed under slates or tiles. See *Eaves.*

Sash. The separate lighter frame to a window, carrying the glass. It may be fixed, hinged, pivoted, or sliding. See *Casement.*

Sash and Frame. A boxed frame with vertically sliding sashes. The boxes at the sides, in which the balance weights move, are built up of a *pulley stile*, *inner* and *outer* linings, and back lining. The outer (top) sash slides between the outer lining and the *parting* beads, and the inner (bottom) sash, between the parting bead and guide bead. Cast-iron or lead balance weights are attached to *sash cords*, that pass over pulleys, and are nailed to the edges of the sash.

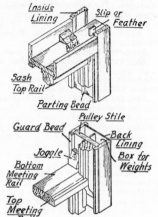

SASH AND FRAME

Sash Balance. See *Pullman Balance.*

Sash Bar. See *Bar.*

Sash Beads. 1. Beads forming the joint between a pivoted sash and the frame. 2. Beads forming a stop or guide for a sash.

Sash Centres. The centres of the pivots for a pivoted sash. They are placed a little above the middle of the height of the sash, so that it closes by its own weight.

Sash Chisel. One with a wide and very thin blade for cutting the pockets in the pulley stile for a sash and frame.

Sash Cords, or **Lines.** See *Sash and Frame.*

Sash Door. A door with the top panels of glass.

Sash Fastener. An appliance for securing balanced sashes. It is fixed on the meeting rails.

Sash Stuff. The material of which sashes are made. It is usually 2 in. × 2 in. and 3 in. × 2 in. yellow deal, and is stocked by most builders. The meeting rails are usually of $1\frac{1}{2}$ in. or $1\frac{3}{4}$ in. stuff. See *Sash and Frame.*

Sash Tool. A round brush used by painters. It is now superseded by the flat brush for oil, but large single-knot brushes are used for distemper.

Sash Weights. Weights used to counterbalance the sashes. See *Sash and Frame.*

SASH FASTENER

Satinay. A light-brown, lustrous hardwood from Queensland. It is very durable, strong, and resists insects. Fumes and polishes well. Used for superior joinery and fittings, flooring, etc. Weight about 48 lb. per c. ft.

Satinwood. The name given to numerous woods that are golden yellow, variegated, lustrous, and fragrant. The chief supplies are species of *Chloroxylon* from Asia, but species of *Zanthoxylon* from Africa and the West Indies are also important. It is esteemed for superior joinery and decorative work generally.

Save Stone. A wedge-shaped stone placed over a stone lintel to relieve the lintel from the weight above.

Savestone. A registered stainless steel sink.

Saving Arch. See *Relieving Arch.*

Savory. A bay in a vaulted ceiling.

Saw Block. A low form of trestle on which timber rests whilst being sawn.

Saw Chops. An appliance for holding saws whilst being sharpened.

Sawder. Solder, *q.v.*

Saw Pit. A pit used for the conversion of heavy timber where machinery is not available. The wood is cut longitudinally by a top and bottom sawyer, the latter standing in the pit.

Saws. Thin blades, discs, or bands of steel with toothed edges, for cutting wood, metal, etc. Joiner's hand-saws are : rip, cross-cut, panel, tenon, dovetail, bow, compass, keyhole, and coping saws. Machine saws are circular, continuous band, oscillating blades, jig, or fret saws. The saws used in masonry are : frame saw, with either plain or corrugated oscillating blades, and circular saws, with either carborundum or diamond saws. The latter have socketed diamonds set round the perimeter. Also see *Hack* and *Pit Saw.*

Saw-tooth Roof. A North Light truss. See *Roofs.*

Saxon. The architecture prevailing in England before the Norman Conquest (A.D. 500-1066). It was of a Keltic type with traces of Romanesque influence. Narrow openings with semicircular or triangle heads, massive stumpy columns, and roughly-shaped simple ornamentation, were characteristics. The best existing example is Earl's Barton Tower.

S.B.C. Abbreviation for *skewback cutting.*

Scabbled, Scabbling, or **Scappling.** 1. Masonry roughly faced with a pick. 2. Dressing with a broad chisel or pick, preparatory to finer dressing.

Scabs. 1. Pieces of superfluous metal on a casting due to a defective mould. 2. A wound in a tree covered by later growth.

Scafco. A registered steel scaffolding.

Scaffixer. A patent chain device for lashing scaffold timbers, instead of using ropes.

Scaffold Boards. The working platform of a scaffold. The boards are 1½ in. by 9 in. and up to 12 ft. long. See *Bracket Scaffold.*

Scaffold Cradle. See *Suspended Scaffold.*

Scaffolding. Temporary platforms, with their supports, for carrying men and materials during the erection of buildings. There are many types : bricklayer's, mason's, saddle, flying, suspended, travelling, etc. They consist of standards, ledgers, braces, and putlogs, all securely lashed together. Scaffold boards rest on the putlogs, and guards and fans are also used. Fir poles are used but are being superseded by weldless steel tubes (tubular scaffolding) which are fixed together by the turn of a set-screw. The tubes are about 2 in. diam., and are light and easily erected. See *Bracket* and *Bricklayer's Scaffold.*

Scaffold Trestle. See *Trestles.*

Scaglia. Italian reddish limestone.

Scagliola. Coloured plasterwork to imitate marble. The required colouring-matter is dissolved in isinglass and mixed with Keene's cement. Fragments of coloured cement or alabaster are interspersed. It is prepared like ordinary plasterwork, then polished with snakestone and finally linseed oil.

Scale. 1. The ratio representing the difference between the dimensions of a drawing and the dimensions of the object. A scale of ½ in = 1 ft., or $\frac{1}{2\frac{1}{2}}$, means that ½ in. on the drawing represents 1 ft. on the object. 2. A thin flake on a surface. 3. Oxide of iron formed on ferrous metal. 4. Encrustations in pipes and boilers due to precipitation of lime salts in the water.

Scalebuoy. An electrically charged, hermetically sealed, glass buoy. A minute electrical discharge precipitates the salts in the water, to prevent scale and corrosion in pipes and boilers.

Scalene. A triangle with three unequal sides.

Scale of Chords. A scale giving the length of the chords for angles up to 90°, and used instead of a protractor for setting-out angles. See *Chords*.

Scale of Slope. An indexed double line used in geometry to indicate the inclination of a plane.

Scaling. Applied to the deposit in boilers and pipes due to hard water.

Scallage. A lychgate, *q.v.*

Scalloped. The edge of thin material cut into a series of curves.

Scamilli. Sub-plinths, or plain blocks, to elevate a column or statue.

Scano. A registered prefabricated wood house.

Scantle. A gauge for regulating slates to the correct length, and holing.

Scantling. 1 Miscellaneous timber of small section. The dimensions are usually over 8 ft. long, and from 2 in. to 4½ in. wide and 2 in. to 4 in. thick. The waste material left by breaking down a log. 2. Stone over 6 ft. long.

Scape. The part of a column where the shaft springs from the base. The Apophyge. See *Mouldings*.

Scappling. See *Scabbled*.

Scarab. A thermo-plastic urea-formaldehyde plastic. See *Urea*.

Scarcement. 1. A local term for the footings of stone walls. 2. The edge to an offset.

Scarf. A joint for connecting timbers lengthways, without increasing the cross-sectional area. See *Table Joint*.

Scarfing. Forming a scarfed joint.

Scarifier. A machine to replace hand-picking, when re-metalling roads.

Scarp. A steep earthwork slope, or embankment.

Scd. Abbreviation for *screwed*.

Scenography. Perspective drawing.

Scented Guarea. A pale mahogany coloured hardwood from Nigeria. Also called Obobonufua and Pearwood. Fairly hard and heavy. Stiff, strong and stable, and polishes well. Used for superior joinery and fittings, flooring, etc. Weight about 41 lb. per c. ft.

Scent Test. Testing drains by pouring a scent with a pungent odour, such as oil of peppermint, into the drain, to locate any escape.

Schedule of Dilapidations. A list of the repairs to be executed by the tenant at the expiration of a lease. The list is prepared by an architect or surveyor on behalf of the landlord.

Schedule of Prices. Estimating for unmeasured work. The quantities are measured on completion, and the job is priced according to the schedule.

Scheme Arch. A segmental arch with the joints of the voussoirs radiating to the centre of the springing line. The bed joints are not normal to the curve. See *Arches*.

Schists. Rocks similar to granite but with planes of foliation so that they split easily into thin laminae.

Schlage Lock. A patent barrel lock operated by pressing a button in the knobs.

Schori. A proprietary anti-corrosive zinc spray.

Sciagraph. The sectional drawing of a building showing the interior.

Scimmer Scoop. Used for mechanical excavating up to a depth of about 6 ft., and for general shallow digging.

Scintled. Bricks *hacked* diagonally, with spaces between, for quick drying.

Sciography, or Sciagraphy. The art of *light and shade* drawing.

Scissors Truss. The simplest form of truss for an open roof for a public building. It is a development of the king-post truss, so that the strut and lower chord form one member. See *Roofs*.

Sclattie. A fine-grain grey granite from Aberdeen. It is used for decorative work. Weight 160 lb. per c. ft. Crushing strength 850 tons per sq. ft.

Scleroscope. An instrument for measuring the hardness of metal by the rebound of a small cylindrical drop-hammer.

Scoinson. See *Sconcheon*.

Scoinson Arch. See *Squinch Arch*.

Scolloped. See *Scalloped*.

Sconce. 1. A bracket carrying a light or reflector. 2. A protecting partition or screen. 3. See *Squinch*. 4. An inglenook, or chimney seat.

Sconcheon, or Scontion. A splayed inside jamb. The side of an opening from the recess to the interior face The interior edge of a window jamb.

Scoot. An American grading term for inferior hardwood lumber, and for scantlings.

Score. To mark with a cut or scratch.

Scotch, or Scutch. A cutting tool used by the bricklayer, consisting of stock, blade, and wedge.

Scotch Bond. English garden wall bond.

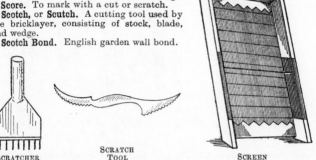

SCRATCHER   SCRATCH TOOL   SCREEN

Scotch Bracketing. Short pieces of laths bedded in screeds of gauged plaster, and placed across the angle between wall and ceiling to carry a moderately sized cornice.

Scotch Fir. See *Northern Pine*.

Scotia. A concave moulding. See *Trochilus* and *Mouldings*.

Scott's Cement. Selenitic lime.

Scottwood. A prefabricated dwelling house. The walls, floors, and roof are on the stressed-skin principle, and made of laminated Oregon Pine.

Scouring. Alternately sprinkling with water and working up a plaster surface with a cross-grained float, to obtain a hard, even texture.

Scraper. 1. A thin steel blade used on hardwoods to remove plane marks before glass-papering. 2. A mechanical excavator for digging, hauling, and spreading. See *Mechanical Shovel*.

Scrappling. The preliminary working of stone.

Scratcher. A plasterer's tool for scratching cement or plaster surfaces to form a key.

Scratching. Applying the first coat of plaster on laths and roughening the surface for a succeeding coat. 2. Shaping, or scraping, a small moulding round a curve or on hardwood with a scratch stock.

Scratch Tools. Small tools of various shapes for cleaning up enrichments in plaster work. They are usually provided with file edges.

Scratch Work. See *Sgraffito*.

Screeds. Narrow bands of plaster used as a guide for the rule when filling in the bays between the screeds. Guides for running a plane surface. An alternative name for grounds.

Screen. 1. A low partition to give protection or privacy. 2. A large rectangular sieve for grading, or sifting, sand, aggregates, etc.

Screen Door. A dwarf door, *q.v.*

Screenings. The refuse after sifting.

Screw. See *Mechanical Powers* and *Screws*.

Screw-down Tap. See *Bib Tap*. The flow of water is regulated by screwing down a washer on to its seating.

Screwdriver. A tool for turning wood screws. There are many types and sizes: brace-bit, London pattern, spindle or cabinet pattern, pump ratchet, spiral and wheel, also screwing machines.

Screwing. Inserting wood screws. The preparation for a screw depends upon the wood, hard or soft, the importance of the work, and the screw,

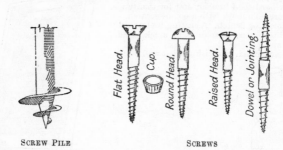

SCREW PILE   SCREWS

whether iron or soft metal. The piece to be fixed should be bored, with bit or drill, for free insertion of the screw, and countersunk if required. The receiving piece should have a small hole at least to give the point of the screw a start. Part-driving with the hammer often breaks the head, also the screw is less efficient. A smear of grease is an advantage. To remove a corroded screw, clean head and slot, heat the head with red-hot iron, put drops of oil on head, and leave for a time, tap end of screwdriver smartly in slot before turning.

Screw Jack. See *Jack*.

Screw Nail. A spiral nail that turns like a screw when driven with the hammer. There are several patent types.

Screw Pile. A cylindrical pile with a screw point. It is revolved so that it bores its way into the earth.

Screws. Cylindrical metal fastenings having a helical thread cut in the metal to engage with the material to which it is screwed. The ordinary type has a head slotted for a screwdriver. They are made of iron, brass copper, or gun-metal, and may be japanned, galvanized, oxidized, chromium- or nickel-plated. The head may be flat, round, or raised. Wood screws are from ¼ in. to 7 in. long, and in gauges from No. 0 to No. 30. Types: wood, coach- or lag-, dowel-, grub-, gutter, self-tapping, and set-screw. See *Screw Nails*.

Screw Shackle. A long cylindrical nut with left- and right-hand threads. It is used to join two rods together to adjust the total length. See *Tension Sleeve*.

Screw Threads. The various types used on bolts and machine parts are: Whitworth, square, buttress, Seller's, knuckle, acme.

Scribbled. Applied to hammer-dressed beds and joints of masonry with marginal chisel drafts.

Scribed Joint. A joint between two intersecting moulded members, in which the moulding on one piece is cut to fit on to the moulding on the other piece.

Scriber. A pointed piece of steel for setting-out on metal.

Scribing. 1. The act of shaping the end, or edge, of a piece of material to fit an irregular surface. 2. Making an incised line on hard material as a guiding line for a tool.

Scrim or **Scrimp.** Coarse canvas used for reinforcing plaster work, such as fibrous plaster and plaster wall-board. Any woven fabric for reinforcing plasterwork or for covering the joints of wall-boards.

Scrimble. See *Scrim*.

Scroll. A carved spiral ornament, the basis of which is a continuous curve built up of circular arcs of increasing radii. See *Handrail Scroll* and *Mouldings*.

Scroll Step. A curtail step, *q.v.*

Scuffled. Applied to roughly sawn surface with fibres removed, but still showing a sawn surface.

Scullery. An adjunct to a kitchen, where the rough work is done, such as cleaning utensils, etc.

Sculpture. The art of carving figures in stone, wood, etc., as practised by a sculptor.

Scum. Lime crystals on new cement work. They are removed by scrubbing with soft soap and water. See *Laitance*. 2. A permanent, or semi-permanent, stain on bricks due to the salts evaporating on the surface before or during burning.

Scumbling. A form of graining. A bright finishing coat of paint applied and partially removed to expose the previous coat, by means of a rag or sponge, to give a marbled, cloudy, or softening, effect.

Scuntion. See *Sconcheon*.

Scurcheon. A Yorkshire term used in stone slating.

Scutch. See *Scotch*.

Scutcheon. See *Escutcheon*.

Scuttle. A small aperture in a roof or ceiling. A trap door.

S.E. Abbreviation for *stopped end*, and *square edged*.

Seal. 1. The water in a trap for a drain. 2. The arrangement for making the joints to drain-pipes air-tight. 3. See *Manhole Cover*.

Sealed Cover. An air-tight cover to a manhole, *q.v.*

Sealing. 1. Closing a joint or an opening. 2. Cementing a strip of slate over the joint of the tiles in tile creasing. 3. Fixing an object to a wall by some form of binding material, as plaster, cement, lead, etc.

Sealocrete. Proprietary waterproofers, hardeners, and colourings for cement, etc.

Sealontone. Proprietary liquid colours and hardeners for cement, etc.

Seam Roll. A hollow, lead roll.

Seams, or Welts. Joints for sheet lead in which the two edges are folded over together. Copper tingles are placed at intervals to stiffen the seam.

Seasoning. Reducing the moisture content of wood to suitable proportions which may be by *natural, wet,* or *kiln* seasoning. Kiln seasoning is the most economical and the best for most woods as it is under control. *Second seasoning* is the further drying of small timbers, or of framing before it is wedged up. See *Moisture Content, Compartment,* and *Progressive Kilns*. 2. Also applied to stone. See *Quarry Sap.*

Seatings. 1. Fireclay blocks on which boilers rest. 2. A bed on which anything rests. A pad or template on which a structural member, as a beam, rests. 3. A collective term for the seats in a public place.

Seaweed. Used as a non-conductor of sound. It is treated and compressed for pugging. See *Cabot's Quilt*.

Seco. A prefabricated house. It has panels of asbestos cement with woodwool-cement core, and plywood structural elements.

Secomastic. A proprietary jointing compound for bedding metal windows, etc. It is a plastic sealing medium and can be applied with the Seco caulking gun.

Secondary Beam. The steel beams carried by the main beams, and in turn carrying the filler

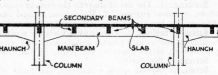

REINFORCED CONCRETE FLOOR

joists. In reinforced concrete they carry the floor slabs. A binder in a wood framed floor.

Secondary Circulation. An auxiliary supply of flow and return pipes to a hot-water system to carry the hot water as close as possible to the taps.

Secondary Clays. Those moved from the source of formation by natural means and combined with other substances. They are classed as: marine (sea deposits), fluviatile (river deposits), and lacustrine (lake deposits).

Secondary Flow. The flow pipe from the top of the cylinder in a Low-pressure System, *q.v.* It also serves as an expansion pipe.

Second Fixing. Joinery that is fixed after plastering: skirtings, picture rails, architraves, etc.

Seconds. 1. The second gauging of plaster for casting. 2. Applied to the quality of some building materials.

Second Seasoning. See *Seasoning.*

Secos. A sacred enclosure.

Secret Dovetail. A dovetailed joint having the appearance of a mitred joint, owing to the pins being buried in the material. See *Dovetails.*

Secret Fixing. The fixing of hardwood joinery without the method of fixing showing on the face. The usual method is by slot screws on the back, which is often adopted for stained and polished work.

Secret Gutter. A valley gutter hidden by the mitred slates or tiles, or one down a gable or the side of a chimney hidden by slates.

Secret Nailing. Inserting nails so that the nail holes are not seen in the finished work. One method for tongued and grooved boards is to skew the nails through the tongues. Another method is to lift a thin layer, with a chisel or special tool, and, when the nail is driven, to glue the layer back in position.

Secret Screw. A woodwork joint in which screw heads are left projecting about $\frac{3}{8}$ in. in one piece, inserted in corresponding holes in the other piece, and driven along parallel slots, thus forming a secure and hidden fastening. It is used to strengthen butt joints. Also see *Soldiers.*

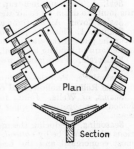

Plan

Section

SECRET GUTTER

Secret Tack. Strips of sheet lead soldered on the back of a sheet of lead, passed through a slot in the boards, and secured on the back of the boards. It is used to secure large vertical sheets of lead, and still allow for expansion and contraction.

Sectioning. Section lines drawn across the portions of an object cut by a section plane, as shown on a drawing. The lines are slightly thinner than the lines of the drawing, and different kinds and arrangements of lines may be used to denote different materials.

Section Modulus. See *Modulus of Section.*

Section Mould. A templet of zinc, cardboard, or plywood, cut to the shape of the section of a member. The templet is marked on the ends of the member, which is shaped accordingly. See *Moulds.*

Sections. Drawings representing the interior arrangements of an object. They are obtained by cutting the object by imaginary *section* planes and then depicting the surface formed by the intersection of the plane with the object.

Sector. 1. A part of a circle bounded by two radii and an arc. See *Circle.* 2. A mathe-

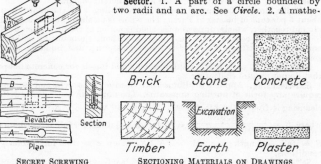

SECRET SCREWING   SECTIONING MATERIALS ON DRAWINGS

matical instrument graduated with sines, tangents, etc., etc., and used for finding the fourth proportional.

Sectroid. The curved surface between two adjacent groins in vaulting.

Securex. A patent joint for copper tubing.

Sedilia. Recessed seats for the priests in the choir of a church.

Sedimentary Rocks. Those formed by the depositing of materials in water, as sandstones and limestones. The variation in pressure, during formation, and chemical combination, cause great variation in the stones. See *Secondary Clays.*

Sedimentation Tank. A tank for sewage to pass through at a speed to allow the solid matter to settle and be removed at intervals.

See. A seat of authority in a cathedral.

Seed-lac. Shellac, *q.v.*

Seekay Wax. A chlorinated naphthalene for fireproofing rubber in buildings.

Seelastic. A proprietary plastic compound for pointing frames, jointing building materials, etc. It is available in several colours, and can be applied cold.

Seepage. To trickle or ooze out. Leakage.

Seger Cones. Pyramids of mixtures of clay and fluxes. They are used to register the result of heat action in kilns, to decide the required temperature.

Segl. A. Abbreviation for *segmental arch.*

Segment. 1. A part of a circle bounded by a chord and an arc. Area of small segment $= \frac{2}{3} hc$, where $c =$ chord and $h =$ height; radius of the circle $= \left[\frac{1}{2} \frac{(\frac{1}{2}c)^2}{h} + h. \right]$ See *Circle.* 2. A part of a geometrical figure cut off by a line or plane. The surface area of part of a sphere = π $(r^2 + h^2)$, where r is the radius of the plane surface; and volume = $\frac{1}{6}\pi h(h^2 + 3r^2)$. 3. Pieces building up circular work.

Segmental Arch. An arch formed of a circular arc less than a semi-circle. See *Arches*.

Segmental Gothic. See *Arches*.

Seize. 1. To stick, or jam, or bind. 2. To lash together with ropes.

Selected Merchantable. A superior grade to Merchantable, in N. American softwoods.

Selek. A proprietary powder, mixed with boiled linseed oil, for jointing metal pipes.

Selenalyte. A registered glass made in six different colours.

Selenitic Lime, or Cement. A feebly-hydraulic lime with the addition of about 5 per cent of plaster of Paris. The whole is mechanically mixed and ground together, which suppresses the slaking action of the lime when used for plaster. It can be used for *rendering* without the addition of hair, and is quick setting. Used for two-coat plastering, and building work generally.

Self-centering. Applied to reinforcements and lintels for concrete floors that do not require shuttering.

Self-cleansing. Applied to drains that cleanse themselves and do not require any form of flushing apparatus. The minimum velocity for self-cleansing is from $2\frac{1}{2}$ to 4 ft. per sec. Vel. = $140 \sqrt{RS} - \sqrt[3]{RS} \times 11$, where $R = \dfrac{\text{sect. area}}{\text{wetted perimeter}}$ = hyd. rad. = $\frac{1}{4}$ diam. for full or half-bore; S = fall $\div$ length; or vel. = $c\sqrt{RS}$, where c = 100 for ordinary conditions.

Self-faced. Applied to laminated stone, as flagstone, that does not require the surface to be dressed.

Selfix. A patent accelerating motor for hot-water systems for public buildings.

Self-supporting Partition. A partition supported by the walls only, and not dependent on the floor for any support.

Sel. Struct. A grading term for pitch pine, meaning *Select Structural*.

Semastic. A proprietary flooring of synthetic resin and fibrous mineral filling, and pigments. It is in the form of tiles.

Semi-arch. A semicircular arch.

Semi-bungalow. A bungalow, but with one or more bedrooms arranged in the roof, forming a two-storied building.

Semmentum. A proprietary fire-resisting cement.

Semprax. A registered name for stainless steel fittings, door furniture, etc.

Semtex. A registered smooth non-slip jointless flooring.

Separate Systems. Applied to drainage systems in which separate drains are used for waste matter, and for rain, or storm, water.

Separators. Strong distance pieces placed between two beams bolted together. They are used to give rigidity and to keep the correct distance. When exposed to the weather, cast-iron spools are used in heavy structural work, because of weathering.

Sepia. A brown pigment.

Sept. A railing.

Septic Tank. A large tank for the collection of sewage. The inlet and outlet are submerged, and no light or air admitted, which decomposes the sewage. The effluent is purified by aerobic action on filters. The system is now out of favour by many authorities.

Sequoia. The standard name for Californian Pine, *q.v.*

Seraya, or **Seriah.** *Shorea spp.* Also called Borneo cedar or mahogany, meranti, etc. The wood is distinguished as *red* and *white* seraya. The red variety is very varied in properties, but generally harder and stronger than white. The wood works readily and polishes well and is used as a substitute for mahogany for superior joinery, etc.

Serpentine. 1. *Gabro.* A varied coloured stone very much like marble, and used extensively for interior decorative work. The chief constituents are hydrated silica of magnesia and magnesia. The veins and spots are supposed to resemble a serpent's skin. 2. Applied in woodwork to a curve of contraflexure.

Serrated. Notched like a saw.

Serval. Asbestos fibre saturated with bitumen and used as sarking felt.

Servant Key. A key that only operates the lock for which it is intended, in a building.

Servery. A place for preparing trays of food for serving in a restaurant.

Service Box. A small reservoir to the supply pipe of a W.C., between the pan and flushing cistern. It holds sufficient water to fill the pan after flushing.

Service Flats. Residential flats for which a staff of servants is engaged to keep the rooms clean, and to prepare food for the tenants, if required.

Service Pipe. A pipe directly subject to water pressure from the service main of the Water Board. The pipe between the service main and the premises receiving the water. The same term is used for a gas supply pipe.

Serving Hatch. A small aperture, with door, in the division wall between two rooms, usually kitchen and dining-room, for the quick serving of dishes, etc. The door may slide or revolve, or it may be hinged or pivoted.

Set. 1. The projection to alternate sides of saw teeth, so that the blade will work freely in the saw-cut. 2. A nail punch. 3. An assortment of the same kind of things, varying in size. 4. See *Setting.* 5. See *Setting Coat.* 6. See *Permanent Set.*

Set-off. See *Offset.*

Set Pot. See *Copper* (1).

Sets. Small blocks of stone, usually granite, used for paving roads.

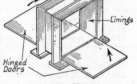

SERVING HATCH

Set Screw. A screw with parallel shank for engaging with metal which is tapped to receive the screw. The head may be square or hexagonal.

Set Squares. Triangular drawing instruments for vertical and inclined lines. They are in pairs (45°, 60°) giving angles of 90°, 60°, 45°, and 30°.

Sett. 1. A cutting tool used by the blacksmith. There are two kinds, *hot* and *cold,* according to the condition of the metal they have to cut. 2. See *Sets.*

Setter-out. See *Setting-out.*

Setting. 1. The quality of hardening in plastic materials. The *initial* set of Portland cement is due to the rapid hydration and crystallization of the aluminates, and the *final* set, or hardening, is due to the chemical action of the silicates, which is much slower. The time for the initial set is 30 minutes, and for final set 10 hours. 2. The laying or fixing of stones and bricks.

Setting Bar. A bar used by masons when setting stone.

Setting Coat. The final coat of plaster, or the skimming coat. It is about ⅛ in. thick, and usually consists of lime putty and very fine sand called setting stuff. Keene's, Parian, or any of the patent plasters may be used, but the background must be of such a nature that it agrees with the setting coat.

Setting Out. Preparing full-size drawings on rods, and transferring the dimensions to the stuff, preparatory to construction. The person responsible for this work is called a *setter-out*.

Setting Stuff. See *Setting Coat*.

Setting-up. Same as *erecting*.

Settlements. The failure of buildings due to defective foundations.

Settling Tanks. Large tanks used in sewage disposal.

Set Up. 1. To turn up the edge of sheet lead. 2. To caulk a lead joint.

Severy. A compartment in vaulting. Infilling.

Sewer Gas. The foul air and vapours from sewers.

Sewer Pill. A large framed skeleton wood sphere. It scours the walls of the sewer as it drifts down the sewer.

Sewers. The pipes, channels, or subways for conveying water and sewage. Usually the controlling local authority is responsible for the sewers whilst the landlord is responsible for the drains. Sewers, in most districts, convey the sewage finally to a *sewage farm* for disposal by scientific methods.

Sexpartite. Applied to Gothic vaulted roofs in which an intermediate pier forms six intersections.

Sextant. 1. A surveying instrument for the measurement of angles. 2. The sixth part of a circle.

Seyssel. Claridge's patent asphalt, obtained from the French Jura mountains.

S.F. Abbreviation for *sunk face*.

S.G. Abbreviation for *specific gravity*. A trade mark for second quality plate glass, meaning *selected glazing*.

Sgraffito. Ornament scratched on plasterwork; or formed by coloured layers of plaster, each successive layer being cut away to expose portions of previous layers, to give the required design.

Shack. A rough hut.

Shackle. A steel link, or loop and bolt, used as a fastening or to secure several pieces together. A lifting shackle is one in which the shackle is used for the hook of a crane. See *Lewis* and *Screw Shackle*.

Shadbolt. A patent fanlight opener and stay.

Shade. 1. A thin lamina of stone used under the joints in stone slating. 2. An arrangement to give protection from the sun.

Shading. 1. Representing shadows on drawings. 2. Sometimes used as an alternative term for *Sectioning*.

Shaff. A patent water tap, bib, pillar, or inclined.

Shaft. 1. The cylindrical portion, or body, of a column, between base and capital. See *Mouldings*. 2. The portion of chimney above the roof. 3. Anything long and slender, as the shaft of a mine or striking tool, or a spire or tall chimney. 4. A conduit or ventilating pipe.

Shafted Impost. One with horizontal mouldings, but with the arch mouldings differing from those of the shaft, or pier, below.

Shakes. A cleavage in the annual rings of timber. A split in the timber. They may be compound, cross, cup or ring, heart, radial, shell, or star shakes, and they are caused by lack of nutrition, frost, and wind, during growth, or by faulty felling and seasoning. See *Timber*. 2. Hand riven ½ in. shingles, longer than normal.

Shale. 1. Laminated rocks. There is great variation. Many are clay shales, and contain carbonaceous matter so that they require little fuel for brickmaking. 2. Sometimes applied to shingle.

Shamah-Duplex. A prefabricated house consisting of steel framework and plastic panels.

Sham Beam. A built up projection on a ceiling to give a panelled effect and to imitate a beam.

Shamrock. A dark-grey sandstone from C. Clare, Ireland. It is hard and compact and used for landings, paving, etc. Weight 167 lb. per c. ft. Crushing strength 1,902 tons per sq. ft.

Shank. 1. The shaft of a column. 2. Long connecting part of an appliance.

Shanty. A temporary wooden building. A hut or mean dwelling.

Shaped Work. Same as curved work.

Shap Fell. A reddish-brown granite from Westmorland. Used for decorative work, columns, etc. Weight 160 lb. per c. ft. Crushing strength 1,200 tons per sq. ft.

Shavehooks. Sharp tools of various shapes used bʸ the plumber for shaving lead before soldering.

Shaw Guard. A patent guard for spindle machines.

Sheal or Shealing. A hut or shelter.

Shea Oak. Tasmanian oak. Several species of hard, heavy, tough, hardwoods, with beautiful figure for superior joinery, veneers, etc. Weight about 50 lb. per c. ft.

Shear. 1. Force that is tangential to a section. In a beam "the shear force, at any section, is the algebraic sum of the forces acting on one side of the section." A tendency for one part to slide over another part, either vertically or horizontally. See *Bending Moment*, and page 366. 2. The curvature of a ship's deck in a fore-and-aft direction.

Shear Legs. Two or three poles spread apart at the feet and secured together at the top, from which pulley tackle is suspended for lifting weights. Sometimes a rope takes the place of the third pole.

Shear Modulus. *G.* Shear stress ÷ shear strain.

Shear Plate. See *Timber Joint Connectors.*

Shears. A machine for cutting steel.

Sheathing. 1. Close boarding on studs or spars used as a base for

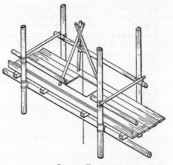

SHEAR LEGS

the coverings. 2. Drawn bronze on a hardwood core. See *Kalamein.* 3. Metal coverings, for timber and concrete, used in under-water work as a protection from marine insects. See *Elio.*

Sheathing Paper. See *Building Paper.*

Sheave. The wheel of a pulley. A pulley block is distinguished as a two-sheaved, three-sheaved, etc., according to the number of grooved wheels. See *Hoisting.* A sliding escutcheon to a keyhole.

Shed. A shelter, used for storing materials or for a workshop. It may be open all round.

Shed Roof. See *Lean-to.*

Sheet Glass. See *Glass.*

Sheeting. 1. Close timbering, running horizontally, for excavations in loose earth. 2. Shuttering or decking.

Sheet Piling. Close piling forming a sheet, or panel, between large piles. Timber,

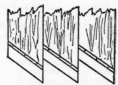

SHEET PILES WITH IRON-COVERED POINTS

concrete, or steel piles forming a continuous sheet to protect the enclosure. They are thin, and pointed for easy driving.

Shelf. A flat, horizontal board, slate, or stone, for supporting articles.

Shelf Nog. A projecting piece of wood built into a brick wall to carry a shelf, instead of using a bracket.

Shell. 1. A semi-cylindrical brace bit. 2. Same as carcass, *q.v.*

Shellac. A resinous gum obtained by melting seed-lac in cotton bags and straining. It is in small thin cakes, and is used with methylated spirits for french polish. It is also used to provide a good hard surface for varnish, and for knotting. It originates from the exudation of insects that feed on the Indian lac trees. It is soluble in alcohol but not in turpentine.

Shellac and White Lead. Used for bedding bricks, for carving.

Shelter. 1. Any form of shed, cabin, or hut, to provide protection against the weather. A decorative form, with seats, is provided in parks, on promenades, and at stopping places for public conveyances. 2. A strong structure to provide protection from air-raids.

Shelving. 1. The fixing of shelves ; or a number of shelves. 2. To slope or incline.

Shepwood. A registered name for a number of patented shapes of glazed bricks and blocks.

Sherardizing. Applying a proprietary rust-proofing compound to iron-work ; especially used for fixings built in walls.

Sheriff. A patent opener and stay for fanlights.

Sheringham Valve. A ventilator consisting of a hinged flap to a cast-iron frame built in the wall. The flap is adjusted as required.

Shet. A joint.

Shides. Shingles.

Shield. 1. A machine working in advance of the brickwork, in tunnel construction. 2. Escutcheon. 3. Anything that protects or defends.

Shieling. Same as *sheal*.

Shilf. Broken slate.

Shims. 1. Packing pieces used to level the surface of battens to carry wall boarding. 2. Plugs or patches for plywood repairs.

Shiner. 1. A thin wide stone placed on edge, and breaking through two or more courses in a rubble wall. 2. A stone laid with the natural bed as the face.

Shingle. 1. Cleft pieces of timber used in place of tiles for roof coverings. Now chiefly of sawn Western red cedar. 2. Round water-worn pebbles, or gravel.

Shingling. 1. A method of producing wrought iron, by squeezing out the refuse material from a bloom of puddled iron. 2. Covering a roof with shingles.

Ship-lap. Rebated weather boards, *q.v.*

Shippers. Hard, sound, stock bricks, imperfect in shape, and so named because of their use for ship's ballast.

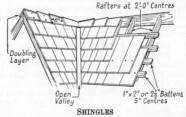

SHINGLES

Shipping Marks. The distinguishing marks of foreign timber-exporting firms. The marks are stencilled, stamped, or hammer-branded on the ends of the timber, and, if the firm is reputable, denote the quality.

Shipping Ton. Implies 42 c. ft. of timber.

Shippon. A cowhouse, or stable.

Ship Spikes. Heavy fastenings forged from square iron bars, with a wedge shaped point, and used for fixing heavy structural timbers.

Ship Worm. See *Teredo Navalis*.

Shivers. See *Spalls* (2).

Shoddy. 1. Square granite under 1 ft. thick. 2. Applied to inferior work or materials.

Shoe. 1. An iron point for the foot of a pile. 2. An iron socket for the foot of a post or rafter. See *Door Frame.* 3. A bent pipe at the foot of a stack to turn the water away from the wall. 4. The seating for the heel of a swing door. See *Floor Spring.* 5. A metal slipper for a sliding door or sash.

Shoot. 1. To straighten the edge of a board with a plane. 2. An inclined trough down which materials are passed from a higher to a lower level. A *Chute.*

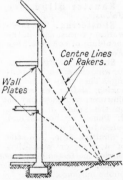

Centre Lines of Rakers.

Wall Plates

RAKING SHORES

Shooting Board. A framed board for steadying a piece of timber whilst shooting the edge.

Shooting Box. A small country house used during the shooting season.

Shore. A wood or iron prop, or support. See *Mar-Shor* and *Roo-shor.*

Shoring. The erection of heavy temporary timbering to support defective buildings, or those undergoing alterations. 2. A collection, or *system*, of shores. See *Dead, Flying,* and *Raking Shores.*

Short Column. Applied in structural design to columns or struts that would fail by direct crushing, to distinguish from one in which bending would also take place. In reinforced concrete a column is "short" if the height is less than 18 × least diam., and for brickwork 8 × diam. See *Columns.*

Shortleaf Pine. Marketed in this country as pitch pine. It is similar to longleaf pine but lighter, softer, and not so durable or strong.

Short Working. A term applied to plaster that is difficult to work and that will not carry the usual proportion of sand.

Shoulder. A projection to give strength and support, as the shoulder of a tenon.

Shouldered Tenon. A tusk tenon, *q.v.*

Shouldered Voussoir. Used in terra-cotta construction for flat arches, as a form of joggle. Also see *Voussoirs.*

Shouldering. Bedding the heads of slates in hair mortar, for roofs in exposed positions.

Show Board. An inside stall board for a shop window. The window bottom.

Show Case. A glazed air-tight case for displaying goods. See *Silent Salesman.*

Shread Head. See *Jerkin Head.*

Shreddings. Short light timbers used as bearers under the roof, in old buildings.

Shrine. 1. A case in which sacred things are deposited. 2. The mausoleum of a saint.

Shrinkage. The contraction of timber and other materials due to the evaporation of moisture. See *Contraction.* Timber contracts chiefly in the direction of the annual rings, as the medullary rays tend to resist radial contraction. This influences the conversion and seasoning of timber. There is a great variation in different woods, but tangential shrinkage is about twice the radial, and the longitudinal shrinkage is negligible in practice except in very few woods.

Shrouds. Same as crypt, *q.v.*

Shuffs. See *Chuffs*.

Shuting. An eaves gutter.

Shutter Bar. A long, pivoted bar for securing folding shutters.

Shutter Blind. See *Jalousie*.

Shuttering. The false-work, or forms, used in the moulding of concrete. See *Formwork*.

Shutters. 1. Protections for windows and other openings. They may be *hinged, boxing, sliding,* or *rolling* shutters; and they may be on the inside or outside. 2. Panels of boards and ledges for formwork.

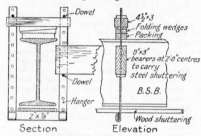

Section — Elevation

SHUTTERING (SUSPENDED)

Shutting Stile. The stile of a door or sash that carries the fastenings; opposite to the hanging stile.

Shy. A protection to a re-entrant angle of a building, to prevent a public nuisance.

Siberian Pine. See *Yellow Pine* and *Redwood*.

Sided Pipes. Cast-iron pipes in which the thickness varies due to a faulty position of the core. The modern rotating core eliminates this defect.

Side Flights. Applied to a double-return stairs where they branch to right and left at the top of the main flight.

Side Gutter. A small lead gutter down the slope of a roof, as to a chimney, dormer, or gable.

Side Hook. See *Bench Hook*.

Side Lights. Same as margin lights for doors, and wing lights for frames.

Side Posts. Secondary queen posts in a truss; also called *princess posts*.

Side Wavers. A local term for *purlins*.

Sidings. 1. Weather boards on a vertical surface. There are a number of market names for varying shapes: feather edge, ship-lap, novelty, rebated, bevel, bungalow, drop, and rustic. 2. The slabs left after squaring a log.

Sidol. A proprietary timber preservative.

Siege. A local term for a mason's *banker*.

Siegwart. 1. A patent fire-resisting floor, consisting of hollow reinforced concrete beams. No centering is required. 2. Also patent blocks for stairs.

Siemens Steel. Steel produced by the Siemens, or open hearth, process.

Sienna. An earth pigment, yellow in colour, and with a strong staining power. When burnt it is a low-toned reddish colour. See *Pigments*.

Sieve. A vessel with a woven wire, or perforated bottom, for sieving loose materials.

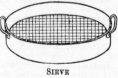

SIEVE

Sieving. Grading loose materials to obtain the finer particles or those of a uniform size by means of a screen, sieve, or riddle.

Sight Rails. Rails nailed to two fixed posts for obtaining levels and gradients. They are used for trench excavations.

Sigmoid, or Sigmoidal Curve. Shaped like the letter **S**.

Sign. A board, etc., to point out the direction and distance of a particular place; or to indicate a particular business or occupation.

Sign Board. A wide fascia board over a shop front describing the business and ownership.

Sika. A patent waterproofing compound for cement, in several forms.

SIGHT RAILS

Sikrex. A patent draught excluder for doors.

Silentium. A patent method of deafening floors with peat and porous tiles.

Silent Salesman. A shop-floor case, about 6 ft. high, for the display of goods.

Silex. A blue-grey sandstone from Halifax, Yorkshire. It is hard, compact, and durable, and used for landings, paving, etc. Weight 168 lb. per c. ft. Crushing strength 2,200 tons per sq. ft.

Silica. SiO_2. Oxide of silicon, an ingredient of many minerals. In crystalline form it is the chief constituent of many building stones. Quartz, flint, rock crystal, are nearly pure silica. It forms about 60 per cent of good brick earth.

Silica Bricks. Firebricks to withstand very high temperatures. They contain over 90 per cent silica.

Silicate Cotton. See *Slag Wool.*

Silicate of Soda. Waterglass. Combined silica and sodium soluble in water and used for waterproofing  SILENT SALESMAN bricks, plaster, stone, etc. It is also used for cleaning down old paintwork. Used alternatively with limewash to coat wooden structures, to make the timber fire-resisting, to some extent.

Silicic Acid. The parent substance from which silicates are derived. The cementing material for sandstones.

Silicon Ester. Used as a preservative for brickwork and stone. The evaporation of the alcohol leaves a layer of hydrated silica that forms a binding material on the face of the stone.

Silky Oak. A Queensland hardwood, light brown, lustrous with beautiful figure and striking silver grain, but it is not an oak. The wood is hard, heavy, strong, and fairly durable, polishes well, and esteemed for superior joinery, bent work, plywood, etc. Weight about 37 lb. per c. ft.

Sill, or **Cill.** The horizontal member at the bottom of framing and openings. Also see *Dead Shores.*

WINDOW SILL

Silo. 1. A tall tower, or battery of towers, for the storage of grain, etc. 2. Air-tight structures in which green crops are preserved for fodder.

Silt. Fine mud and grit.

Silt Box. A loose iron box at the bottom of a gulley for collecting mud and grit, or silt. A handle is provided for easy removal and emptying.

Silt Well. A soakage pit.

Silurian Rocks. The lowest division of the palaeozoic strata. They are post-Cambrian and pre-Devonian, and consist of upper and lower Silurani. See *Slates*.

Silver. A metal wedge or packing inserted between the legs of a chain lewis when hoisting blocks of masonry.

Silver Ash. A white lustrous hardwood from Queensland. Similar to Maple Silkwood and also called Cudgerie. Fairly hard, heavy, and strong, with attractive grain. Easy to work and polishes well. Used for flooring, bentwork, joinery, veneers, etc. Weight about 39 lb. per c. ft.

Silver Beech. New Zealand Beech.

Silver Grain, or Felt. The beautiful figure obtained by cutting oak beech, etc., along the medullary rays.

Silver Greywood. See *Chuglum*.

Silvering. Coating glass for mirrors. The glass is coated with a transparent film of tin on which is deposited a layer of silver, which is protected by paint or composition.

Silver Soldering. Brazing with silver solder instead of spelter.

Silver Spruce. See *Spruce*.

Simplex. A patent glazing for roofs.

Simplex In situ Piles. A hollow metal core is driven, into which concrete is poured and the core withdrawn; the concrete may be reinforced by a metal skeleton which is left in position when the outer core is lifted.

Simplified Plumbing. See *One Pipe System*.

Simpson's Rule. A mathematical formula for obtaining irregular areas and volumes. The area is divided into an even number of equal parallel spaces (s) giving an uneven number of ordinates (h). For eight spaces, area $= \frac{1}{3}s\{h_1 + h_9 + 2(h_3 + h_5 + h_7) + 4(h_2 + h_4 + h_6 + h_8)\}$. For the volume of a solid, sectional areas are substituted for the ordinates.

Single Floor. A floor consisting only of common joists and floor boards.

Single-lap Tiles. Plain tiles.

Single Lath. A $\frac{3}{16}$ in. plasterer's lath.

Singles. Small slates, about 12 in. × 8 in.

Single Shear. When the resistance to shear is on one section plane only. Safe single shear $= 5\frac{1}{2}$ tons per sq. in. for mild steel. For double shear allow $1\frac{3}{4}$ × single shear.

Sink. A receptacle for receiving dirty water, as for kitchen waste, etc. It is emptied, by a trapped waste pipe and gulley, to the drains. 2. A trap door for the scenery, in a theatre stage.

Sinking. A part sunk below the surrounding surface. A recess on a surface.

Sinking Square. An adjustable square for testing the depth of a sinking.

Sinter. Calcareous or siliceous rock formed by spring deposit.

Siphon. 1. See *Syphonage*. 2. An appliance to produce siphonage. A receptacle for collecting the water that may accumulate in gas pipes.

Siphonic Closet. A W.C. in which the contents of the basin are removed by siphonic action. It has a double seal, between which is a long siphonic leg where the action takes place. See *Syphonage*.

Sirapite. A patent plaster made from gypsum impregnated with petroleum. It sets and dries rapidly, and is harder than lime plaster. It is used for two-coat work.

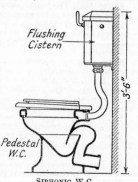

SIPHONIC W.C.

Siris. Two species of dark-brown variegated hardwoods from India, Black and White Siris. They are both excellent decorative woods for superior joinery and for purposes where strength and elasticity are required. Yellow Siris from Australia is similar to White Siris. Weight from 28 to 40 lb. per c. ft.

Sisal. Used in plaster as a substitute for hair.

Sisalcraft. A proprietary *building paper* (*q.v.*).

Sissing. See *Cissing*.

Sissoo. A strong and durable, dark, reddish-brown hardwood from Bengal. Used for structural work. Similar to rosewood.

Sister. Applied to numerous objects that are similar to, but of less importance than, another object.

Sister Hook. A safety clip hook. A pair of hooks fitting close together on the same axis, and used for hoisting purposes.

Site. A plot of ground for building on.

Site Concrete. Surface concrete, *q.v.*

Sitka. The largest and most esteemed spruce, from Sitka Islands, W. Canada. It is very strong for its weight, and used for nearly all purposes where strength, lightness, and clean white appearance are required. Weight about 24 lb. per c. ft. See *Spruce*.

Sitting Room. A room, in a small house, for relaxation. A parlour.

Situ. See *In situ*.

Sitz Bath. A hip bath. One in which a bath is taken in a sitting posture. See *Bidet*.

Sixpartite. See *Sexpartite*.

Size. A powdered animal glue. It is sometimes sold in a jelly form. See *Size Water*.

Size Stick. A gauge used by slaters.

Size Water. Size dissolved in hot water and used as a groundwork to prevent absorption of distemper, paste, etc., and mixed with plaster to retard the setting. It is sometimes used to prevent the raising of the grain of wood with woolly or interlocked grain.

S.J. Abbreviation for *soldered joint*.

Skeeling. Plastering on the slate battens and between the rafters, instead of lathing on the underside of the rafters.

Skeleton Flashings. See *Step Flashings*.

Skeleton Framing. 1. Framed linings for door openings. 2. Any light type of framing to be covered by boards or other material. 3. See *Framed Construction*.

Skeleton Key. A specially-shaped key that will operate a number of different locks.

Skeleton Linings. See *Skeleton Framing* (1) and *Framed Linings*.

Skeleton Steps. Treads without risers.

Skep. A basket for hoisting granular materials, etc.

Skerries. An Irish limestone from Dublin. Grey to black in colour.

Skerry. Pieces of lime in brick earth.

Skew. "On the skew" means in an oblique position. Out of square or twisted.

Skew Arch, or **Bridge.** One running obliquely to its faces. The axis of the arch is not at right angles to the face.

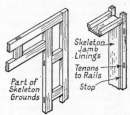

SKELETON GROUNDS AND LININGS

Skewback. The bricks, or stones, forming the abutments for a segmental arch. The surface that is worked radially to form the bed for the first voussoir. See *Centering* and *Arches*.

Skew-corbel, or **Skew-butt.** The footstone for a gable coping. See *Gable* and *Kneeler.*

Skew Fillet. A tilting fillet along a gable wall to raise the slates and so throw the water on to the roof and away from the wall.

Skew Flashing. The lead flashings down a gable wall projecting above the roof.

Skew Nailing. Nails driven obliquely to the surface to give greater security. Alternate nails are often driven in opposite directions.

Skewput. A kneeler, *q.v.*

Skews. Stone slates used for swept valleys.

Skew-table. A kneeler. See *Skew-corbel.*

Skids. Short, small pieces of wood used for packing to a plane surface. Also called sticks or stickers.

Skiffling. See *Knobbling.*

Skilting. Same as skirting, *q.v.*

Skim, To. To apply or remove a thin coating.

Skimming Coat. See *Setting Coat.*

Skinoff. A patent retarding agent for cement. It is applied to the shuttering.

Skin Stress. The maximum, or outer, stress acting in a structural member.

Skintling. See *Scintled.*

Skip. See *Skep.*

Skirting. A projecting finish between wall and floor. It may be of wood, metal, or cement.

Skirting Block. A plinth block, *q.v.*

Skirting Board. A plinth with moulded edge. A moulded baseboard. Also called a *washboard* or *base-plate.* See *Finishings* and *Soldiers.* Also *Metal Trims.*

Skirts. Projecting eaves.

Skylight. A glazed frame running parallel with the roof surface.

Skylux. A patent multiple gear for continuous lights.

Skymeter. An instrument for determining sill ratios for isophotal diagrams in problems concerning ancient lights.

Skyscraper. A tall building of many stories characteristic of American architecture. A recent example in New York is 1,240 ft. high, having 86 stories plus an observation tower 200 ft. high. See *Height of Buildings.*

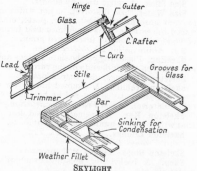

SKYLIGHT

S.L. Abbreviation for *short lengths.*

Slab and Girder. Applied to reinforced concrete floors, consisting of a thin floor slab supported by secondary beams. The floor slab is considered as the top flange of the beam, resisting compression. See *Secondary Beam.*

Slabbing. Squaring a log.

Slab Canvas. Coarse scrim used in plaster slabs.

Slabs. 1. The outer pieces left after squaring a log. 2. See *Slab and Girder.* 3. Thin rectangular blocks of stone, marble, concrete, etc.

Slack Blocks. The supporting wedges used for centres in bridge construction.

Slacking, or **Slacking Off.** Easing the strain on the ropes of lifting tackle as the suspended weight rests in its position. 2. See *Slaking*.

Slade. A sloping path.

Slag. The waste products from a blast furnace, or the scoria from a volcano. The former is used for cement, as an aggregate, for slag wool, and for artificial pumice. Owing to the presence of sulphur, vitrified metal and glass, fluxes, etc., it requires care in use.

Slag Bricks. Bricks made from blast furnace slag. The slag is ground with hydrated lime. They are regular in shape and durable.

Slag Wool. Mineral wool, produced by blowing steam through molten furnace slag. It is one of the best non-conductors, or insulators.

Slaking. Mixing water with quicklime. See *Lime*.

Slamming Strip. 1. An edge strip to the shutting stile of a flush door. 2. A planted door stop to form the rebate.

Slap Dash. A local term for rough-casting.

Slash. Rough or random cutting with tools.

Slash Sawn. Timber cut tangentially to the annual rings. See *Rift Sawn*.

Slat. 1. A stone slate. 2. See *Slats*.

Slate-and-a-half Slates. Specially wide slates, used for valleys, hips, and at the verge of a roof to avoid the use of a narrow piece for forming the bond.

Slate Boarding. See *Sarking*.

Slate Cramp. A dovetailed piece of slate used at the joint between two stones to prevent lateral movement.

Slate Hung. A slated vertical surface.

Slate Nails. The best are copper and composition. The latter are made of 63 per cent copper, 33 per cent zinc, and 4 per cent tin. Inferior nails are: malleable, galvanized, and zinc. For chemical works, chrome-iron alloy or lead pegs are used and bent round steel angles. The nails are from $1\frac{1}{4}$ to $2\frac{1}{2}$ in. long, with flat, thin heads.

Slate Ridge. Consists of a roll and two leaves, or wings. See *Ridge Covering*.

Slate Rule. A wood rod for measuring the courses in stone slates.

Slates. Thin laminae of fine-grained metamorphic clay rock from the Cambrian and Silurian formations. They are grey to blue-black, or green, in colour. The slate is quarried and reduced to small blocks (about 3 in. thick), sawn square, and then split with a wide chisel and mallet to the required thickness, $\frac{3}{16}$ in. to $\frac{1}{4}$ in. The best slates are from N. Wales, Westmorland, Lancashire, S. Wales, Cornwall. Slates were graded as Smalls, Doubles, Headers, Ladies, Viscountesses, Countesses, Marchionesses, Duchesses, Princesses, Empresses, Imperials, Rags, Queens, Random, Ton. It is now more usual to quote sizes. See *Stone or Greystone Slates*.

Slatex. A proprietary bituminous waterproofer.

Slats. Laths. Long narrow slips of wood, as used in Venetian blinds. 2. Cotswold slates.

Slatter. A mason who prepares stone slates.

Sleeper Plate. A wall plate on a sleeper wall.

Sleepers. 1. Strong horizontal timbers resting on the ground as a support for other timbers, rails, etc. 2. Valley boards resting on spars of main roof to carry the feet of the jack rafters. Used instead of valley rafters.

Sleeper Wall. A dwarf honey-combed wall supporting ground floor joists.

Slenderness Ratio. A term used in the design of columns and struts, being the ratio of the effective length to the *radius of gyration*, $l \div K$. For a wood strut it is the ratio of length to thickness.

Sleeve. A coupling for pipes, shafts, etc. A thimble. See *Ferrule*.

Slicing. A method of preparing decorative veneers. It is more economical than sawing. The logs are quartered, cut to length into flitches, and steamed. A long knife slices off the shavings, like a huge plane, from 4 to 140 sheets per inch of thickness. The usual thickness is about $\frac{1}{28}$ in.

Slide. An inclined plane for transporting goods.

Slide Rule. A mathematical calculating instrument, graduated according to the logarithms of the numbers, and also giving trigonometrical values for angles.

Sliding Door. A door that slides in grooves or on pulleys. See *Coburn Fittings*.

Sliding Door Lock. A special type of lock in which the bolt shoots upwards behind the striking plate.

Sliding Sashes. Sashes that slide horizontally on runners. Sashes sliding vertically are more correctly termed *balanced sashes* to distinguish the two types. See *Sash and Frame* and *Yorkshire Light*.

SLIDING DOOR LOCK

Sliding Shutters. Balanced shutters that operate like a *sash and frame, q.v.*

Slings. 1. Ropes or chains used for securing an object when hoisting. 2. Hangers. A piece of wood or metal by means of which another member is suspended.

Slip. 1. See *Slurry*. 2. See *Glaze*. 3. See *Parting Slip*.

Slip Feather. A tongue cut diagonally across the grain to obtain the maximum strength for a grooved joint. Plywood serves the same purpose.

Slip Hook. One that is secure against accidental removal but can be freed easily.

Slip Mortise. 1. An open mortise. 2. A chase mortise (*q.v.*).

Slipper. 1. An extended top to a gully so that the waste pipes do not discharge immediately over the gully. 2. An appliance on the sole of anything to assist sliding. See *Horse* (3).

Slipway. A long sloping way on which a cradle runs from which ships are launched.

Sliver. A splinter. A piece of wood torn away.

Slope of Grain. The angle between the axis of a piece of wood and the direction of the grain. It is stated as the tangent of the angle the slope makes with the edge. The strength of the timber decreases with increase of the angle, over 1 in 40, to as much as 30 per cent for a slope of 1 in 10.

Slop Moulding. Wetting the mould with water, when moulding bricks, to prevent the clay from sticking to the mould. See *Sand Faced*.

Slop Sink. A sink for filthy water used by the chambermaids in hotels and institutions.

Slop W.C. Water closets in which only waste water is used. A pivoted tipper, or tippler, gradually fills, which moves the centre of gravity so that it rotates and flushes the basin. The type is now out of favour, although extensively used 50 years ago.

Slote. A stage trap-door.

Slot Screwed. A method of fixing screws so that the shank is free to move in a slot to allow for expansion and contraction of the wood.

Sludge. Mud. The more solid sewage when separated from the liquid matter.

Sluice. A valve or shutter for controlling the flow of water.

SLOT SCREW

Sluice Gate. A sliding gate or partition used as a dam. A flood gate.

Slump Test. A test of the consistency of concrete. A metal truncated cone, tapering from 8 in. to 4 in. diam. and 12 in. high, is filled with concrete and periodically punned at 3 in. layers. When filled the cone is removed and the subsidence of the concrete measured. A slump of 3 in. for foundation and road concrete is recommended, and 2 in. for cement mortar.

Slurry. A semi-fluid mixture. The term is especially applied to the mixture of materials for the manufacture of Portland cement.

Slype. See *Tresaunte*.

Sm. Abbreviation for surface measure.

Small Knot. One between $\frac{1}{2}$ in. and $\frac{3}{4}$ in. diameter.

Small Tool. A plasterer's steel tool, of varying shapes, for making good to mitres, etc.

Smalls. Slates about 12 in. × 6 in., or 10 in. × 8 in. About 530 per square at 3 in. lap.

Smalt. A blue pigment.

Smell Test. See *Scent Test*.

Smelting. Fusing iron ore with a flux in a furnace to separate the iron, which is run off in a fluid state.

Smith and Founder. Applied in Quantities to forged and cast metal work.

Smith Floor. A patent fire-resisting floor with two-way reinforcement and pre-cast blocks.

Smithing. The craft of working, or forging, iron. The craftsman is a *smith*, or blacksmith.

Smith's. 1. A patent interlocking tile. 2. Proprietary waterproofers.

Smithy. A blacksmith's workshop.

Smoke Shelf. A ledge at the back and over a fireplace to modify down-draught.

Smoke Test. Injecting dense smoke into drains, after stopping the ends, to test for a leakage. See *Drain Testing*.

Smudge. 1. A black mixture used by the plumber for painting lead so that solder will not adhere. It is also called *soil*, and consists of lamp-black and glue size; or proprietary mixtures may be obtained. 2. Paint residues mixed together and used for painting the insides of gutters, etc.

Snack Bar. A counter at which light refreshments may be obtained.

Snag. Applied to an obstruction or drawback that holds up working processes.

Snail. A short curly decorative figure in walnut, etc.

Snake Stone. Ammonite. Used for polishing the surface of imitation marble, etc.

Snakewood. A decorative hardwood from Guiana. It is warm hazel in colour with black spots.

Snape. An oblique bevel to the end of a timber to fit a splayed surface.

Snaped. Logs with the ends tapered for easy removal from the forest.

Snap Head. See *Rivets*.

Snapped Flints. Flints broken across the middle of their lengths and used for facing flint walls.

Snapped Headers. See *Blind Header*.

Snapping the Line. Using the *chalk line* (*q.v.*).

Snatch Block. A pulley block with a hinged side strap which is moved so that the rope can be placed in the pulley groove. See *Mast*.

Sneck. 1. The small stone that levels up the riser (*q.v.*) in snecked rubble. See p. 300. 2. A local term for a door latch, or fall bar.

Sneck Head. The catch for a latch.

Snecked Rubble. Masonry walls built of uncoursed roughly squared stones, irregular in size, but forming a unit, of riser, leveller, and sneck.

Sniff Hole. An air-hole in the dome of a flushing cistern to terminate the flush.

Snifting Valve. Used in water mains to allow air to escape where a pipe line rises above the hydraulic gradient.

Snipe's Bill. A wooden plane for working sunk mouldings under projecting members.

Snips. Shears for cutting tin or zinc.

Snow Boards. 1. Long strips of wood, or lags, about 3 in. × 1 in., nailed to 3 in. × 2 in. bearers and placed over roof gutters. The lags are spaced

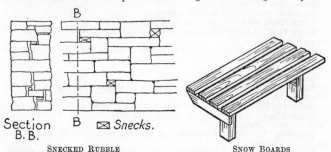

Section B.B.   B   ⊠ *Snecks.*

SNECKED RUBBLE   SNOW BOARDS

about 2 in. apart so that as the snow melts it drops through into the gutter. Also called Snow Slats and Snow Ducts. 2. A long board, on edge, in front of an eaves gutter.

Snowcem. A proprietary waterproof cement paint for brickwork and concrete.

Snowcrete. A registered name for a white or cream Portland cement prepared by the Cement Marketing Co., Ltd.

Snow Guards. Horizontal boards placed on edge, and about 2 in. above an eaves gutter, to check the fall of snow from the roof.

Snug. Applied to anything that is a good fit. Also see *Lug.*

S.O. Abbreviation for *supplied only.* Not fixed.

Soakage. Water passing through by saturation.

Soakaway. A pit of broken stones, etc., to collect the drainage from land drains.

Soakers. Pieces of lead formed into right angles and bonded with slates to make a watertight joint between a wall and the roof. Stepped flashings are used over the soakers.

Soap. A solution of soap and common alum used as a surface waterproofer for cement, and known as the Sylvester process.

Soaps. An alternative name for queen closers.

Soapstone. See *Talc.*

Socket. 1. The eye of a dovetail joint. See *Dovetail.* 2. A cavity or opening into which anything is fitted endwise, as the faucet to a pipe. 3. Also see *Pivot.*

Socle. A plinth. A plain block to a pedestal.

Sodium Acetate. Used for fireproofing timber.

Sodium Aluminate. Used for water softening.

Sodium Chloride. Common salt.

Soff. Abbreviation for *soffit.*

Soffit. 1. The horizontal lining at the head of an opening. 2. The underside of an arch, stairs, etc. See *Eaves, Treads,* and *Arch.*

Soft Pine. Applied to several N. American and Canadian pines, especially to Ponderosa pine.

Softwoods. Applied to woods from coniferous trees. The cellular structure is simpler than that of hardwoods. The softwoods used in this country include: Cedar, Chil, Cypress, numerous species of Fir, Hemlock, Kauri, Larch, Matai, several species of Pine, Pitch Pine, Podo, Rimu, Sequoia, several species of Spruce, Tamarack, Totara, Yellowwood, Yew.

Soil. 1. See *Smudge*. 2. Sifted ashes or cinders, used, when making London stock bricks, to reduce contraction and give colour and fuel. 3. The upper stratum of earth; mould or loam. 4. The contents of a cesspit.

Soiling. Placing a layer of sandy loam or cinder dust over the clay in the settling tanks, when brick-making.

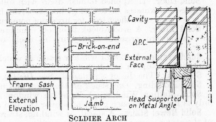

SOLDIER ARCH

Soil Pipes. Vertical pipes from W.C.s, etc., for conveying waste matter to the drains. They are of heavy cast iron or lead, and should be air-tight, except for the out-let vent above the eaves.

Sol. A proprietary hard-gloss paint.

Solar. See *Sollar*.

Solarium. A room at the top of a building designed to obtain maximum sunshine, and often equipped with artificial sunlight apparatus.

Solcheck. A patent tile to protect the asphalt on flat roofs.

Solder. To unite metals with a metallic substance in a state of fusion which hardens on cooling and makes a secure joint. See *Brazing*, *Plumber's Solder*, and *Flux*.

Soldered Dot. Used for fixing large vertical surfaces of sheet lead. A spherical hollow is scooped out of the boards and the lead dressed into it and secured by a screw and washer. The hollow is then filled with solder.

Soldering Iron. See *Copper Bit*.

Soldier Arch. A flat arch constructed of uncut bricks on end. The joints are recessed for a cement joggle. Wire hangers are fixed in the joggles and hooked round the reinforcement in the *in situ* concrete lintel. Many other methods are used to give support.

Soldier Course. A course of bricks on end.

Soldiers. 1. Short vertical grounds (S), to which joinery work is fixed, especially skirting boards. 2. Heavy vertical timbers placed across several walings and strutted. This is done in stages, to remove the lower struts for a deep excavation, as the wall is built. Also see *Puncheon* (3). 3. Upright members carrying vertical boarding for formwork. 4. Profiles.

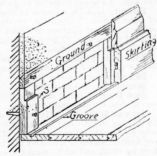

SOLDIERS FOR SKIRTING BOARD

Sole. See *Sill*. The bottom surface of an object, as a plane.

Sole Piece, or **Plate.** The sleeper carrying the feet of *raking shores, q.v.* Also see *Sleeper*.

Solid Bossed. Angles, etc., in sheet lead, bossed to the required shape, instead of soldering the projections. See *Bossing*.

Solid Bridging, or **Strutting.** Short pieces of board driven tightly between floor joists to stiffen them. See *Herring-bone Strutting*.

Solidcor. A registered construction for flush doors.

Solid Door. 1. Applied to flush doors with solid core. 2. A door built up of three thicknesses of t. and g. boards with the inside layer horizontal. It is covered with sheet iron for fireproof construction.

Solid Floor. A wood-block floor on concrete. Wood floors in which the joists are pressed tightly up to each other, for a fire-resisting floor.

Solid Frame. A door or window frame in which the stiles, etc., are solid, and not built-up, or cased.

Solidium. The die of a pedestal. See *Die* (1).

Solid Moulding. One stuck on the solid, and not *planted*.

Solid Mullion Window. A Venetian window in which the centre sash only is hung. The sash cords pass over the side sashes.

Solid Newel Stairs. A spiral, or winding, stairs, in which the inside end of the stone steps are worked into a cylinder, forming a continuous cylindrical newel. See *Winding Stairs*.

Solid Panel. A flush panel (*q.v.*).

Solid Partition. A partition without a cavity. The term is usually confined to partitions formed by plastering on both sides of some type of metal lathing. Temporary studs are erected against the metal lathing whilst the opposite side is plastered, unless it is strengthened by other means, such as T-irons. See *Expanded Metal*.

Solids. A geometrical term for cone, cylinder, cube, prism, pyramid, prismoid, sphere, spheroid, ellipsoid. See *Polyhedra*.

Solid Steps. Rectangular stone steps. Solid, built-up, wooden steps are used for resisting fire.

Soligen Driers. A quick-drying agent mixed with an organic acid and used for paints, etc.

Solignum. A proprietary preservative for timber. It has the qualities of creosote without the offensive odour.

Solite. A registered construction for flush doors.

Sollar. An open gallery, at the top of a building, exposed to the sun. An upper chamber or garret.

Solvent. 1. A liquid used to make the mixing and applying of paint easier. Turpentine, or its substitutes, is the common solvent. 2. Any mixture or liquid used for removing old paint, instead of burning-off by a blow-lamp. Any substance that will dissolve another substance.

Somerville Floor. A patent fire-resisting floor consisting of hollow reinforced concrete tubes supported by secondary beams.

Sommer. See *Summer Beam*.

Sommering. The radiating joints of a flat, brick arch.

Soot Door. A small cast-iron door fixed at a sharp bend in a smoke flue, for cleaning purposes.

Sopwith Staff. A levelling staff used in surveying. It is usually 14 ft. long and divided into three sections so that the parts slide over each other when not in use. The feet are marked by large red figures, tenths of a foot by black figures, and hundredths of a foot by black and white marks alternately.

Sorel Cement. Magnesite cement. It is one of the strongest

SOPWITH
STAFF

cements, and used in composition floors. It consists of magnesia paste, MgO, and magnesium chloride solution, $MgCl_2$.

Sorting. Same as *grading*.

Soss Hinge. A patent invisible hinge. It is mortised into the centre of the edge so that no part is seen when the door is closed.

Soufflet. A form of quatrefoil in Flamboyant tracery.

Sough (suff.). A gully grating. A drain.

Sound Boarding. Boards between joists to carry pugging.

Sounding Board. A canopy to a pulpit used where the acoustic properties of the building are not good. The best results are given by one of parabolic form.

Sound-proofing. Insulating floors and walls to prevent the passage of sound. One principle of sound insulation is to trap still air in small spaces or pores. Hence porous or cellular materials are effective in varying degrees. Usually other qualities are required in addition, such as fire and vermin proof. The most efficient materials are : slagwool or silicate cotton, Cabot's quilt, pulp wall-boards, eel-grass, cellular plaster, granulated cork, hair felt, seaweed, etc.

Souring. Storing clay paste for several days to allow the decomposition of vegetable matter to increase the plasticity.

Southern Pine. Pitch pine, *q.v.*

Southland Beech. New Zealand beech.

Southwood. A patent reversible window for easy cleaning and replacing of glass.

Sow. See *Pig Iron*.

Sowdal. See *Saddle Bars*.

S.P. Abbreviation for *soil pipe*.

Spaced Slating. See *Open Slating*.

Space Frame. The name given to a structural frame in mechanics when the members are not all in one plane.

Spader. A small tool for consolidating reinforced concrete.

Spall. To break away the edges of stone.

Spalls. Pieces of broken stone used to fill the spaces between irregular stones in rubble walls.

Span. The distance between the supports for a beam, truss, or arch. See *Effective Span* and *Arch*.

Spandrel, or **Spandril.** Any triangular filling to bring to a level line. The triangular framing, or boards, under the outer string of a stairs. The filling to the right and left of an arch to the level of the crown. See *Inserted Tenon*.

Spandrel Steps. Stone steps triangular in section, either to give more headroom or for ornamentation. The underside may be moulded, or worked to a flush or to a broken soffit. See *Tread*.

Spanish Blind. An outside blind working on spring rollers and with a sliding hood.

Spanish Tiles. These consist of *under* and *over* tiles, each tapering and half-round in section. The under-tile is laid between vertical battens. They are 14 in. long with the under-tile tapering from 9 in. to $7\frac{3}{4}$ in. diam., and the over-tile $7\frac{1}{4}$ in. to $5\frac{3}{4}$ in.

Spanner. A lever for turning a nut. Also called a *key* unless it is adjustable.

Span Piece. A local term for a collar beam (*q.v.*).

Span Roof. The most common type of roof, having equal pitch on each side. The term is used to distinguish from a lean-to.

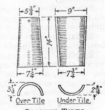

SPANISH TILES

Sparge Pipe. A perforated water pipe for flushing a urinal.

Spars. 1. Common rafters. 2. Small scaffold poles. 3. See *Fluor Spar.*

Spathic Ores. The more pure ferrous carbonate ores from which most British pig iron is smelted.

Spearpoint. A patent clip for fixing wood joists to concrete.

Specification. A statement describing the quality of the work and materials for a particular piece of work

Specific Gravity. *S. G.* The relative weight of a substance to that of an equal volume of water. Water weighs approximately 62·5 lb. per c. ft., hence if the S.G. of a certain stone is 2·2, the weight is 62·5 × 2·2 lb. per c. ft.

Specific Heat. The specific heat, s, of a substance is the number of calories required to raise the temperature of one gram through one degree centigrade. See *Heat.*

Specky. A blown sheet glass similar to Cathedral but with added imperfections to give an antique effect.

Spectrum. A coloured band produced by analysing light through a prism or spectroscope.

Speculum. A metallic reflector used in surveying instruments.

Spelter. 1, Hard solder. See *Brazing.* 2. Zinc in ingots.

Spent. Applied to materials that have lost their essential qualities, as cement that has been exposed and lost its setting power.

Spere. A screen or shelter to an entrance.

Spers. Same as spars, *q.v.*

Sperver. The frame at the top of a canopy.

Sphere. A ball, or globe. A solid, formed by the revolution of a semi-circle about its diameter. Vol. = $\frac{4}{3}\pi R^3$. Surface area = $4\pi R^2$.

Spherical. Having the form of a sphere; or part of a sphere.

Spheroid. A solid formed by an ellipse revolving about one of its axes. When about the major axis it is *prolate*, and when about its minor axis it is *oblate*. See *Ellipsoid.*

Spheroidal. Having the form of a spheroid, or part of a spheroid.

Spider Gauge. A grasshopper gauge.

Spiegeleisen. Cast iron, containing manganese and carbon, used in the production of steel by the Bessemer process.

Spigot. The plain end of a pipe that fits into the faucet, or socket, end of another pipe.

Spike Grid. See *Timber Joint Connectors.*

Spikes. 1. Strong nails over 4 in. long. 2. Any long pointed iron bar, used as a fastening in timber construction.

Spile. 1. A splinter of wood. 2. A plug. 3. A wood pile. 4. Slender rods used for fencing.

Spiling. Scribing equal distances from an uneven surface by means of a rule or light rod, instead of by compasses.

Spills. *Splinters* on metal bars due to impurities: phosphorus and sulphur.

Spindle. 1. A moulding machine in which the cutters are carried by a vertical spindle, above the table top. 2. A steel bar to carry the knobs for a lock. 3. Any long slender pin forming an axis of rotation.

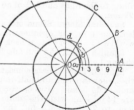

ARCHIMEDIAN SPIRAL

Spinner. A short piece of wood used for twisting wires in fencing.

Spiral. 1. Having the form of a screw, or a helix. 2. A gradually expanding curve. A curve that continually recedes from the centre about which it revolves.

Spiral Grain. Due to the fibres being spirally aligned in the growing tree, which usually makes the wood difficult to work.

Spiral Stairs. A circular staircase, with the treads consisting of winders only, round a central newel. See *Winding Stairs.*

Spire. A tall pyramidal roof to a tower.

Spirelet. A spire springing from a roof instead of from a tower. A small spire.

Spirit Level. A long piece of hard wood containing a glass tube nearly filled with spirit. The tube is slightly bent and fixed with the round side uppermost so that the air-bubble in the spirit is at the centre when the level is perfectly horizontal. See *Level.*

Spirit Stain. A stain for wood made of methylated spirits and the required colouring matter. It does not raise the grain like a water-stain.

Spit. A spade-depth deep. A term used when removing the top earth from a site.

Spitter. A short pipe or outlet, from a cesspool or eaves gutter, to the downpipe.

Splash Board. A board placed on its edge against the wall, to prevent the rain from splashing the dust from the scaffold on to the wall. A weather board to an external door.

Splats. Cover strips for the joints of wall-boards.

Splay. A bevel or slope.

Splayed Grounds. Those with a bevelled edge to form a key for the plaster.

Splayed Heading. A bevelled heading joint for floor boards, instead of a butt joint.

Splayed Jambs. Jambs, or reveals, not at right angles to the face of the wall.

Splayed Joint. A joint in which the two pieces are bevelled and overlap, without increasing the sectional area.

Splayed Skirting. One with the top edge bevelled instead of moulded.

Splay Knot. A spike knot, one cut lengthwise of the branch.

Splice. A lengthening joint; but more especially applied to ropes, when the joint is made by interweaving the strands of the two pieces. A splayed joint.

Spline. 1. A thin *feather* placed obliquely into a mitred joint to strengthen it. A saw-cut is made, and the spline glued and inserted in the saw-cut. 2. A flexible rod.

Split Bill. A metal fixing, leaded or cemented into the wall, and having two projecting sharp spurs, or points, that are turned outwards to provide the fixing.

Split Course. A course of bricks less than the ordinary thickness of the bricks. The bricks are cut lengthways with bolster or trowel.

Split Heads. Supports for plasterer's scaffolds. They consist of two pieces of floor boards splayed outwards at the feet and cleated. A notch at the top carries a bearer that in turn carries the scaffold boards.

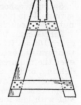

SPLIT HEAD

Split Pin. A divided iron pin that, when inserted in a hole, has the ends opened out to keep it in position.

Split Pipe. A drain pipe cut lengthwise. A channel.

Splits. Bricks less than the ordinary thickness, as used in a split course.

Spokeshave. A woodworking tool for shaping curved surfaces.

S-polygon. A diagram derived when finding the extreme stresses produced by unsymmetrical bending, in the design of structural members.

Spools. Cast-iron separators placed between two timbers bolted together, to prevent the moisture from collecting between the timbers. They are used between the heads of the sticks for pile trestles, in timber bridge construction.

Spot Board. A plasterer's board, about 3 ft. square. It rests on a stand about 2 ft. 3 in. high, and is used for working up the stuff before applying it to wall or ceiling.

Spot Levelling. A term used in surveying. The site is divided into imaginary squares and the levels of the corners taken.

Spouting. A northern term for eaves gutters. They are made of long lengths of red deal hollowed out and moulded on the front.

Spouts. 1. Troughs or pipes for conveying liquids. 2. Shafts or ducts in flour mills, distinguished as dog-legged, breeches, diamond shaped, etc.

Spraying. Applying paint by a special appliance that sprays the paint evenly over the surface. The spray gun is worked by compressed air, and is very speedy for large surfaces with liquid preservatives, waterproofers, lacquer, varnish, etc.

Spreaders. Horizontal timbers of small section fixed between layers of shuttering to keep them the required distance apart when pouring the concrete between.

Sprig. A small nail with very small head. Glaziers' sprigs, or points, are square in section and without head. See *Nails*.

Sprigging. Temporary nailing of anything in position.

Spring. 1. A thin piece of steel so tempered that it springs back to its original shape when pressure is removed. 2. See *Apophyge*. 3. A pipe bend, for a turn less than 90°. 4. See *Sprung*.

SPRING BOWS

Spring Beam. A tie beam. A beam with considerable deflection.

Spring Bows. Compasses for small circles, with a screw for fine adjustment.

Springer. 1. The lowest voussoir of an arch. See *Arches*. 2. See *Footstone*. 3. A triangular fillet nailed to a trimmer as an abutment for the trimmer arch.

Spring Floor. A specially constructed dance floor, *q.v.*

Spring Hinge. See *Floor Spring*.

Springing. The line, or level, from which an arch springs, or commences to turn. See *Arch*.

Spring Latch. See *Rim Latch* and *Night Latch*.

Spring Snib. A sash fastener.

Springtree. A carriage to a flight of stairs.

Spring Washer. A washer in the form of a spring. It is placed under the nut to lock the nut. See *Lock Nuts*.

SPRINGER FOR VAULT

Sprinklers. A system of pipes with a sprinkler head, or drencher, to every 100 sq. ft. of floor area. A soft solder connection is melted when the temperature exceeds 150° F., which opens a valve and the water sprays over the required area. It is the most efficient method of automatic fire fighting, and is usually embodied in fire-resisting construction.

Sprocketed Eaves. Projecting eaves formed by sprocket pieces.

Sprocket Pieces. Short rafters fixed at the foot of the common rafters and with less inclination to make a break in the surface of the roof and to form projecting eaves.

Sprockets. Teeth on a wheel that engage with the links of a chain.

Sprowl. A registered chimney scaffold.

Spruce. *Picea spp.* White deal; but more especially applied to Silver Spruce, Sitka, or *Picea sitchensis*. It is clean, white, uniform, and easily wrought. It weathers badly but is very elastic and strong for its weight. Used for constructional work, joinery, flooring, shuttering, etc. Weight from 24 to 34 lb. per c. ft. See *Fir* and *Whitewood*.

Sprung. A term applied to a distorted curved member, and to a flexible piece that is bent to force it into position.

Spud. A dowel in the foot of a door post. See *Dowel*.

Spun Glass. 1. See *Crown Glass*. 2. Similar to slag wool and used for insulating.

Spur. 1. A strut. 2. A sharp projection. 3. The marking point to a gauge. 4. A spur-wheel is one with projecting teeth on the rim. See *Pinion*, *Rack*, and *Winch*.

Sq. Abbreviation for *square*.

S.Q. Abbreviation for *squint quoin*, and for *silvering quality*, the first quality of plate glass.

Square. 1. A tool for testing right angles. 2. A measurement of area containing 100 sq. ft. 3. See *Steel square*. 4. A pane of glass.

Square Billet. Ornamentation characteristic of Norman architecture. See *Billet*.

Square Cut. A system of preparing the wreaths of handrails, as distinguished from the bevel cut system.

Squared Rubble. Walling built of irregular stones, but having rectangular faces.

Square Head. Applied to a boundary wall built up to a building without bonding.

Square Joint. A butt joint.

Square Panel. A panel without mouldings.

Square Shoot. 1. A wood down-pipe for rain-water. 2. See *Shoot*.

Square Staff. A square fillet used, instead of a bead, for an angle-staff.

Square-to Roof. A North-light roof truss, *q.v.*

Square Turning. Ornamenting balusters, newels, etc., on all four sides, instead of by circular turning. The process may be done on a machine.

Squaring. 1. Working out the sizes of the different items on the dimensions paper into superficial areas, etc., when preparing a bill of quantities. 2. Correcting framing so that it is rectangular.

Squeegee. A roller appliance for spreading plastic material.

Squeeze. Obtaining the profile of a moulding or enrichment by pressing plaster into the members, in the form of a block. The squeeze, when set, is used to mark the zinc for the running mould, or to form a mould for a gelatine mould, in repair work.

Squinch Arch. A pendentive arch across an angle to carry the masonry above, as when a square tower is brought to an octagonal one.

Squint. An oblique opening in a wall. In a church it is directed towards the altar.

Squint Quoin. A quoin not forming a right angle.

Squints. Specially moulded bricks for squint quoins.

St. Abbreviation for *straight* and *struck*.

S.T. Abbreviation for *surface trenches*.

Stabil. A patent steel clip for securing concrete reinforcement in position.

Stability. 1. The resistance against being overthrown. The term is especially applied to walls and masonry structures. The three forces acting for equilibrium, or stability, are : weight of wall, pressure on wall (wind, water, earth, etc.), and upward reaction of ground. For overturning : moment of weight = moment of pressure. 2. See *Moisture Content*.

Stable. 1. A building for horses or cattle. **2.** A property of certain woods that enables them to resist change of volume with change of moisture content.

Stack. 1. See *Chimney Stack*. **2.** A rain-water pipe from eaves to ground. See *Down Pipes*. **3.** A pile of timbers as arranged for seasoning. See *Air Seasoning*.

Stadia Points. Hair lines on the diaphragm of a theodolite or level used in surveying. They are used to calculate the horizontal distance, D, by means of the interception, x, of the hair lines on the staff. Then $D = 100x + C$ where C is a constant that varies from 1 ft. to 2 ft. according to conditions. This method of calculating the distance is called *tacheometry*.

Stadium. A running track surrounded by stands. The name implies athletics, but modern usage has extended it to cycle and dog racing.

Staff. 1. A stick or pole. **2.** See *Sopwith Staff*. **3.** A flagstaff.

Staffa. A registered bender for steel reinforcement for concrete.

Staff Bead. See *Angle Bead* and *Guard Bead*.

Staffordshire Blues. Bricks used in engineering work, foundations, and in damp positions. They are dark blue or nearly black in colour, due to oxide of iron. They are hard, dense, and nearly impervious, and may be obtained as wire-cuts, or pressed.

Stage. 1. An elevated platform, as in a theatre. A scaffold. A story, step, or floor. **2.** The divisions between the offsets in Gothic buttresses. **3.** The horizontal divisions in windows.

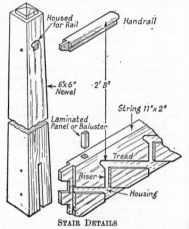

STAIR DETAILS

Stage Door. A back entrance to a theatre for staff, actors, etc.

Staggered. 1. Studs placed irregularly in a partition so that the two faces are independent, to make it non-conducting. **2.** Not in a straight line, as zigzag riveting. See *Hollow Partitions*.

Staging. A scaffold or temporary platform.

Stain. A dye to give the required colour to a material. For timber it may be water, oil, or spirit stain.

Stained Glass. Windows glazed with pieces of differently-coloured glass to build up a particular design. They are common in churches and public buildings, and called *lead lights*. See *Cames*. The glass is stained by metallic oxide in production, or flashed, or painted.

Stainless Steel. Extensively used for shop fronts because it does not oxidize, or rust, and retains its bright polished appearance in all weathers. See *Rustless Steel*. Contains about 20% chromium, 8% nickel, and a small percentage of one or more of the following elements: silicon tungsten, molybdenum, titanium, and copper, according to special requirements.

Staircase. The apartment containing the stairs. The term includes all the materials used in the stairs and finishings.

Stair Horse. A carriage piece, *q.v.*

Stairs. A succession of steps connecting two points at different levels.

They may be of wood, stone, concrete, or metal. The various types are bracketed, close newel, continuous, dog-leg, double-return, half-turn, geometrical, helical, newel, open newel, quarter turn, solid newel, spiral, straight, and winding. See *Step* and *Tread*.

Stairway. The well of a stairs.

Staith. An elevated wharf for shipping coal, etc.

Stakes. Pieces of wood that are pointed for driving into the ground.

Staking Driving stakes at the required positions and levels.

Staking Out. Setting out a plot of ground for building purposes, by driving stakes in salient positions.

Stalk. 1. The vertical part of a reinforced concrete retaining wall. See *Retaining Wall*. 2. A chimney stack.

Stall. 1. A partly enclosed seat in the choir of a church. 2. A seat at the front of the pit in a theatre. 3. A division in a stable for one animal. 4. A table for the display of goods. 5. A bay, furnished with bench, between the book presses in a library.

Stall Board. The stout sill on which the window sash rests in a shop front.

Stall-board Lights. A window under the stall board for a shop front to light a basement.

Stall Riser. The riser between the pavement and stall board to a shop front. It may be of marble, granite, tiles, wood, etc.

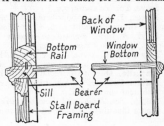

STALL BOARD

Stanchion. An iron or steel post forming an intermediate support for a beam. They are usually encased in fire-resisting materials. An upright support. 2. Sometimes applied to the mullions of a window. 3. A pair of bars for confining cattle in a stall.

Stand. 1. A table, shelf, rack, etc., on which things are placed: hat, umbrella, music, lamp, etc. 2. A frame or support. 2. A tier of seats for spectators, usually out-of-doors, or temporary.

Standard. 1. A temporary wooden post, as to a scaffold. See *Derrick Tower*. 2. A measure of timber for soft woods. See *St. Petersburg* and *London Standards*. 3. A bench end, *q.v.* 4. A local term for a door post and for quarterings. 5. A patent roof glazing.

Standard Grey. See *Granite*.

Standard Measurement. An agreed method of measuring building work, published by the Surveyors' Institution.

Standard Mix. Concrete in the proportion of 1·2·4, giving a safe crushing strength of 750 lb. per sq. in. See *Concrete*.

Standard Notation. The symbols used in structural design as agreed upon by the professional associations concerned.

Standard Steel Sections. See *British Standard Sections*.

Standard Wire Gauge. A standard, legal gauge for wire, ranging from 7/0 to 50. Also called the Imperial or British S.W.G.

Stand Sheet. A glazed sash without a frame.

Stanford Joint. A patent joint for drain-pipes. A tapered bituminous collar on the spigot ensures the pipes being in alignment.

Stanley. A registered name for several joiner's tools, a steel square, and for drawing and surveying instruments.

Staple. A metal loop to receive a hook, pin, bolt, or padlock. See *Hasp* and *Box Staple*.

Star Drill. See *Drills and Chisels*.

Starlings. Piles used as a protection for bridge piers.

Star Shake. Radiating shakes at the heart of a balk or log.

Starter Frame. A shallow 2-in. box used in shuttering to allow for the base of a wall or column to be cast with the mass concrete or floor.

Startex. A proprietary wood-fibre wall-board.

Statics. A branch of mechanics that deals with the action of forces in maintaining equilibrium. See *Graphic Statics*.

Station Poles. See *Ranging Poles*.

Station Roof. A roof with central support only as used on island platforms in railway stations. See *Roofs*.

Statuary Marble. A white marble from Italy used for statues, etc. The first quality is uniform and free from veins and blemishes.

Statuesque. Ornamentation having the character of a statue.

Stave, To. To caulk, *q.v.*

Staves. 1. Narrow boards used to build up a curved surface. See *Jacket*. 2. The rungs of a ladder. 3. Cylindrical bars in a rack.

Stay. A brace or tie to steady a post or framing. See *Derrick Crane*. Also see *Casement Stay*.

Stay Bar. 1. A metal bar across the tops of the mullions to tracery windows. 2. The bars to strengthen lead lights.

Staybrite. One of the registered makes of stainless steels (*q.v.*).

Stayset Damper. A patent damper for regulating the flow of air in mechanical ventilation.

Steam. See *High-* and *Low-pressure Systems*.

Steam Cleaning. A method of cleaning the stonework of existing buildings. See *Water Jet*.

Steaming. Placing wood in a steam bath to soften it and make it more plastic for bending, slicing, or peeling.

Steam Navvy. A mechanical excavator.

Steane House. A prefabricated house of steel, concrete, briquette panels, etc.

Stearates. Compounds of stearic acids used as admixtures for natural cements.

Steel. See *Mild Steel*.

Steel Casements. Casement windows made of steel. A large variety of standard shapes and sizes is obtainable, with or without wood surrounds. See *Metal Casements*.

Steel Frame Construction. See *Framed Structure*.

Steel Lathing. Any form of expanded metal, etc., used for reinforcing plastic materials. See *Solid Partitions* and *Expanded Metal*.

Steel Sections. See *British Standard Sections*.

Steel Square. A large graduated square used by the carpenter for obtaining lengths and bevels in timber framing. The blade is about 24 in. × 2 in. and the tongue about 18 in. × 1½ in. A " take down " square has the two arms loose jointed for easy packing. The graduations are based on trigonometry.

Steel Wool. A bunch of fine steel threads used for cleaning paint off metal, etc.

Steening, or Steining. The brick, stone, or concrete lining to a well or cesspool.

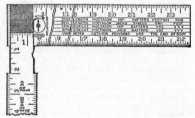

STEEL SQUARE

Steeple. A pyramidal cover to a tower. It is not so slender as a spire.

Steer. A stack of timber standing vertically, for seasoning.

Stelai. The pillar-like monuments to early Grecian tombs.

Stellar Vaulting. See *Lierne Vaulting.*

Stemming Piece. 1. A short piece of joist between chimney breast and trimmer. It is used instead of a furring piece, to carry the ends of the floor boards. 2. Solid strutting to joists.

Stemple. A supporting timber to a platform.

Stench Trap. See *Trap* (1).

Stencil Work. Painted ornamentation formed by painting over a thin plate cut out to the required design. It is the usual method for repetition work.

Step. A unit in a stairs. A rise from one level to another by means of one tread and riser. The proportions of rise and going may be obtained from: Rise × Going = 66 in. The various kinds are: balanced, bull-nose, commode, curtail, dancing, drum end, flier, return, solid, winder. See *Going* and *Tread.*

Step Irons. Strong iron staples built into the joints of a manhole for the descent of workmen.

Step Joint. A structural notched joint for two timbers making an angle with each other. A double step is used to prevent weakening the member unduly.

STEP IRON

Step Ladder. Steep stairs with narrow strings and treads and no risers. It has a framed stay hinged to the back for domestic use.

Stepped Flashings. Lead or zinc flashings let into brickwork joints to make the joint between wall and roof water-tight. They are in short lengths,

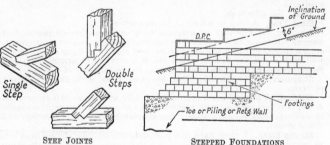

STEP JOINTS   STEPPED FOUNDATIONS

and step down the wall to suit the inclination of the roof. See *Soakers.*

Stepped Foundation. See *Benched Foundation.*

Stepped String. A cut string, *q.v.*

Stepped Voussoirs. The stones of an arch square-shouldered at the top to suit the horizontal courses of bricks or stones. See *Voussoirs.*

Stepping. See *Benching* (1).

Step Turner. A plumber's tool made of wood, used for turning sheet lead at right angles, as for stepped flashings.

Stere. A metric cubic measure of 35·3 c. ft.

Stereobate. See *Stylobate.*

Stereotomy. Geometry dealing with the sections of solids.

Sterilizer. For the purification of water. There are many methods, such as the "De Clor" filtration plant, the "R.U.V." process, the "Ozonair" process, chloramine, etc. Sand filtration under scientific supervision is considered satisfactory.

Sterling. See *Starling*.

Steuben. A registered moulded decorative glass. Claimed to be the clearest crystal glass.

Stewart's Piles. *In situ* piles formed by driving a cylindrical shell, into which the concrete is poured as the shell is withdrawn.

Stick and Rag Work. Fibrous plaster (*q.v.*).

Stickers. Thin pieces of softwood used for packing between boards whilst seasoning, to allow air to circulate round the boards. Also called *skids* and *sticks*.

Sticking. The act of shaping mouldings with a plane or machine.

Sticking Board. A framed board for steadying small pieces whilst being moulded, or *stuck*.

Sticks. An American term for timbers circular in section and used for posts in trestle bridge construction; or for any single piece of timber. See *Stickers*.

Stiff. Applied to a beam able to withstand bending under a load. Rigid, or the reverse of flexible.

Stiffeners. Angle or T-irons used to stiffen the web of a built-up plate girder.

Stiles. The vertical members on the outer edges of a piece of framing.

Still Room. 1. A room for distilling liquors. 2. A housekeeper's pantry.

Stilted. Applied to an arch that rises vertically above the impost for a short distance before it is *turned*. See *Arches* and *Vaulting*.

Stilting. Raising the transverse ribs to suit the diagonal ribs in unequal intersecting vaults. Raising the springing of one of a pair of intersecting vaults.

Stilts. See *Starlings*.

Stink Trap. A trap to a drain. An *interceptor*.

Stippling. 1. Forming an irregular or granular surface on plaster and cement surfaces with a brush. The method is applied to paint, distemper, etc.; a special brush, a *stippler*, is used, and applied at right angles to the wall with a *stabbing* action, which leaves a mottled or granular surface. Special plates or cylinders of rubber may be used instead of a brush.

Stippolyte. A registered rolled obscured glass with small irregular pattern.

Stirrups. A piece of wood or iron from which another piece is suspended. A tie. Vertical straps tying together the top and bottom reinforcement in concrete, and resisting shear. See *Continuity Stirrups, Centering,* and *Flitch Beam*.

Stirrup Strap. A strap fixed to a post and carrying a horizontal member, as between king post and tie beam in a roof truss. In this case the strap is fixed to the post by gibs and cotters. See *Cottered Joint*.

Stoa. A detached collonade or a portico to ancient Greek buildings.

Stobs. Rough uprights for fencing, or fixings for the feet of posts.

Stock. 1. The principal supporting parts of tools, as the stock of a plane, etc. 2. The timbers supporting a ship during building. 3. The frame carrying the dies for screw-cutting. 4. See *London Stock*.

Stockade. A palisade. A protective enclosure formed by stakes.

Stock Board. A board, with a *kick* to form the frog, placed at the bottom of the mould, when brick moulding by hand.

Stock Brush. A plasterer's brush for damping surfaces with water.

Stockholm Tar. A timber preservative obtained by destructive distillation of the waste from pines and firs after conversion from tree to timber.

Stock Lock. A heavy wooden casing containing a bolt operated by a key. A dead lock.

Stocks. Clamp-burnt bricks. See *London Stocks*.

Stock Sizes. Sizes of materials in common demand and kept in stock by building firms and timber yards.

Stockyard. A large yard with pens, stables, sheds, etc., for cattle awaiting slaughter or sale.

Stoep. A terraced verandah in front of a house (South Africa). See *Stoop*.

Stoke Ground. A light-brown bathstone from Somerset. It is fine-grained

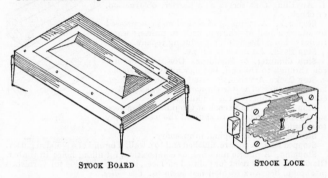

STOCK BOARD   STOCK LOCK

and used for interior carving and selected stone for exterior work. Weight 126 lb. per c. ft. Crushing strength 107 tons per sq. ft.

Stones. Those used in building and constructional work are known as granites, limestones, and sandstones. Subdivisions are: syenites, diorites, basalts, whinstones, gritstones, traps, etc. They are also classified as igneous, sedimentary, and metamorphic, according to the formation; and as silicious, calcareous, and argillaceous, according to the chemical composition.

Stone Tongs. An appliance consisting of two bent levers loosely riveted together, and used for lifting blocks of stone.

Stone, or **Greystone, Slates.** Thin laminated stones used for roof coverings such as Collyweston slates. They are named according to the sizes, as slates, and are recorded on a special rule used by the workmen; bachelors, becks, cocks, cussomes, followers, muffities, skews, tants, wivets, etc.

Stoneware. Sanitary ware made from plastic clays of the Lias formation; approx. 75 per cent silica, 25 per cent alumina. Ground flint, sharp sand, decomposed granite, etc., are added to diminish shrinkage. Usually salt glazed.

Stonhard. A registered name for a number of proprietary materials: waterproofers, fillers, hardeners, preservatives, etc.

Stoniflex. A proprietary waterproof material for roofs.

Stonite. A proprietary range of wall-finishing materials.

Stool. 1. A framed support. 2. A seat without a back. 3. A foot rest. 4. A low bench for kneeling upon. 5. A base for a cupboard, etc. 6. A concrete block foundation for a timber building. 7. See *Stooling*.

Stooling, or **Stool.** A flat seating, as on a weathered sill, for a jamb or mullion. See *Sill*. Also see *Boxing* (3). 2. The stump of a tree.

Stool Valve. A lever valve for admitting water to a W.C. A weighted lever closes the valve which is opened by pulling a handle.

Stoop. 1. A basin for holy water. See *Stoup*. 2. The landing to a flight of stairs leading to an outside door. 3. A tapered piece spiked to a post, for strength.

Stoothing. Battening on walls.

Stop. 1. An ornamental termination to a moulding stuck on the solid. 2. A projection on a bench top to steady the stuff. 3. The projections on a door casing, or lining, against which the door closes. See *Finishings*. 4. Anything that serves as a barrier, obstruction, or check.

Stopback. A stopping of soaked and compressed bread, or a ball of whiting, used to prevent water from trickling through a pipe during repairs.

Stop Bead. An inner bead, *q.v.*

Stop Chamfer, or **Moulding.** One that does not run through to the end, but dies away on a *stop*.

Stop Cock. A valve, operated by hand, for regulating the flow of water through a pipe, as used to cut off the supply to the building.

Stop Ferrule. A stop cock at the junction between supply pipe and main, to shut off the water supply to a building.

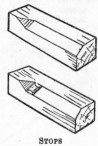

STOPS

Stoping. Same as coping, in masonry.

Stopped End. A square, finished end to a wall. The end of a wooden gutter.

Stopping. A plastic material, or composition, for filling holes in timber. Glazier's putty is used for painted surfaces, litharge and glue for varnished pitchpine, beeswax for polished surfaces. See *Filler*.

Stoppon. A proprietary substitute for clearcole.

Stop Shutters. Temporary timbers to limit the amount of work concreted in one shift. They are usually provided with fillets to form a key for later work.

Stopstara. A proprietary putty for glazing metal windows.

Stop Stone. A stone fixed in the ground against which the meeting stiles of a pair of gates close. It is more usual to have a hinged metal stop to avoid the obstruction.

Stop Valve. See *Stop Cock*.

Storage Cistern. A tank for containing water supplied by the Water Board as distinct from a hot-water and flushing cistern.

Storey, or **Story.** A part of a building between two consecutive floors or between top floor and roof. See *Floors*.

Storey Posts. Wooden posts supporting a floor or a beam carrying a wall. A newel extending from floor to floor.

Storey Rod. A long strip of wood on which the heights of the steps between two floors are set out. It is used when setting-out stairs.

Storm Chain. 1. Steel chain supports to a sunblind to safeguard against the material leaving the rail or roller. 2. A chain to a door or casement as a safeguard against the wind.

Stormproof Windows. The name applied to numerous registered wood casement windows. In some cases they are only provided with numerous throatings, but in others there is a projecting lip on the face of the casements which necessitates the use of cranked hinges. Hood mouldings are also used over the tops of the casements as additional protection.

Storm Water. Rain-water collected by the drains of a building and the adjacent ground. See *Separate Systems*.

Storm, or **Double, Window.** 1. A frame with two sets of balanced sashes or casements, with a space between, to prevent outside noise from penetrating into the building. 2. A vertical window in a sloping roof. It is the opposite of a dormer window, and is entirely within the roof.

Story. Same as storey, *q.v.*

Stoup. A basin for holy water. It is placed in the wall, and near the door of a church.

Stove. 1. Usually a cylindrical cast-iron arrangement for slow combustion; the fuel is coke or anthracite coal. Many types provide hot water. Oil and gas stoves are also commonly used, but the latter are for cooking. 2. A chamber specially heated for quick drying. See *Heating, Direct,* and *Indirect.*

Stoving. Seasoning timber, or drying enamelled goods, in specially heated chambers.

Stovis. A mobile air compressor for spraying.

Straddle Poles. Scaffold poles lying astride a roof to carry a saddle

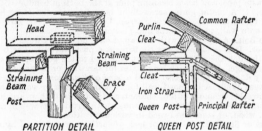

PARTITION DETAIL   QUEEN POST DETAIL
STRAINING BEAMS

scaffold. The two poles cross each other, and are lashed together at the ridge, and to the tops of the standards.

Straddle Scaffold. See *Saddle Scaffold.*

Straight Arch. An arch with a level soffit, or with just sufficient rise to counteract the appearance of sagging. Also called a camber arch, or flat arch, or a lintel. See *Arches.*

Straightedge. A long piece of seasoned yellow pine with parallel straight edges for testing levels, etc. The dimensions depend upon the requirements, but it is sufficiently strong to remain rigid when placed on edge.

Straight Flight. A flight of stairs consisting of fliers only.

Straight Grain. Timber with straight fibres parallel to the axis of the stuff. It is the best and strongest for constructional work.

Straight Joint. 1. A plain butt joint. 2. Vertical joints in brickwork falling together in adjacent courses. 3. Opposite to *breaking joint.*

Straight Lock. One simply screwed on the face of the door, without cutting the woodwork. Opposite to cut lock, *q.v.*

Strain. The alteration of shape of a structural member due to the stress. It is measured as the extension or contraction per inch of length, therefore strain = extension ÷ length. See *Modulus of Elasticity.*

Strainer. 1. A hollow perforated globe or box at the foot of a suction pipe to a pump. 2. Perforated metal or woven mesh for sieving liquids.

Straining Beam. A horizontal timber between the heads of the queen posts, in a roof truss. A horizontal strut in structural framing. See *Roofs* and *Centering.*

Straining Piece. A horizontal piece receiving the thrust from the struts in a system of flying shores.

Straining Sill. A horizontal timber between the feet of queen posts. See *Roofs.*

Straining Standards. Iron posts for tightening up the wire in wire fences. Also see *Bricanion*.

Strakes. The sections of plates for building up a tall steel chimney.

Stramit. A proprietary insulating wallboard consisting of compressed straw faced with kraft paper. Standard size is 8 ft. by 4 ft. by 2 in., but larger sizes are available. It is light, strong, fire-resisting and easily cut and fixed.

Stranded Caisson. A box-like structure with sides and bottom but no top. It is built on land, launched, and towed to the site, which is prepared previously. The pier is built in the caisson which sinks, or is stranded, under the weight.

S-trap. A sanitary trap shaped like the letter ∪. It is generally used for a waste pipe, and has a cleaning plug at the bottom bend.

Strap. 1. A metal plate, screwed or bolted across the joint between two, or more, timbers. Also see *Stirrup*, *Centering*, and *Heel Strap*. 2. A batten on a wall, to carry laths and plaster.

Strap Bolt. A fastening with a strap at one end and a thread and nut at the other. It is used for heavy timbers, as a post and rail. The bolt part passes through the rail and the strap is bolted to the post, or *vice versa*.

Strap Hinge. Hinges with long plates for screwing or bolting to the face of heavy doors or gates. Also called a *band and gudgeon*, *q.v.*

Strapped Wall. A battened wall for lath and plaster or wallboard.

Strapwork. Ornamentation, characteristic of Elizabethan architecture, imitating interlacing straps, or bands, with rivets. It was formed in plaster, wood, or stone, and used in various positions.

Strata or **Stratum.** A layer, or bed, of earth or rock formed by natural causes.

Stress. The internal force acting in a structural member, per square inch of section, resisting the action of the external forces; $=$ load $\div$ area $= f \div a$. See f and *Modulus of Elasticity*.

Stress Diagram. A diagram drawn parallel to the members of a structural frame, and showing the magnitude of the stresses acting in the various members of the frame. It is also called a reciprocal diagram. See *Graphic Statics*.

Stressed Skin Construction. Structural plywood construction consisting of outer and inner skins, to resist the external forces, which are connected by an inside core of wood ribs. The whole is resin-bonded into a prefabricated unit for walls, floors, and roofs of houses. The manufacture requires expensive equipment and can only be used in mass production.

Stress Graded. Applied to certain woods that are graded according to their strength value, for constructional work.

Stretcher. 1. Bricks, or stones, laid lengthways with their faces showing on the face of the wall. 2. A long tie only able to resist tension.

Stretcher Bond. Brick walls showing stretchers only on the face.

STRETCHER BOND

Stretcher, or **Stretching, Course.** A course of stretchers. See *English Bond*.

Stretch-out. A development. Especially applied to the development of the cylindrical string for a geometrical stair.

Stria. 1. Thread-like lines, or veins, on marble. 2. A flute to a column.

Striations, Striated. Faint lines in glass, or on the surface. 2. Small thread-like parallel lines or channels. See *Stria*.

Striking. Removing a centre, or other temporary timbering, when the work is completed.

Striking an Arc. Describing a circular arc of large radius, as in full size setting-out.

Striking-off Lines. The wall and ceiling lines to which the stock is cut, when preparing a bench mould for fibrous plaster work.

Striking Plate. A plate, screwed to the rebate, against which the latch bolt of a mortise lock strikes when the door closes, and with which the bolts engage.

Striking Points. The centres for describing circular arcs.

String Course. Any ornamental horizontal band on the face of a building to emphasize the horizontal divisions. They are often projecting, and usually continue the line of the window sills.

Stringer Beams, or Stringers. 1. Long horizontal members in a trussed bridge and in steel-framed buildings (see *Framed Construction*); or timbers tying together the heads of trestles in trestle bridges. 2. A B.S.B. under a flight of stone stairs as a precaution against fire. 3. Walings.

String Piece. The horizontal member of a Belfast roof truss, *q.v.*

Strings. The inclined supports carrying the ends of the steps of a stairs. They are distinguished as: bracketed, close, cut, outside, wall, or wreathed. The slabs at the free end of concrete or faced steps. A slender tie holding the ends of a curved member.

Stringybark. Applied to several species of Eucalyptus, *q.v.* They are mostly excellent woods and used for all purposes from structural work to superior joinery and flooring. Also called Tasmanian Oak, Messmate, etc. Weight about 54 lb. per c. ft.

Strip. A grading term for wood less than 4 in. wide and less than 2 in. thick. A narrow slender piece of wood.

Strip Flooring. Pieces of narrow thin hardwoods for facing worn or inferior floorings.

Stripping. 1. Removing old wall-paper or paint, preparatory to applying new. 2. See *Manger's Soap*, *Quickstryp*, and *Burning Off*. 3. Destroying the thread of a screw of bolt. 4. Removing shuttering to concrete.

Strip Slates. Ruberoid roofing imitation slates.

Stript. A proprietary paint remover.

Strix. The fillet between the flutes of a column.

Stroked. Ashlar with regular chisel marks across the face of the stone forming a series of small flutes.

Strong Clay. Clay free from other substances, such as sand, etc.

Strong Room. A fire- and burglar-proof chamber for storing valuables.

Struck Joint. See *Pointing*.

Struck Joint Work. Brickwork in which the joints are pointed with the point of a trowel as the work proceeds.

Structural Glass. 1. Glass units used for building non-bearing walls. Glass bricks. See *Pavement Lights*.

Structural Steel. See *British Standard Sections*.

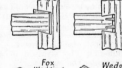

Strut. An inclined compressional member of a frame. See *Centering*, *Timbering*, and *Roofs*.

Strutting. Short timbers placed between joists to stiffen them and to prevent them from canting. See *Herring-bone*, *Keyed*, and *Solid Strutting*.

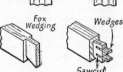

Strutting Piece. Same as straining piece, *q.v.*

Strypt. A registered fire-resisting composite door.

Stub. See *Nib*.

STUB TENONS WITH SECRET WEDGING

Stub Tenon. A short tenon not passing through the material in which the mortise is formed.

Stuc. Plasterwork to imitate stone.

Stucco, or **Stuke.** A general term for plastered surfaces, but especially applied to cement surfaces to imitate stone. *Common* or *trowelled stucco* is ordinary plastering set with greystone-lime putty, and rather coarse sand, and finished with a cork or felt-faced float to give it a rough surface.

Stuck Moulds. Mouldings stuck on the solid. See *Sticking*.

Studding, or **Studs.** 1. The quarterings, or rough vertical timbers, used in a partition. 2. A *stud* is also a form of set screw in which a nut takes the place of the head.

Studio. The work-place of an artist. Applied to painting, sculpture, music, film work, etc.

Stuff. 1. A joiner's term for converted timber. 2. See *Coarse Stuff* and *Fine Stuff*.

Stuffing Box. Any arrangement for packing round a rod to make it steam- or water-tight.

Stugged. Applied to stone that is roughly dressed with a pointed chisel.

Stuke. See *Stucco*.

Stump Tenon. A short tenon increased in thickness at the root, for strength. It is often confused with a stub tenon (*q.v.*). An alternative method is to taper the tenon to increase the thickness at the root.

Stunt Head. Temporary timbering built across a trench to form the boundary for the day's work, when pouring concrete for the base of a wall, etc. Also see *Stop Shutters*.

Stupa. See *Tope*.

Stupid. A *running-out* machine for hollow block making. The machine is horizontal, and a plunger presses the clay through the mouthpiece.

Sty. An enclosure for pigs.

Style. A distinctive type of architecture or design. Also STUMP TENON see *Stile*.

Stylobate. A continuous unbroken pedestal for a range, or row, of columns. The term is sometimes applied to a pedestal, which is actually a fragment of a stylobate.

Sub-Contracts. Work sublet by the chief contractor to specialist firms.

Subgrade. The existing road bed for tar-macadam roads.

Sub-plinth. A second lower plinth, placed under the principal one, to a pedestal.

Subsellia. Small hinged seats in choir stalls that can be raised or lowered as required.

Subsidence. Sinking or settling down due to faulty foundations.

Sub-sill. The sill, or stall board, under the bottom rail of a shop front.

Sub-soil. The stratum of earth immediately under the surface.

Sub-soil Drains. Drains with open joints, for removing water from saturated ground. Semicircular pipes, and coarse broken stone, brick, etc., are sometimes used. See *Land Drain*.

Substitution. A term used in Graphic Statics when two members are temporarily replaced by one member, for the purpose of obtaining a graphical solution on the force diagram. See *French Truss*.

SUBSELLIA

Subway. An underground passage.

Suction. The term for the adhesion of mortar to the bricks.

Suction Plant. Vacuum cleaning plant.

Sucupira. A lustrous tan-brown hardwood, with darker stripes from C. America. It is very hard, heavy, durable, and strong. It is difficult to work and used where strength and durability are of primary importance. Weight about 54 lb. per c. ft.

Suffolk Bricks. Made from a mixture of chalk and clay, producing a pale yellow to white brick. Also called Suffolk Whites or Gault bricks.

Suffolk Latch. See *Norfolk Latch.*

Sugar. Retards the setting of Portland cement and plaster of paris.

Sugar of Lead. See *Lead Acetate.*

Sugar Pine. One of the best soft pines, from W.N. America. Large sizes free from defects. Used for the same purposes as White Pine, very resonant. Weight about 24 lb. per c. ft.

Sugar Soap. Used for cleaning and stripping paint. See *Manger's Soap.*

Suite. A set of rooms. A series, or set, of things. The furniture suitable for a particular room.

Sulphur Cement. A mixture of equal parts of pitch and sulphur, used for fixing iron work.

Summer Beam or **Summer Tree.** A large beam, or one used as a lintel for dead load only.

Summer House. A small structure in a garden for use in summer.

Summer Stone. 1. A large stone bridging columns. 2. A skew corbel.

Sump. A hole for collecting the water in trenches or mines to keep them dry. A pump is used to empty the sump, or sumph.

Sumping. Soaking crusted or dry clay in a shallow pit, or sump.

Sun Blind. See *Blind* and *Rolling Shutter.*

Sundeala. A registered wood-fibre wall-board.

Sundri or **Sunderwood.** A dark reddish-brown hardwood, with fine black markings from India. It is very hard, heavy, strong, and durable, and used for all purposes from structural work to furniture. Weight about 57 lb. per c. ft.

Sunglow. A patent electric heater in the form of a lighting bowl. The bowl is of silica and glows red with radiant heat.

Sunk Bead. See *Mouldings.*

Sunk Draft. A margin sunk below the face, when dressing masonry.

Sunk Face. Applied to a stone with a sunk panel.

Sunk Gutter. A secret gutter sunk below the plane of the roof surface, as down the side of a chimney.

Sunk Mouldings. Mouldings below the surface of the framing.

Sunk Panels. Panels sunk in the solid material, wood or stone, as in pilasters, bench ends, newels, etc.

Sunk Shelf. A plate shelf. The top is recessed so that plates will stand against the wall.

Sunray. A patent vapour tube lighting.

Supa. A yellowish-bronze to dark-brown hardwood from the Philippines. Rather hard, heavy, and durable, but not strong. Difficult to work, polishes well. Used for superior joinery, flooring, etc. Weight about 39 lb. per c. ft.

Super. 1. Abbreviation for superficial. Square measure. 2. A proprietary wood-fibre wall-board.

Super Cement. A superior form of Portland cement, described as *tannocatalysed.* It produces a harder, stronger, and denser concrete than does ordinary Portland, and is better able to withstand destructive agents, such as chemically-charged water.

Superimposed Carving. A separate carved piece placed over a deeply sunk hollow. The method is common in large Gothic mouldings.

Superimposed Load. All loads other than dead load, in structural design, under the L.C.C. Code of Practice.

Superposition. See *Whipple Murphy Truss*.

Superstructure. The structure above the base and footings, or foundations.

Supertex. A proprietary jointless flooring consisting of rubber latex, with cement, aggregate, and pigments.

Supply Cistern. A cold-water storage tank for a hot-water system. See *Low-pressure System*. It also provides a reserve supply in case of main failure.

Supply Pipe. Any service water pipe not serving as a communication pipe.

Surbase. The top mouldings of a pedestal. The crowning moulding of a dado or dado framing.

Surbased. Applied to an arch or vault when the rise is less than half the span.

Surcharged Wall. A retaining wall to an embankment. The earth rises from the top of the wall. See *Angle of Repose*.

Surface Box. A cover to a valve for an underground pipe, as a water main.

Surface Coatings. Membranous, or film, waterproofers, sprayed or brushed on the surface. There are many proprietary materials for the purpose. Fluosilicate solutions, hydrochloric acid solution followed by oxalate of ammonia, cellulose, oils, bitumen, waxes, chlorinated rubber, paints, etc., are all surface coatings.

Surface Concrete. A layer of concrete, at least 4 in. thick, covering the ground within the walls of a building to prevent ground gases from rising.

Surfaced. Same as planed.

Surface of Operation. A plane surface prepared for applying square, templets, etc., so as to work the rest of the stone.

Surfacer. A woodworking planing machine.

Surface Water. Rain-water collected on ground surface as distinct from roof surface.

Surfex. A proprietary plastic decorative material. It can be applied in any thickness and to nearly any material. Surfex is a powder, and is mixed with water, and applied with a brush.

Surmounted. A vault with the rise greater than half the span. Stilted.

Surround. A decorative border round anything, as the tiles to a register grate.

Survey. An inspection of land, buildings, etc., by a qualified person, to assess its value for a particular purpose.

Suspended Ceiling. A ceiling with the joists hanging from the roof timbers or from the bottom member of a steel truss. It is also formed below floors to provide sound insulation, space for ducts, pipes, etc., or to give a level ceiling surface to cover any projecting floor members.

Suspended Scaffold. A self-contained scaffold suspended by steel ropes from two outrigger beams. Four winches, fixed at the corners, can wind the scaffold to any required height.

Suspended Shuttering. Formwork for floors and beams suspended from the steelwork of the floor. Steel hangers, on each side of a secondary beam, are drilled to receive the dowels that in turn carry the bearers. The shuttering, or decking, is usually of pressed steel plates, about 2 ft. square. See *Shuttering*.

Suspended Tread Floor. See *Truscon Floor*.

Suspension Bridge. One in which the load is carried by chains, or cables, anchored at the ends and passing over elevated towers.

Suspensol. A proprietary graphite, waterproof paint. It is made in several colours.

Sussex Bond. A *garden-wall bond*, consisting of three stretchers and one header throughout each course. The header is placed over the centre of the middle stretcher in the course below.

Swag. A decorative carved loop of leaves, husks, etc. A festoon. It is a feature of Adam architecture.

Swage. A blacksmith s tool for shaping hot metal. It is the reverse of the required shape and usually consists of top and bottom sets.

Swager. A machine for setting bandsaws.

Swage Saw. A thick circular saw, ground on one side to a thin circumference, for deep cutting wide stuff into thin stuff that can curl away from the saw.

Swallow Tail. A dovetail.

Swamp Gum. A species of Eucalyptus similar to Stringybark, *q.v.*

Swan. A patent electric heater, or booster, for raising the temperature of hot water to boiling point.

Swan-neck Chisel. A strong, long, curved chisel used for levering out the core of the mortise for a mortise lock.

Swan Reflector. A rolled glass with one side covered with small hemispherical projections to reflect the light from any angle. Used for signs, etc.

Swan's Neck. A combination of ramp and knee, to give a quick rise to a handrail. A pipe, of contra-flexure, from gutter to down pipe, for overhanging eaves or a parapet gutter. See *Cornice* and *Knee*.

Swarf. Composition for *rust joints* (*q.v.*). Iron filings and sal-ammoniac.

Sway Rods. Diagonal bracing to resist the wind on structures.

S.W.D. Abbreviation for *stoneware drain*.

Sweating. 1. Water oozing through a porous pipe. 2. Condensation.

Sweat-out. The reverse of dry-out, *q.v.* The plaster appears damp and soft after set has taken place, due to clay or loam in the sand, no air circulation, damp foggy weather, or no porosity in the background.

Sweep. 1. A curve of large radius. 2. A freehand symmetrical curve. 3. A bundle of shavings, or rags, to keep a flue clear during construction.

Sweetbark. A lustrous pinkish-rose hardwood, with faint flecks, from Queensland. Fairly hard and heavy but not durable. Fumes and polishes well. Hard on cutting tools. Used for superior joinery, bentwork, flooring, etc. Weight about 41 lb. per c. ft.

Swelling. See *Entasis*.

Swept Valley. One in which a "tile-and-a-half" tile is cut in the angle to give the appearance of a continuous course of tiles. The same method is used with random slates.

S.W.G. Abbreviation for *Standard Wire Gauge*. See *Birmingham Gauge*.

Swing Bar. A pivoted bar for securing a pair of gates.

Swing Bridge. A bridge that is counterbalanced and rotates on a pivot.

Swing Door. A door without stops to the frame, so that it opens in both directions. Special hinges (helical) or floor springs, are required for this type of door.

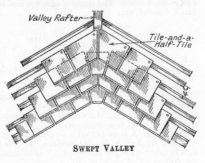

SWEPT VALLEY

Swinging Arches. The operation of *easing* the centering for large arches, especially continuous ones forming a bridge.

Swing Post. The hanging post for a gate.

Swing Sash. A pivoted sash.

Swing Saw. A pendulum cross-cut saw.

Swirl. Irregular figure in wood as at the intersection of branch and stem or around a knot.

Switch. A device to break or make a circuit, or to change the course of an electric current.

Switchboard. A collection of electrical switch-gear on one board.

Swivel. A joint formed so that one part rotates freely round the other part, and commonly used in gas fittings, etc.

Sword. A piece of thin iron used for hacking concrete to form a key.

Sx Board. A registered insulating wall-board.

Sycamore. A hard white timber with even grain, used for kitchen utensils and furniture. Radial cutting produces a beautiful figure valuable for veneer. It is sometimes dyed and called Greywood. Weight about 48 lb. per c. ft.

Syenite. Rock composed of hornblende and felspar.

Sylvadure. A proprietary preservative for timber.

Sylvester Process. See *Soap*.

Syncore. A proprietary plyboard with a core of compressed wood segments.

Synthaprufe. A liquid tar-rubber composition for damp-proofing. It is applied cold, brushed or squeegeed, and used for fixing wood blocks.

Synthetic. Applied to manufactured products that are intended to imitate natural products.

Synthetic Ochres. See *Pigments*.

Synthetic Paints. Applied to paint and varnish in which linseed oil is replaced by artificial substances, because of the present shortage of oil. It has a different chemical reaction when applied.

Syphonage. Causing liquid to flow up a pipe by first creating a vacuum in the pipe; the atmospheric pressure forces the liquid up the pipe. The height cannot exceed 34 ft., for water, as this column balances atmospheric pressure. The maximum height of siphonage in practice is 26 ft. owing to friction. See *Siphonic Closet*. A sudden rush of water may cause a partial vacuum and unseal a sanitary trap. See *Anti-siphonage*.

Systyle. A range of columns two diams. apart. See *Columns*.

Szerelmey Stone Liquid. A proprietary waterproofer for brick, stone, and cement walls. The name is also applied to a cleaning process for stone, brick, faience, etc.

T

t. Abbreviation for *tensile stress*.

Tabby. Lime mixed with gravel and shells, and made plastic to make building blocks.

Tabern. A cellar.

Tabernacle. A receptacle for anything holy or consecrated. A niche for a saint.

Tabernacle Work. Ornamental open-work in churches, as often applied to the tops of niches, stalls, etc.

Tablature. A painting on a wall or ceiling.

Table. 1. Any wide, flat surface. 2. The scarfings for a tabled joint. 3. The cross-piece, or flange, of a T-iron section. 4. A horizontal band of mouldings. 5. A sheet of crown glass.

Table Joint. 1. An engineering masonry joint between large flat surfaces. A long rectangular
projection on one piece fits
into a corresponding recess
in the other piece, to prevent
lateral movement. Now seldom used. 2. A scarfed joint
for heavy constructional

TABLE OR SCARFED JOINT

timbers, in which the fitted surfaces are parallel to the edges of the timber
The joint is strengthened by hardwood folding wedges or keys.

Tablet. A thin sheet of wood, stone, etc., for an inscription

Tabling. Same as *Table Joint*.

Tablinum. A space opening on to the atrium of ancient Roman buildings.

Tabular. The flat plane part of anything. Like a table.

Tace. Same as sally, *q.v.*

Tacheometry. See *Stadia Points*.

Tachymeter. A surveyor's instrument for finding positions rapidly.

Tacking. Fixing a thing temporarily or not securely.

Tacking Rivets. Rivets connecting flange plates together only, and not
subject to calculated stress.

Tackle. See *Pulley Blocks*.

Tacks. 1. Short nails with flat heads. 2. See *Tingle*.

Tacky. Paint or varnish not quite dry. Sticky.

Tænia. See *Tenia*.

Taft Joint. See *Blown Joint*.

Taggers. Thin sheet iron or tin-plate.

Tail. The rear end of anything. See *Retaining Wall*. The lower part of
a slate.

Tail Bay. The last bay in a roof or floor.

Tail Gate. The lower gate of a canal lock.

Tailing. Fixing the end of a projecting member securely into the
wall. Also *tailing in* or *tailing down*. See *Reinforced Concrete Cornice*.

Tailing Irons. A Standard Steel Section built across the wall end of a
projecting member, for tailing down.

Tailings. Refuse or dregs. Especially applied to the washing of ores.

Tail Joist. A trimmed joist. See *Trimmer*.

Tail Pipe. A suction pipe to a pump.

Tail Race. The water channel below a mill-wheel.

Tail Trimmer. A trimmer against a wall to avoid building the ends of
the joists into the wall.

Take-down Square. See *Steel Square*.

Take-up, To. Applied to a leaky joint that stops leaking due to natural
causes.

Taking a Squeeze. See *Squeeze*.

Taking Off. Taking the dimensions from the drawings when preparing
bills of quantities. The sizes and abbreviated description are transferred
to the *dimension paper*.

Talc. A soft stone that feels greasy to the touch, in which silicate of
magnesia predominates. French chalk. See *Mica*.

Tall-boy. 1. A contrivance or tall funnel to a chimney to prevent
down-draught. 2. A tall chest of drawers.

Tallow Wood. A yellow-brown species of Eucalyptus, *q.v.*, from Queensland and N.S. Wales. Excellent wood and used for nearly all purposes.
Weight about 60 lb. per c. ft.

Tallus Wall. A wall battered on one face only.

Tally. 1. A piece of wood notched or marked to register a number of
objects. 2. See *Chain*. 3. Slates sold by *count*, per 1000 slates.

Talon. The French term for an ogee moulding.

Talon Fixing. Patent sheradized steel fastenings for insulating boards.

Talus. An inclination or slope.

Tamalin. A reddish-brown hardwood, with darker lines, also called Burma Tulipwood. It is very hard, heavy, durable and strong, with beautiful figure. Difficult to work, polishes well. Used for superior joinery, bentwood, etc. Weight about 60 lb. per c. ft.

Tamarack. Larch, *q.v.*

Tambour. 1. Anything drum-shaped. 2. The wall supporting a dome. 3. A short stumpy pillar. 4. The *vase, campana,* or *bell,* of the Corinthian and Composite capitals. 5. A wooden screen or vestibule to a church porch. 6. A palisaded defence to a gate.

Tambours. Narrow strips mounted on steel strips or canvas for a flexible shutter, as for rolling shutters.

Tampin. A cone-shaped piece of boxwood used by the plumber to open the ends of lead pipes.

Tamping, or **Tamp.** Consolidating concrete. See *Punning.*

Tanalized Timber. Wood impregnated with a preservative.

Tanalith. A grade of Wolman salts, *q.v.* It is for exposed wood, and gives protection against fungi and fire.

Tang. The projecting part of a tool to which the handle is fixed.

Tangent. A line touching a curve but which, when produced, does not cut the curve. See *Circle* and *Ellipse.*

Tangent Plane. A plane surface in contact with a curved surface.

Tangent System. Used in wreath construction for handrails. The application of tangential planes to cylindrical surfaces.

Tangile or Tanguile. A species of Red Lauan, *q.v.* Excellent wood and used as mahogany for superior joinery, etc. Weight about 36 lb. per c. ft.

Tank. A large cistern, usually rectangular, for holding liquid.

Tanking. A method of making the walls and floor of a basement in waterlogged soil waterproof. A lining of mastic asphalt is built in the walls and floor. Also called a tanked basement.

Tank System. A hot-water system in which the hot water flows from the boiler to a tank placed just below the cold storage cistern. The cylinder system is preferred because of the loss of heat through radiation from the long circulation pipes in the tank system. See *Low-pressure System.*

Tannsteel. A steel alloy impervious to mechanical tools and with a very high melting point. A blow-pipe produces a protective layer that checks further action.

Tant. See *Stone Slates.*

Tap. 1. See *Bib-tap.* 2. A pipe, or plug, for controlling the flow of liquid. 3. A tool for forming the reverse thread in a drilled hole to engage with the screw-thread of a bolt or screw.

Tap Bolt. A set-screw.

Taper Pipe. A conical-shaped piece for joining two pipes of different diameters. See *Diminishing Piece.*

Tapering Gutters. Parapet and V-gutters that vary in width, to obtain the necessary fall.

Tapestry Bricks. Rustic bricks (*q.v.*).

Tap House. An inn. A place where liquors are served.

Tappet. A cam to give intermittent motion to some part of a machine.

Tapping. Preparing a drilled hole with a *tap,* to receive a screw.

Tar. A dark viscous fluid obtained by the destructive distillation of coniferous woods and coal. It is used as a preservative for wood and iron, and to form a waterproof surface to walls.

Tarmacadam. Roads formed of macadam and bitumen, or tar. See *Macadam.*

Tarpaulin. Large waterproofed canvas sheets used as protective coverings.

Tarran. A registered prefabricated house with panels of lignocrete slabs; also a plumbing unit and other prefabricated building units.

Tarus. See *Curb Roll.*

Tarviated. Macadam roads bound with tar.

Tas-de-charge. The bottom courses from which the ribs spring, in vaulting. Springing stones with horizontal bed joints, in Gothic vaults.

Tasimeter. Apparatus measuring temperature and humidity variations.

Task Work. Same as *piece work.*

Tasmanian Oak. *Eucalyptus obliqua.* An excellent wood that resembles oak when seasoned but without ray figure. It may be used for nearly all purposes, including structural work and superior joinery. Other species are now marketed as Tasmanian Oak, and also called Alpine, Mountain, and Red Ash. They are similar woods. Weight about 48 lb. per c. ft.

Tassels. Same as Torsels, *q.v.*

Tavern. An inn, *q.v.*

T-beams. Beams in reinforced concrete floors, where part of the floor slab is calculated as a compressional flange for the beams. See *Secondary Beam.*

Teagle. A hoist or lift. Lifting tackle.

Teak. *Tectona grandis.* A hard, strong, durable timber, that resists the attack of insects and fire more than any other timber. Difficult to work when seasoned, due to the hardening of an aromatic secretion. The colour is yellowish-green when cut, but it quickly turns to dark brown. Polishes badly but oils well. Obtained from Burma and Siam. Weight about 53 lb. per c. ft.

Tears. Same as runs, or curtaining, *q.v.*

Teasing. Working out defects in varnished surfaces. Straightening out fibrous materials.

Teasel or **Teazle Post.** A corner post for a timber framed building.

Teaze Tenons. Tenons reduced in width so that they may cross each other at right angles for timbers on the same level. Often confused with box tenon, *q.v.* See page 326.

Technicrete. A proprietary waterproofer, anti-freeze, and cement hardener in liquid form.

Teco. Patent timber connectors, *q.v.*

Tecuta. Thin bronze sheets for flat roofs. They are applied with a bituminous paste, and may be plain or patterned.

Tee Hinge. A long strap hinge, shaped like a T and used for batten doors. The long arm is tapered and from 9 in. to 2 ft. long, and the cross-piece from 4 in. to 6 in. long. They are screwed on the face of the door and frame. See *T-Hinge.*

Tee Iron. See *British Standard Sections.*

Teeth. 1. Broken, or serrated, edges to the clay column as it leaves the mouthpiece when brick moulding. The fault is due to the middle of the column travelling too fast. 2. The serrated edges of saws.

Tegula. The flat part, or under tile, of old Roman tiles.

Telamones. See *Atlantes.*

Telemeter Level. A surveying instrument for obtaining distances, etc.

Telescopic Centering. A rectangular sheet-steel bearer in three sections. The end sections are fish-tailed at one end to rest on supports, while the other end slides into the middle section for the required length.

Telescopic Scaffold Boards. Patent extending scaffold boards.

Telford. A prefabricated house with structural steel panels and felt insulation.

Tellard. A proprietary stain for cement.

Temenos. A sacred precinct to a Greek temple.

Tempera. Distemper, *q.v.*

Temperac Quilt. An insulating sheet consisting of eel grass enclosed in waterproof paper. It is obtained in rolls 28 yd. long and up to $1\frac{3}{8}$ in. thick.

Temper, or **Tempering.** 1. To mix plastic materials in the correct proportions, or to correct some defect in the materials. 2. Making the prepared brick earth into a homogeneous mass for the moulder. 3. Bringing tool steel to the correct degree of hardness or toughness. The steel is first hardened and then reheated. The required temper is judged by the colour of the film of oxide that appears on the surface.

Temperature. The degree of "hotness" of a body, as measured by a thermometer. The scales adopted for measuring are: Fahrenheit, Centigrade, and Réaumur. See *Heat* and *Expansion.*

Tempinis. A yellowish-red hardwood, with darker markings, from Malay. It is an excellent wood where strength and durability are required, and a substitute or hard mahogany. Weight about 56 lb. per c. ft.

Templaster. A registered name for several building products: fire cement, fluxes, fillers, etc.

Template. A stone slab bedded in a wall to distribute the pressure from a concentrated load, as the end of a beam or truss. See *Templet.*

Temple. 1. A place of worship, or a building dedicated to some worthy object. 2. The "Inns of Court" at the London Law Courts.

Templet. A pattern, or mould, made of thin material and used to mark the material to the required shape. Also the *square* and *mitre* templets used by the joiner for setting out and scribing. See *Template.*

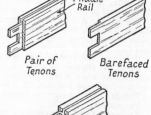

PARTHENON ATHENS
PERIPTERAL OCTASTYLE

TEMPLE

Temporary Timbering. Supporting timbers used to allow permanent work to proceed: centering, shoring, scaffolding, piling, formwork, and timbering to excavations.

Temse. A sieve or strainer.

Tender. An offer to execute a piece of work at a certain cost. See *Estimate.*

Tenement. A set of apartments, or a flat. A house used by one family. Anything held by tenancy.

Tenia. The fillet separating the Doric frieze from the architrave.

Tenon. The end of a piece of wood reduced in section so that it can

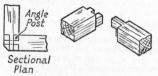

TEASE TENON

TENONS

be inserted into a recess, or *mortise,* in another piece. Tenons may be haunched, franked, stub, stump, box, double, twin, barefaced, housed, forked, chase, shouldered, dovetailed, teaze, tusk, loose, or hammer headed. The width of a tenon should be three to five times the thickness, and the mortise, *q.v.*, is usually one-third the thickness of the stuff.

Tenon Bar Splice. An end to end joint in heavy bearing timbers. Two steel bars are mortised through the timbers at the correct distance from the ends and bolts passed through the bars, on each side of the timbers, to pull the ends together.

Tenon Saw. A woodworker's saw with a thin blade stiffened by a metal back, and used for sawing tenons, etc.

Tension, or Tensile Stress. A stress in the material tending to pull the fibres apart lengthways. See *f*.

Tension Rods. Structural members that are resisting tensile stress only. They are usually round or rectangular in section.

Tension Sleeve. See *Screw Shackle*. Used on wide gates to lift the nose, and in long tension rods.

Tensovic. A proprietary hardened wood. It is impregnated by synthetic resin which increases the tensile strength and stability considerably.

Tenter Hooks. Square hooks, sharpened at both ends. They are driven into the tops of fences as a protection, forming "chevaux de frise."

Tentest. A proprietary fibre insulating wall-board.

Tepidarium. A room of moderate temperature in ancient Roman Baths.

Terebene. A liquid *driers* for paints. It consists of a drying agent reduced in oil of turpentine and spirits.

Teredo Navalis. The ship-worm. A marine insect very destructive to timber.

Terminal. A finial, *q.v.*

Terminous. The upper part of a pillar carved as a bust or statue.

Termites. Insects, similar to ants, very destructive to timber.

Terne Plate. Steel sheets coated with an alloy of lead and tin, and used for roofing, etc.

Terrace. A raised platform or space adjoining a building, usually with balustrade and steps. A level path on a sloping site. A flat roof to a house. A *terrace house* is one of a number of attached houses in a row.

Terra Cotta. Blocks used in building, made from specially prepared brick earth moulded to the required shape and burnt to vitrification. It varies in colour, like bricks, according to the chemical composition. Ground glass, pottery, and sand are sometimes added to prevent excessive shrinkage, and the blocks are made hollow with walls and webs, or diaphragms, about 1 in. thick for the same reason. The blocks are filled with coke breeze concrete before fixing.

BACK OF TERRA-COTTA BLOCK

Terrawode-Brickwood. Patent partition blocks that will take nails, etc., for fixing purposes.

Terrazzo. Venetian mosaic. See *Mosaic*.

Terantum. A grey fragrant hardwood from Malay. Hard and durable. Used for structural work, piling, flooring, etc. Weight about 38 lb. per c. ft.

Tessellated. Paving, with different colours of tiles, laid in geometrical patterns.

Tesserae. Same as *Tesselated*, but formed of small cubes instead of tiles. Mosaic work. A tessera is a small cube of marble, glass, pottery, etc. See *Mosaic*.

Tester. A flat canopy over a pulpit, tomb, or bed.

Testing Drains. See *Drain Testing*.

Tests. The scientific tests applied to wood to prove its resistance and suitability for its purpose include : compression, bending, tension, crushing, shear, torsion, deflection, elasticity, fast or variable colour, figure, grain, stability, durability, resilience, resonance, resistance to wear, etc. Other factors include specific gravity or weight, plasticity, moisture content, workability. See *Resistance.*

Tetrahedron. A solid bounded by four equilateral triangles. One of the five regular solids. Surface area = sq. of linear side × 1·732. Volume = cube of linear side × 0·118.

Tetrastoon. A courtyard with porticoes on all sides (classical architecture).

Tetrastyle. A portico having four columns in front.

Texture Bricks. See *Rustic Bricks.*

Textured. Plastic surfaces worked into some form of decorative treatment with a stippling brush, or special tools. See *Swirl* and *Travertine.*

t.g.b. Abbreviation for *tongued, grooved,* and *beaded.*

Thatchboard. Proprietary insulating boards formed of straw. They are made in large sizes and up to 2 in. thick, and may be faced with plaster or cement. See *Building Boards.*

Thatching. Roofs of country buildings formed of bundles of straw, or reeds, laced together with withes.

Theatre. 1. A building for dramatic performances. A playhouse. 2. An operating room in a hospital.

Theodolite. A telescopic instrument used in surveying for measuring angles, either vertically or horizontally.

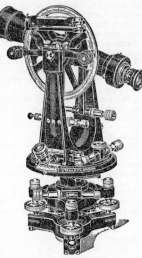

THEODOLITE *(Stanley)*

Therm. The unit of measurement of gas consumption, used by gas companies, and consisting of 100,000 B.Th.U. It depends upon the calorific value of the gas, but it usually equals 200 c. ft.

Thermacoust. A proprietary wood-wool building slab.

Thermae. The baths of ancient Rome and Greece.

Thermal. See *Heat.*

Thermatox. A proprietary fumigator against insects.

Thermex. A proprietary anti-actinic, or heat-excluding glass.

Therming. Square turning, *q.v.*

Thermo-. Applied to any subject in which heat is of importance, as thermo-dynamics, which is the science that deals with the conversion of heat to energy.

Thermolite. A patent flat-roofing system insulated against heat from the sun.

Thermolux. Glass silk sandwiched between sheets of drawn glass. It is used for light diffusing.

Thermometer. A graduated instrument for determining temperature. See *Temperature.*

Thermopil. A proprietary acoustical material.

Thermopile. Apparatus for comparing the radiative values of surfaces.

Thermoplastic Resin. Vinyl resin. It can be made plastic by re-heating.

Thermoscope. A bar used instead of a Seger cone, *q.v.*

Thermostat. An automatic apparatus for regulating temperature.

Thermosteel. A proprietary insulating steel decking for roofs, etc.

Thermotex. A proprietary insulating board.

Thermotile. Proprietary white tiles for flat roofs. They are 1 ft. square and protect the asphalt against extremes of temperature.

Thicknesser. A woodworking machine for planing stuff to a regular thickness.

Thickness Moulding. A bed-moulding, *q.v.*

Thimble. See *Ferrule.*

Thimble Joint. A loose cast iron box, used to join together two cast-iron columns at the intersection with a floor. The columns and the box are flanged, and *centred* by means of a turned projection fitting into a corresponding recess. Also see *Ferrule* (2).

T-hinge. A hinge for batten doors. See *Tee Hinge.*

Thinners. A liquid added to paint to make it work freely and to thin the paint. Turpentine is generally used; or turpentine substitute, which is a rectified mineral spirit.

T-HINGE

Thinwin. A variegated chocolate-brown hardwood from India and Burma. It is very hard, tough, and durable, with beautiful figure. Used where strength and durability are required, superior joinery, fittings, etc. Weight about 57 lb. per c. ft.

Third Fixings. The final joinery finishings of buildings. It implies the finishings following the initial decorating: hanging doors, locks, furniture, etc. See *Fixings.*

Thistle Board. A gypsum wall-board faced on both sides with brown paper, and made in sheets up to 8 ft. × 4 ft. A retarded hemi-hydrate plaster (Thistle plaster) is supplied to coat the boards if required.

Thitka. A yellowish-brown hardwood, also called Burma mahogany, because of its resemblance in grain and texture. It is used for superior joinery, etc. Weight about 37 lb. per c. ft.

Thitmin. A species of Podocarpus from India and Burma. See *Podo.*

Thole. 1. An obsolete term for a dome, cupola, or circular building. 2. A wooden peg.

Tholobate. The substructure supporting a dome.

Thoroughfare. Literally means a passage between two streets, but usually implies any main street.

Thrawn. Same as warped, *q.v.*

Thread. The prominent spiral part of a screw. See *Screw Thread.*

Three-centred Arch. One shaped like an ellipse but built up of circular arcs struck from three centres. See *Arches.*

Three-coat Work. Plasterwork applied in three separate coats: "render, float, and set," or "lath, plaster, float, and set."

Three-four-five Rule. The application of trigonometry (Pythagoras's Theorem) for obtaining a right-angle. "The square on the hypotenuse of a right-angled triangle is equal to the sum of the squares on the other two sides." Hence, if the sides of a right-angled triangle are 3, 4, and 5, we have $3^2 + 4^2 = 9 + 16 = 25 = 5^2$. This rule is applied when setting out large right-angles by means of a tape measure, or laths, by using 3 ft., 4 ft., and 5 ft. or multiples of those lengths.

Three-light Window. A window with three main divisions as formed by two mullions. See *Windows* and *Venetian Window.*

Three-moment Theorem. See *Clapeyron's Theorem.*

Three-pinned Arch. An arch hinged at the crown and at the abutments. This type of arch allows for an accurate graphical solution, as the line of pressure passes through the three hinges, and the horizontal thrust at the crown can be determined.

Three-quarter Bat. A brick broken to three-quarter length. It is sometimes used to form bond in place of a header and closer.

Three-quarter Header. A header in masonry three-quarters of the thickness of the wall in length.

Three-way Strap. A wrought-iron strap shaped to tie together three members in different directions, as at the apex of a king-post roof truss.

Threshold. The sill to an external door.

Threshold of Audibility. The base on which the sensation scale of noise is measured.

Throat. 1. The minimum thickness of a weld along a line passing through the root. 2. The lower end of a flue where it widens out for the fireplace. 3. The opening in a plane where the shavings escape.

Throating. A small groove on the underside of a projecting member, or between two surfaces nearly in contact, to break the flow of water. See *Sill.* Throatings are necessary on the stiles and mullions of the frame of a casement window, where the casements open, otherwise the rain works through by capillary attraction.

Through Bonder or **Stone.** See *Bonder.*

Throw. 1. The relative vertical movement of strata due to faulting, *q.v.* 2. The height that a man can throw excavated earth. The maximum is usually accepted as 6 ft. in deep excavations for trenches, etc.

Thrust. The horizontal pressure of an arch at the abutments and crown or of a strut. The push exerted by an inclined member in compression.

Thumbat. A wall-hook for fixing sheet lead.

Thumb Latch. See *Norfolk Latch.*

Thumb Moulding. One shaped like a thumb, consisting of a quarter ellipse and a small fillet, and used at the edge of a table, etc. See *Mouldings.*

Thumb Planes. Very small planes used on moulded circular work.

Thumb Screw. A screw turned by hand and used for securing sliding sashes.

THUMB SCREW

Thwacking Frame. See *Horse* (1).

Thwaites. A registered builders' hoist.

Tidal Valve. An arrangement, with copper ball, to prevent back flow.

Tie. Any member used to prevent two other members from separating, or spreading. A tensional member in a frame or truss.

Tie Beam. The horizontal timber tying together the feet of the principal rafters of a roof truss. See *Roofs.*

Tied. Applied to a run for workmen that has cleats fixed under to prevent unequal sagging. See *Bridging Run.*

Tied House or Cottage. A house owned by an employer, firm, or authority for their workers only, who cease to be tenants with loss of employment.

Tie Irons. Used for tying together thin walls. They should be treated to prevent corrosion. See *Cavity Walls* and *Wall Ties.*

Tie Lines. Lines used to check a survey by triangulation.

Tier. One of several stages placed one above the other.

Tiercerons. Intermediate ribs in vaulting to reduce the width of the cells, or compartments.

Tie Rod. A steel, or wrought iron, rod in tension. A steel rod used in place of a tie beam. See *Roofs* and *Camp Sheeting.*

Ti-Foon. A registered stormproof window, *q.v.*

Tige. The shaft of a column.

Tight Knot. A firm sound knot. See *Knots*.

Tight Sheathing. Matched boarding on studs or rafters as a base for the final coverings. The term is especially applied to diagonal boarding under sidings to timber buildings.

Tile-and-a-Half. Specially made wide tiles for forming the bond at the verge of the roof, instead of using half-tiles, or for a swept valley (*q.v.*). Also see *Open Valley.*

Tile Battens. The laths or which roof tiles are hung. They vary in size but are usually about 2 in. by 1 in.

Tile Creasing. See *Creasing*.

Tile Hanging. Fixing roofing tiles on walls, as a protection against the weather.

Tile Pins. Oak pegs used instead of nails. They are used in chemical works.

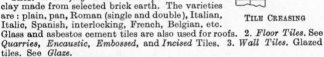

Tiles. 1. *Roofing Tiles.* Thin plates of burnt clay made from selected brick earth. The varieties are : plain, pan, Roman (single and double), Italian, Italic, Spanish, interlocking, French, Belgian, etc.

TILE CREASING

Glass and asbestos cement tiles are also used for roofs. 2. *Floor Tiles.* See *Quarries, Encaustic, Embossed,* and *Incised* Tiles. 3. *Wall Tiles.* Glazed tiles. See *Glaze.*

Tile Stones. See *Stone Slates*.

Tilia. The lime tree.

Till. 1. A receptacle for money, usually a drawer with bowl-shaped compartments. 2. See *Boulder Clay*.

Tiller. The top handle of a pit saw, the other end is called the box.

Tilted. In a sloping position. Not vertical or horizontal.

Tilting Fillet. A triangular strip of wood used at the eaves to tilt the bottom course of slates or tiles, so that the succeeding courses bed with a close joint.

Tiltman Blocks. A patent solid partition block, keyed together with double V-joints.

Timber. Wood suitable for building and structual purposes, whether as logs, converted stuff, or as standing trees. Usually the term is applied to stuff of large section or in bulk. Small stuff is usually called wood, especially when in the hands of the craftsman or machinist. There are over 4000 woods in common use throughout the world, and the number of those in use in this country is continually increasing. The best known softwoods used in buildings and for structural work in this country include : Redwood, Whitewood, Douglas Fir, Hoop Pine, White

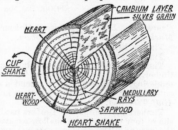

SECTION OF TIMBER TREE

Pine, Red Pine, Podo, Pitch Pine, Sequoia, Sitka, Spruce, Kauri, Rimu, Larch, Hemlock, Red Cedar. The best known hardwoods include : American White Wood or Canary Wood, Ash, Bagac, Balsa, Beech, Birch, Black Bean, Blackwood, Camphor Wood, Cedar, Chestnut, Chuglum, Crabwood, Danta, Ebony, Elm, Eng, Gaboon, Greenheart, Guarea, Haldu, Iroko, Jarrah, Karri, Kokko, Lauan, Mahogany, Makore, Maple, Maple Silkwood,

Merante, Mora, Myrtle, Oak, Obeche, Padauk, Purpleheart, Pyinma, Pyin-kado, Rosewood, Sapele, Sycamore, Satin Walnut, Seraya, Teak, Thitka, Toon, Walnut. The market forms are balk, battens, boards, casewood, deals, die square, ends, flitch, log, mast, pitwood, planchette, plank, planking, poles, quarterings, scantlings, thick stuff. Wood is sold by board or cubic measure, standard, float, fathom, square, load, or by weight. A large number of other important woods are described alphabetically. Also see

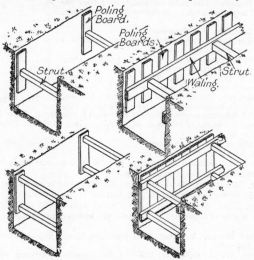

TIMBERING FOR TRENCHES

Conversion, Defects, Diseases, Moisture Content, Preservation, Shipping Marks, and *Fire Resisting Timbers.*

Timber Bricks. See *Nog.*

Timber Dogs. See *Dog.*

Timber Framed. Applied to a building in which the structural part is of timber, except the foundations. The panels, which may be brick, cement rendering, or other materials, intended only to keep out the weather. The outside may be covered with sidings and the inside with wallboards. See *Half Timbered.*

Timbering. Temporary timbers used for excavations. The illustration shows four methods, for shallow excavations, according to the firmness of the ground. The requirements depend upon the depth, the pressure on surrounding areas, and the firmness of the earth. In some cases, especially of site excavation, heavy structural timbering is required designed by engineers.

Timber Joint Connectors. Patent metal connectors for heavy construction joints. There are several types, such as the Alligator, Bulldog, and Teco. They all serve the same purpose, to lessen the number of bolts and outer plates and to prevent lateral movement. The Teco types include toothed ring, split ring, and the shear plate which is for jointing timber to steel. The toothed ring is up to 4 in. diameter and the split ring up to 8 in. diameter. Other types include bat, claw plates, and spike grids. They all allow for higher loads on the structure than ordinary bolted joints.

Time Sheet. A specially prepared sheet given to the workman so that he can register the time he spends on any particular job.

Timesing. A column on the dimensions paper, when "taking off," in which is denoted the number of times a particular item is repeated on a job.

Timonex. See *Antimony Oxide*.

Tindalo. A pale-orange to dark-red hardwood from the Philippines. It is

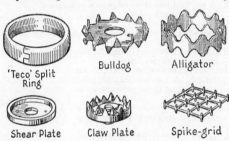

'Teco' Split Ring   Bulldog   Alligator

Shear Plate   Claw Plate   Spike-grid

TIMBER JOINT CONNECTORS

hard, heavy, durable, and strong, and polishes well. It is used for superior joinery, flooring, etc. Weight about 51 lb. per c. ft.

Tingle. Strips of lead or copper used to tie down the edges of sheet lead ; or to stiffen a hollow roll or seam ; or to prevent one sheet of glass from sliding down over another, in roof glazing. 2. Bricks, or a notched plate, used to keep the bricklayer's line tightly stretched.

Tingle-tingle. Stringybark, *q.v.*

Tinned. A soldering iron coated with solder or tin preparatory to making a soldered joint.

Tin Plate. Thin mild steel coated with tin to prevent corrosion.

Tinted Glass. Glass coloured by fusing colours in the molten materials See *Flashed Glass*.

Tintocrete. A cement spray for brickwork, concrete, or stone. It gives a glazed non-absorbent surface.

Tintopal. A registered rolled glass, opal finished, and in 30 different colours.

Tip Cart. A lorry or cart that tips over sideways or backwards to deposit loose materials.

Tipler, or **Tipper.** See *Slop W.C.*

Tips. 1. The projections from the air-holes after pouring the gelatine for a plasterer's jelly mould. 2. Pits provided by local authorities in which dry refuse is deposited. 3. A small brush used in gilding.

T-iron. See *British Standard Sections*.

Titan Cement. Made from the by-products of blast furnaces where titaniferous ore is used. It is cheap to produce, and is very strong and acid-resisting.

Titanium White. A non-poisonous white pigment for paint. It is not affected by acid or sulphur fumes. See *Pigments*.

Title Deeds. Legal documents showing the owner's rights to certain properties.

T.L.B. Rubbers. A well-known brick for gauged work, made by T Laurence & Sons, Berks.

Tobby. A surface box.

Tobin's Tubes. Rectangular tubes placed vertically in or on a wall for ventilation purposes. The air is admitted about 6 ft. above floor level by means of an adjustable flap valve.

Toe. 1. The projection of the base over the face of a reinforced concrete retaining wall. See *Retaining Wall.* 2. The front end of a plane. 3. The bottom of the shutting edge of a door. 4. The foot of a strut.

Toe Board. A scaffold board on edge as a protection.

Toe Joint. A bird's-mouth joint.

To-fall. Same as lean-to, *q.v.*

Toggle. A short bar, or pin, placed through the eye at the end of a rope.

Toggle Bolts. Patent Rawlplug fixings for sheet fixings, hollow blocks, etc. They may be spring or gravity bolts.

Toggle Joint. 1. A knee or elbow joint. 2. An end to end joint. 3. A mechanical link joint used in machines exerting great pressure, as in stone crushers, etc.

Toledo. A patent steel trestle, or horse. It is collapsible and has corrugated jaws to grip the bearer.

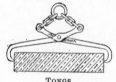

TONGS

Tolerance. 1. An agreed difference in dimensions to allow for unavoidable imperfections of workmanship. 2. See *Limits.* 3. See *Clearance.*

Toll Bar, or Gate. An obstruction across a road or river to prevent passage except on payment of a fee, or toll.

Tondino. An astragal, *q.v.*

Tongs. See *Stone Tongs.*

Tongue and Groove. A joint for boards to build up a plane surface, as floorboards and matching. A tongue on the edge of one board fits in a corresponding recess in the adjacent board.

Tongues. Long thin strips of wood or metal fitted in grooves to strengthen a joint. See *Matching* and *Egg and Tongue.*

Tonk's Fittings. Metal adjustable supports for shelves to cupboards, etc.

Ton Slates. Welsh slates sold by weight.

Tooling, or Tooled. See *Stroked.*

Tool Pad. A woodworker's tool, consisting of a number of interchangeable small bits fitting into a handle by means of chuck and jaws.

Toon. Indian or Moulmein Cedar. Very similar to a soft mahogany. It is excellent wood, polishes well, and used for superior joinery and fittings. Weight about 34 lb. per c. ft.

Tooth. 1. A quality in fillers that allows for cutting down with pumice stone. 2. A point or cog of a saw, rake, gear wheel, etc. See *Teeth, Rack,* and *Pinion.*

Toothing. Leaving the end of a wall unfinished, by allowing alternate courses to project 2¼ in. so that the wall may be continued without cutting. Cutting away alternate courses, or several courses together at intervals, in a wall to provide bond for new work.

Toothing Plane. A plane in which the edge of the cutting iron is serrated like a saw. It is used to roughen a surface to provide a key for glue, as when veneering.

Tooth Ornament. An early Gothic moulding consisting of a series of small pyramids, cut at the sides, and placed in the hollows of large mouldings. See *Dog Tooth.*

Top Beam. A *Collar Beam.*

Top Cap. The strata of earth above the stone in a quarry.

Topes. Early Indian tombs, consisting of a cylindrical body with pilasters and a domical cover.

Top Hung. Applied to casements that are hinged on the top edge and that open outwards by means of a stay or quadrant.

Top Lighting. Any form of lighting from overhead : lantern lights, etc.

Topography. The description of a place : town, district, locality, etc.

Topping. Correcting the points of the teeth of a circular saw. A piece of hard stone is brought up to the saw teeth whilst it is running, to just touch the prominent teeth and level the teeth.

Topping Coat. See *Browning Coat.*

Top Rail. The top horizontal member of a piece of framing.

Torching. Pointing underneath slates or tiles with cement or lime mortar, to exclude rain and wind. It is a questionable advantage, because of capillarity, and unnecessary in a good roof.

Torfoleum. A proprietary insulating board.

Torpedo Splice. A device for joining the ends of the wires in prestressed concrete.

Torque. 1. An ornament carved in the form of a rope. 2. A twisting force. See *Torsion.*

Torsels. 1. A twisted scroll. 2. A template. 3. The pieces that tie under the ends of a mantel-tree.

Torsion. A stress to resist twisting, as in a driving shaft.

Torso. The trunk only of a statue.

Torus. A large semicircular moulding, or bead. It is often used with a cavetto for the moulding of a skirting. See *Mouldings.*

Tosh Nailing. Same as skew nailing, *q.v.*

Totara. A species of *Podocarpus* from New Zealand. See *Podo.*

Tote. 1. The handle to a bench plane. 2. Abbreviation for *totalizator*, an automatic betting machine.

Touch. A plumber's term for tallow.

Touch Stone. A compact black siliceous stone used in the testing of precious metals.

Toughened Plate. Plate glass heated and suddenly chilled, which produces a state of compression on the surfaces and tension in the interior. When the glass is fractured it falls into powder.

Toupie. A solid cutter for shallow mouldings, on a spindle machine.

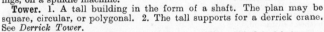

TOWER BOLT

Tower. 1. A tall building in the form of a shaft. The plan may be square, circular, or polygonal. 2. The tall supports for a derrick crane. See *Derrick Tower.*

Tower Bolt. A fastening in the form of a sliding cylindrical bolt, controlled by two or more keepers riveted to a back plate.

Tower Gantry. A derrick tower.

Toxement. A proprietary damp-proof course.

Trabeated. 1. Having an entablature. 2. Combinations of beams or lintels. Construction without arches.

Trabeation. An entablature.

Trabs. A wall plate or beam.

Tracery. The geometrical arrangement of mullions and bars in Gothic windows. The ornamentation is obtained by flowing lines and foliation.

and is also applied to panels in stone and wood. See *Plate Tracery, Bar Tracery,* and *Foils.*

Traces. A geometrical term for the intersection of planes, or of a line with a plane.

Trachelium. The neck of a column.

Trachyte. Volcanic rock similar to syenite.

Tracing. A copy of a drawing obtained by tracing over the original on transparent paper or linen. See *Blue Print.*

Tracitex. A registered tracing cloth.

Track. A trench for wall foundations.

Traffolyte. A proprietary decorative wallboard.

TRACERY PANEL SHOWING
2, 3, 4, AND 5 CUSPS

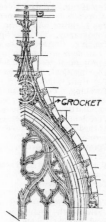

←CROCKET

TRACERY WINDOW

Trafford Tiles. Asbestos cement tiles with tapered corrugations that give close fitting lap joints.

Trail. A continuous ornamental band, representing leaves, tendrils, etc. See *Vignette.*

Trammel. 1. An appliance for drawing elliptical curves. It consists of two grooved timbers halved at right angles to each other. The grooves control a beam compass that describes the ellipse. 2. A beam compass.

Transept. The transverse part of a church built in the form of a cross. The short arms of the cross at right angles to the nave.

Translucent. Not transparent, but transmitting light.

Transom. 1. An intermediate horizontal member of a frame, or opening, between the head and sill. 2. A beam across a sawpit.

Transparent. Like clear glass. Anything that transmits rays of light without diffusion.

Transporter Bridge. One in which the load travels suspended on trolleys.

Transverse Arches. Arches running at right angles to the main arches, in vaulting, etc.

Trap. 1. A bend in a sanitary pipe, so formed that it is always filled with water, to *trap* the air in the drains, and so prevent it from escaping at undesirable points. Traps are named after the letter they represent: D-trap, P-trap, S-trap. Other types are grease trap, gulley trap, and petrol interceptor trap. 2. A scaffold board projecting over a putlog so as to be a source of danger. 3. See *Diorite.*

Trap Door. A door in a horizontal surface, as in a floor or ceiling.

Trapezoid, Trapezium. A quadrilateral with unequal sides but with two opposite sides parallel. A *trapezium* is a quadrilateral with no sides parallel; a kite-shaped figure. These two definitions are often interchanged. See *Quadrilaterals*.

Trap Screw. A screwed plug for a lead trap, for cleaning purposes.

Trass. A volcanic substance, similar to puzzolana, used for hydraulic cement.

Travel. A term used by craftsmen to denote a horizontal distance, especially applied to the two ends of an inclined member or piece of work.

Travellers. The trussed beams carrying a movable winch or travelling crane. They are supported on strong rails and posts, and used for transporting heavy material from one part of the works to another.

Travelling Cradle. A swinging cradle, or scaffold, suspended from a track projecting over the top of the building. It can be raised, lowered, or moved along the track as desired. The popular type is Carter's.

Travelling Crane. See *Travellers*.

Travelling Gantry. One that is mounted on wheels so that it can be moved along rails.

Traverse. 1. A curtained screen for privacy. 2. A steel straightedge fixed in a wood block and used as a seating whilst straightening the edge of a slate with axe or whittle.

Traverser. A movable platform in large garages for manipulating the cars.

Traverse Survey. One that consists of noting the direction of a series of straight lines and measuring the distances. A *closed traverse* returns to the starting point, to check the results.

Traversing. 1. Anything going across from side to side, as planing across the grain of wood, or a gallery across the end of a building. 2. Moving the templet across the face of an arch when obtaining the shape of the voussoirs.

Travertine. A white or creamy limestone, formed by springs containing lime in solution. The structure has a concretionary appearance, with casts of leaves, twigs, and branches of trees.

Travertine Texture. Plastering to imitate travertine. A thick twig is drawn through and up the plastic material so that it falls over, forming uneven lines.

Tray. A flat shallow vessel.

Treacle Moulding. A projecting thumb moulding with a continuous hollow underneath to provide a grip for the fingers.

Tread. The top, or horizontal, surface of a step. See *Step* and *Going*.

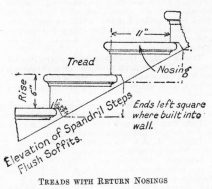

TREADS WITH RETURN NOSINGS

Treenail. A large hardwood pin to secure a tenon, or to fix laminated timbers together. Hardwood pegs used for roof tiling chemical works, etc.

Treetex. An insulating wall-board.

Trefoil. A tracery foliation consisting of three *foils* and *cusps*. See *Tracery*. Also see *Arches*.

Trefoil Arch. An arch consisting of four segments, built on an equilateral triangle. See *Arches*.

Treillage. The framework for an espalier (*q.v.*).

Trellis. Timber framing with the panels formed of thin narrow laths, or wire, crossing each other, usually diagonally, to form lattice work. An *espalier* (*q.v.*).

Trembling Bell. An electric bell that rings continuously whilst the button is pressed.

Tremie. A long box-shaped funnel for depositing plastic concrete below water. The tremie is traversed over the site by a travelling crane and the concrete deposited where required.

Trenail. See *Treenail*.

Trench. A long narrow excavation, usually for drains or foundations. Also see *Trenching* (2).

Trench Drain. A subsoil drain in which the trench is partly filled with broken brick or rubble.

Trenching. 1. Cutting trenches. They should not be wider or deeper than is necessary. The width depends upon whether the wall can be built astride (called "following the wall"), or whether the operatives must work from each side. A clear width of 2 ft. 6 in. on each side is necessary for the latter. The clearance is required on one side only for overhand work. 2. Cutting small recesses in wood to receive the end of another piece.

Trench Timbering. See *Timbering*.

Tresaunte. A passage in a large house between the hall and offices; a part of the cloisters of a monastery.

Trestle. A bearer with splayed legs braced together. They are used for carrying scaffold boards, and are usually about 5 ft. high for work up to ceiling height. A folding trestle consists of two short wide ladders hinged at the top, and with staves at different levels. Also see *Trestle Bridge*.

Trestle Bridge. One whose bed rests on framed trestles. The supports are long, uncut, or round, timbers, called sticks. This type of construction is only used where timber is plentiful and steel scarce.

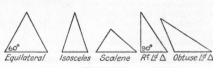

FOLDING TRESTLE

Tretol. A proprietary fibrous bituminous compound for roofs, etc. Also a liquid cement hardener, and other building products.

Trial Pits. Pits sunk to ascertain the nature of the strata, preliminary to designing foundations or tunnelling.

Trianco. 1. A vibrating table for concrete. 2. A registered hollow concrete block for speedy erection of walls.

Triangle. A plane figure with three sides. Triangles are named according to the sides, as equilateral (all equal), isosceles (two equal), scalene (none equal); or according to

Equilateral   Isosceles   Scalene   Rt ld △   Obtuse ld △

TRIANGLES

the angles, as right-angled (one 90°), acute (all less than 90°), obtuse (one more than 90°). Area = $\frac{1}{2}$ × base × perp. height.

Triangle of Forces. "If three forces, acting at a point, are in equilibrium, then the forces can be represented in magnitude and direction as the sides of a triangle, taken in proper order." See *Graphic Statics*.

Triangulation. 1. Surveying by dividing up the site into a number of triangles. 2. Arranging a frame, or truss, in triangular formation to conform to a structural "perfect" frame.

Triapsidal. Having a triple apse.

Tribune. A platform, pulpit, or rostrum, for speech-making.

Tricalcium Aluminate. The main cause of shrinkage in cement work, but contributing largely to the strength for the first 28 days.

Tricalcium Silicate. A constituent of cement that contributes to the strength of concrete.

Triclinium. A dining room in ancient Roman buildings.

Tricosal. Proprietary liquid preparations for waterproofing and hardening cement.

Triddlings. Motor droppings.

Triforium. An arcade or gallery above the arches of the nave of a church. When it has no windows to the open air it is called a blind-story.

Triglyph. A block, with perpendicular grooves, repeated at regular intervals in the Doric frieze. It has two full and two half-grooves, or flutes. See *Doric*.

Trigonometry. Mathematics dealing with the relations between the sides and angles of triangles. The *trig. ratios* (sin, cos, tan, cosec, sec, cot, etc.) are concerned with right-angled triangles.

Trigonometrical Survey. See *Triangulation*.

Trillor Vibrator. A patent electric machine for vibrating concrete to assist consolidation.

Trim. An architrave. The interior finishings to a building : architraves, skirtings, picture rails, cornices, etc.

Trim, or Trimming. 1. Framing an opening in a floor, ceiling, or roof. 2. Straightening or squaring the edges of a frame or the ends of a board. 3. Correcting the mitred end of a moulding. See *Trimming Machine*.

Trimmed Joist. A floor joist carried by a trimmer. See *Hearth*.

Trimmer. The joist supported by two *trimming joists*, and in turn carrying a number of *trimmed joists*. The arrangement is to form, or trim, an opening in a floor, as for a fireplace or staircase. See *Tusk Tenon* and *Hearth*.

Trimmer Arch. An arch, from trimmer to chimney breast, carrying the hearth. See *Hearth*.

Trimmer Joint. See *Tusk Tenon* and *Hearth*.

Trimming Joists. The stronger floor joists carrying the trimmer. See *Hearth*.

Trimming Machine. A *mitring machine* ; it may be operated by a lever handle above or a treadle below. A *guillotine*.

Trinasco. A proprietary asphaltic liquid for sealing and re-conditioning old surfaces and roofs.

Trinazzo. Coloured asphalt with marble chippings. See *Colorphalt*.

Trincomalee. A dark-red hardwood, with darker lines, from India. It is very hard, heavy, strong and durable. Difficult to work, polishes well. Used for superior joinery, fittings, bentwork, etc. Weight about 58 lb. per c. ft.

Tringle. 1. A small fillet over the Doric triglyph. 2. A rod on which curtain rings run.

Triolith. A grade of Wolman salts, *q.v.* It is for interior timbers.

Trip. A mechanism to produce instantaneous release of a machine part.

Tripartite. Divided into three parts.

Tri-pedal. A patent non-slip iron paving for heavy traffic.

Triplex Glass. Laminated glass. See *Safety Glass*.

Tripod. A support consisting of three legs spread apart at the feet and secured together at the top.

Tripod or **Triple Mast.** A very tall three-legged mast. They may be over 100 ft. high, of scarfed timbers, as used for electric power transmission before the present grid-scheme pylons were introduced.

Triptych. As diptych, but with three leaves.

Trivet. A movable bracket attached to the top bar of a firegrate.

Trochilus. A concave moulding. The outline consists of two circular arcs of different radii. A scotia. See *Mouldings*.

Trolley. A small truck, or narrow cart, usually with four low wheels that are sometimes flanged to run on lines. Like a bogey but the body is fixed to the ends of the axles. See *Derrick Crane*.

Trough. A long narrow vessel or channel for water.

Trough Gutter. A box gutter (*q.v.*).

Trowel. A steel hand-tool used for spreading plastic materials.

PLASTERER'S LAYING TROWEL

Trowelled Surface. The surface of plastic materials finished with a trowel, or *ironed*. This brings up the fat and gives a smoother finish.

Trower Cornice. A registered glass cornice with strip lighting behind. It is made in various sections, in 3 ft. lengths, and is screwed to the wall.

T.R.S. Abbreviation for tough, rubber sheathed cable, in electric wiring.

Trueing-up. Preparing the surface of wood by hand.

True Shape Section. A section mould for a radial section, as for the springer to a vault or the voussoir for an arch.

True to Scale. A process of printing (see *Blue Print*) in which the dimensions of the paper are not influenced by immersion in water.

Trumeau. A pier or pier glass.

Trumpet Arch. An arch in a thick wall having the soffit made conical to admit more light. It serves the same purpose as an "arch with orders" (*q.v.*).

Truncated. A solid having the top cut away by a plane. Also see *Frustum*.

Trunk. 1. The shaft of a column. 2. A large rectangular pipe.

Trunnel. See *Treenail*.

Trunnion. The projections on each side of a rotating cylinder that support the cylinder on the carriage.

Truscon Paste. A proprietary waterproofer for cement.

Truscon Precast Units. A sound-resisting system of flooring consisting of a patent U-shaped reinforced concrete lintel which is precast to suit the span. It has a vee groove for continuity bars and *in situ* concrete. The units may be placed either way up for "suspended tread" or suspended ceiling.

Truss. 1. A self-contained, triangulated, rigid frame of wood or iron, arranged and designed to transfer the loads, acting on the frame, to the supports. See *Roofs and Trusses*. 2. The ornamental projecting head to a pilaster for a shop front. The trusses terminate the fascia, sun-blind, etc., and may be of wood or stone. 3. A bracket or console.

Trussed Beam. A compound beam consisting of several members arranged to form an unyielding truss. It usually consists of a stout cast iron saddle under the middle, with rods fixed at the saddle and at the ends of the beam.

Trussed Centre. A framed centre for an arch.

Trussed Partition. A framed partition. It is usually supported at the ends only, and may carry a floor and ceiling. See *Partition*.

Trussed Purlin. A similar arrangement to a trussed beam, but lighter and with wood saddle. The trussing of purlins is more usual in steelwork.

Trussed Rafter Roof. 1. The usual type of steel truss used instead of a king-post truss. It consists of two trussed rafters connected by a horizontal bar. 2. Timber roofs having the collar of each couple stiffened by braces. A scissors truss, *q.v.* See *Roofs.*

Trussed Roof. A roof for spans over 20 ft., consisting of triangulated frames, or trusses, placed at 10 to 12 ft., intervals along the walls. The frames carry the purlins and ridge, and the shape of the trusses depends upon the span and the requirements.

Trussing. Triangulating a frame to make it a self-contained truss.

Trussit. A patent expanded metal, chiefly intended for thin concrete or plaster walls.

Trying Plane. A long plane used by the joiner for straightening edges.

Try-square. A tool for testing right angles.

T-section. One of the standard steel sections.

T-square. An appliance used by a draughtsman for drawing horizontal lines. It consists of stock and blade, which are usually of mahogany, with ebony edges. See *Drawing Appliances.*

Tube Lamp. A patent filament electric lamp, curved and shaped to harmonize with decorative schemes.

Tubular Scaffolding. See *Scaffolding.*

Tuck Pointing. A method adopted for old brickwork. The joints are raked and pointed, and a shallow $\frac{3}{16}$ in. groove formed in the joints. The whole is then coloured as required and the grooves filled with lime putty so that it projects about $\frac{1}{8}$ in. See *Bastard* and *Pointing.*

Tudor. A transitional style of architecture prevailing in the sixteenth century, and sometimes called Late Perpendicular. Some of the characteristics were : fan vaulting, the *four-centred arch,* panelled walls, and shallow mouldings. See *Arches.*

Tudor Flower. A trefoil ornament common in Tudor architecture.

Tufa or Tuff. A porous durable stone, formed by deposits from calcareous waters, or of volcanic origin. See *Travertine* and *Trass.*

Tufboard. A registered wood-fibre wallboard.

Tug. An iron hoop from which lifting tackle is suspended.

Tulipwood. A decorative hardwood. The colour is usually bright red with darker markings, but varies considerably, and loses lustre on exposure.

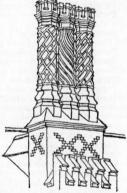

TUDOR CHIMNEYS

Tumbler. The mechanism in a lock that holds the bolt in position, until turned by the key.

Tumbling Bay Junction. A drain-pipe connection for a very steep or vertical branch. It is used for connecting a shallow drain to a deep manhole.

Tumbling Courses. Inclined brickwork courses intersecting horizontal courses, as in weathering to external flues.

Tun. The shaft of a chimney.

Tungcrete. A proprietary textured paint, to imitate stone.

Tung Oil. See *China Wood Oil.* Thinned with turps or white spirits, it is used to prevent the dusting of cement floors, and as a priming coat for cement. Also used for special varnishes, bath enamels, etc.

Tungum. A non-corroding bronze, or copper alloy. Impervious to acids and alkalis. It may be cast, drawn, spun, or extruded.

Tunnelite. A rapid-hardening cement.

Tunnel Kiln. Also called a *draw* or *continuous-running kiln.* A kiln for lime-burning with alternate layers of coal and stone.

Tunnel Vault. A barrel vault (*q.v.*).

Tupeloe. A U.S.A. timber comprising several species of gum, and often sold in this country as American whitewood.

Tuque. A greenish to olive hardwood from Venezuela. Very similar to greenheart and used for the same purposes. Weight about 34 lb. per c. ft.

Turkey Stone. See *Oilstones.*

Turkey Umber. A brown earth pigment for paint.

Turk's Head. A brush for limewashing, etc. It can be fixed to a long shaft to reduce scaffolding to a minimum.

Turnall. Various proprietary asbestos-cement products.

Turnbuckle, or Button. 1. A simple form of catch, for a cupboard door, that swivels on a screw. 2. A similar fastening on the inside of a cupboard door operated by a knob outside. A cupboard turn. 3. A coupling that regulates the tension of the connected parts. See *Screw Shackle.*

Turnery. Articles ornamented by turning in a lathe.

Turn-in. A small horizontal curve to a handrail to fix into a wall or newel.

Turning an Arch. Erecting or building an arch.

Turning Bar. See *Chimney Bar.*

Turning Piece. A solid *centre* for a segmental arch with small rise and span.

Turning Pin. A conical piece of hardwood, used by the plumber for opening the ends of lead pipes.

TURN-BUTTON

Turnpike. Same as toll-gate, *q.v.*, or turnstile.

Turnstile. A short post with horizontal arms that rotate to allow passage at the entrance or exit of a building or enclosure.

Turn Tread. See *Winder.*

Turpentine. 1. A resinous substance from trees, from which oil, or spirits, of turpentine (turps) is distilled. Turps is used as a diluent for paint, and makes it work freely and serves as a drier. 2. A brownish-red, lustrous, hardwood from N.S. Wales. It is very hard, heavy, and durable, and resists fire and *teredo navalis.* Used for marine piling, flooring, structural work, etc. Weight about 62 lb. per c. ft.

Turret. A small tower on a building. It may be used for observation purposes, ventilating, to carry a bell, or simply for ornament.

Turret Steps. Solid newel stairs in a circular chamber. Winding stairs.

Turriculated. Having small turrets.

Tuscan. The simplest of the five orders of architecture. It is the Roman modification of the Doric, but without triglyphs, and flutes in the columns.

Tusk Nailing. Two nails driven at opposite angles to increase the holding properties. Skew nailing.

Tusk Tenon. A combination of tenon and housing joint for *bearing* timbers, at right angles to each other, in constructional work. It is generally used between trimmer and trimming joist. See *Hearth.*

Tusses or Tusks. Stones left projecting to bond with later work. *Toothing stones.*

Tuyere. The pipe conveying the forced draught to a furnace.

Twin Tenons. A divided tenon. See *Gunstock Stile.*

Twintite. A patent plunger valve for steam.

Twist. 1. A form of warp in which the distortion is of a spiral character

as when the ends of a board are twisted permanently in opposite directions.
2. See *Ramp and Twist*.

Twisted Fibres. Cross grain in timber, making it unfit for constructional work. Interlocked or spiral grain, *q.v.*

Twisteel. A registered reinforcement for concrete.

Two-bolt Lock. A lock combining both lock and latch. The former is operated by a key and the latter by either knobs and spindle or a slide.

Two-coat Work. Plastered surfaces completed in two coats. See *Three-coat Work*.

Two-foot. A rule two feet long. It may be two-fold or four-fold.

Two-hinged Arch. An arched rib hinged at the ends. The stresses are statically indeterminate, unless certain conditions are assumed.

Two-light Frame. A window with two main divisions, as formed by one mullion. See *Windows*.

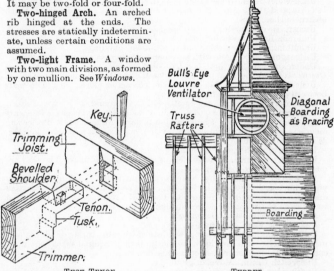

TUSK TENON

TURRET

Two-pipe System. A hot water system in which as many flow and return pipes as required are taken direct to each system of radiators. Also see *One-pipe System*.

Two-to-one. Applied to snecked rubble walling.

Two-way Column. An angle column having a re-entrant angle, to face in two directions.

Tympanum. The triangular or segmental panel of a pediment. The filling between the head of an opening and an arch above.

Type. A canopy over a pulpit. A cupola roof to a turret.

Type Sash. A window made of sash stuff but with some part opening on pivots. All the material is of uniform thickness, as with a *fixed* or *standsheet* sash.

Typhon. A simple form of temple in early Egyptian architecture.

Tyrol. A patent hand machine for applying Cullamix coloured rendering with open textured finish.

U

Ukay. A classification for Columbian timbers, especially Douglas **Fir**.

U.K. Floor. A self-centering fire-resisting floor consisting of reinforced concrete secondary beams supporting hollow concrete tubes.

Ulmus. The elm tree.

Ultimate Strength. Breaking strength. The load that just produces rupture.

Ultramarine. A blue pigment for paint, obtained from *lapis lazuli*; or it may be chemically prepared artificially. The latter should not be used in conjunction with lead pigments.

Ultra-violet Glass. Glass that allows the ultra-violet rays to pass through. It is claimed to be more beneficial to health than ordinary glass which obstructs these rays.

Umber. Brown earth pigments used in paint. See *Pigments*.

Umbrella Roof. A station roof. See *Roofs*.

Unbuttoning. Demolishing old steel-framed buildings.

Undecagon. A polygon with eleven sides. Area = sq. of side × 9·366.

Undercloak. 1. The first layer of lead worked over a roll. Also see *Overcloak*. 2. A course of plain tiles placed under Italian tiles, at the eaves.

Undercoat. A coat of flat paint under enamel or gloss paint. The coats between priming and finishing coat.

Undercroft. An apartment or chapel below ground.

Undercut. 1. A shoulder, to a tenon, cut slightly out of square to ensure a tight joint on the face. Preparing joints hard on the face. 2. Projecting members moulded to form a drip, by hollowing out a horizontal surface, as to a cornice stone.

Underdrawing. Same as torching, *q.v.*

Under-eaves Course. A double eaves course for plain tiles. The tiles are made 7 in. long for the purpose.

Under-feed System. A hot-water system in which the flow and return pipes run together round the building. The connections for the radiators are taken off the mains as required. The chief advantage is that the cooled water from the radiators passes straight to the return pipe.

Underfoot. Same as underpin, *q.v.*

Underlining Felt. Sarking felt.

Underpinning. Strengthening foundations by substituting or adding new materials. The term also includes the timbering required before the work can proceed: dead shores, sleepers, needles, etc.

Underpitch Groin. A Welsh groin, *q.v.*

Under-ridge Tiles. A course of specially made plain tiles, 9 in. long, for finishing the covering at the ridge.

Under Tile. The under tile, or tegula, in double tiles, like Italian or Spanish, *q.v.*

Undulating. An edge or surface with alternating concave and convex curves.

Uniform Load. A load may be uniformly distributed over a beam, or structural frame, so that the reactions are equal; or it may be uniformly increasing from one end to the other. See *Distributed Load*.

Unilac. A registered hardboard with lacquered finish.

Union Clip. A clip for joining together cast-iron gutters.

Union Joint. A pipe joint in which the two ends screw equally into a ferrule, or union.

Unisan. A patent chemical closet.

Uniseco. A prefabricated dwelling-house.

Unit. Anything taken as one. Inches or feet, etc., may be the English units for measurements; lbs., cwts., or tons may be the units of weight.

Unitex. A registered hardboard.

Unit of Bond. The portion of a brickwork course that is repeated continuously throughout the length. See *Double Flemish.* In a 1½ brick wall in English bond it consists of a stretcher and two headers.

Unit Stress. The stress per unit of sectional area. See *Safe Load,* and *E.*

Unity of Form. Designing and relating the various elements of a building to give a unified composition.

Universal. 1. A prefabricated house incorporating asbestos cement. 2. An asbestos-cement sheeting for large spans.

Universal Plane. A patent adjustable plane for ploughing, moulding, rebating, etc. It was also called the 55 plane because of the number of cutters, but blanks are supplied that can be cut to any required contour. It replaces the numerous joiner's moulding planes.

Universal Woodworker. A multiple machine that performs various operations, such as planing, moulding, tenoning, sawing, etc.

Unstable. Applied to wood that varies much with change of moisture content.

Untrimmed Floors. A floor with common, or bridging, joists only.

Unwin's Formulae. The various formulae derived by Professor Unwin and used in structural design.

Unwrought. Timber left from the saw. Not wrought.

Upset. A defect in timber due to a severe blow, as in faulty felling. The fibres are broken across the grain. Also called thunder shake, rupture, and cross fracture.

Upstart. Same as *Stooling.*

Upturn or Upstand. The portion of the lead of a gutter or flat turned up against the wall. See *Flashings.*

Uralite. 1. A mineral having the crystalline form of augite and the cleavage characteristics of hornblende. 2. A proprietary fireproof insulating material, in hard or soft qualities, and in sheets.

Urastone. A trade name for various asbestos-cement products.

Urea. A white crystalline solid, $NH_2CO.NH_2$, obtained by decomposing ammonium carbonate, which is produced by combining dry ammonia and dry carbon dioxide. It is readily soluble in water. Urea is used in synthetic resins, and is an important preservative, making wood nearly immune from insect and fungal attack. It increases the stability of the wood which can be glued or painted. The plasticity of green wood is increased greatly by soaking in a solution of 1 lb. urea to 1 pint of water for several days at ordinary temperatures. The wood is bent and heated in an oven and sets quickly to the required shape. Urea is the important constituent in urea-formaldehyde resins, and in phenolic products of the Bakelite types. See *Glue* and *Plastic Glues.*

Urinettes. Appliances in public conveniences for women. The pan has a movable grid.

Urunday. A reddish-brown hardwood from tropical America. It is very hard, heavy, durable, and strong, and used for piling, structural work, superior joinery, etc. Weight about 63 lb. per c. ft.

V

Val de Travers. A registered name for all kinds of asphaltic work.

Valley. 1. A re-entrant, or recessed, angle in a roof. The intersection of two roof surfaces meeting internally, having an external angle less than 180°. It corresponds to a hip, which is formed when the surfaces meet externally. See *Swept Valley* and *Roofs.* V-shaped channel or gutter.

Valley Boards. Boards running down a valley to carry the lead or zinc.

Valley Rafter. The timber on which the jack rafters intersect to form a valley. See *Roof Terms*.

Valton Floor. A patent hardwood dance floor on springs.

Valtor Girder. A patent girder for dance floors. It rests on springs that can be made solid by a locking device.

Value-cost Contract. One in which the owner bears the actual cost of the work and pays the contractor a proportionate fee.

Valve. A cover to an aperture opening in one direction only. There are many varieties used in building: safety, draw-off cocks, air cocks, mains valves, radiator valves, etc. They are classified as *safety* or *stop* valves (*q.v*).

Valve Closet. The most efficient of water closets. It is expensive because of the mechanism required, and is operated by a lifting handle in the seat top. No flushing cistern is required. The pan fills automatically after use.

Vandyke. Forming ornamentation by indenting.

Vandyke Brown. A brown earth pigment for paint. Also made artificially. It is ground in water for graining. See *Pigments*.

Vandyke Pieces. The pieces of lead left from cutting skeleton flashings.

Vane. A weathercock. An indicator showing the direction of the wind.

Vanishing Points. Points on the picture plane used in radial, or perspective, drawing.

Vanway. A passage, to a building, for lorries.

Variation Order. An order from the architect authorizing some change from the original contract.

Variegated. Marked with irregular stripes or patches of different colours.

Varnene. A proprietary oil varnish stain.

Varnish. Resin dissolved in spirits or oil and used as a finishing coat for painted surfaces. It sets hard and forms a glossy protective coat.

Varved Clay. A glacial-lacustrine deposit. Alternate thin layers of fine sand or silt and clay.

Vase. 1. The body of a Corinthian capital. See *Bell*. 2. An ornamental vessel.

Vat. A large tub or vessel for holding liquids, especially in manufacturing processes.

Vault. An arched roof. 2. An underground chamber. 3. A chamber with a vaulted roof.

Vaulting. Covering chambers with stone or bricks in arched form; or, a number of vaults collectively. The

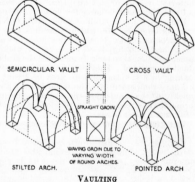

SEMICIRCULAR VAULT   CROSS VAULT

STRAIGHT GROIN

WAVING GROIN DUE TO VARYING WIDTH OF ROUND ARCHES.

STILTED ARCH.   POINTED ARCH

VAULTING

three main divisions are barrel, groined, and ribbed vaults, which are elaborated into lierne, stellar, fan, or conoidal. The outline may be cylindrical, elliptical, Gothic or pointed, conic, spherical. Other descriptions are: simple, compound, single, double, cross, diagonal, surmounted or stilted, surbased, and rampant.

Vaulting Shaft. A small column or shaft from which the ribs spring, or a vaulted roof. The shaft rises from a corbel.

Vectorial Angle. The angle between two radius vectors, as used when drawing a geometrical spiral.

Vector Polygon. The *polygon of forces* as used with the link polygon to obtain the resultant or equilibriant for a system of forces. The construction is based on the theorem : "If any number of forces act at a point and that point is in equilibrium, then the forces can be represented in magnitude and direction as the sides of a polygon taken in proper order."

Vectors. Straight lines denoting direction and magnitude. Directed quantities.

Vee Gutter. See *V-gutter.*

Vee Joint. Chamfers on the edges of match-boarding to break the joints.

Vee Roof. See *V-roof.*

Veevic. A proprietary oil-powder wash and a paint cleaner.

Vegetable Black. See *Black Pigments.*

Vehicle. The fluid used in paints in which the base and pigments are distributed and suspended. It may be oil, spirits, or water, but is chiefly oil. The oil may be boiled or raw linseed, China wood, poppy, or nut oil. Gold size and varnish may also be used as vehicles.

Vein. A streak or stripe of different colour.

Veined Pipes. Cast-iron pipes with grooves due to a defective core when moulded.

Velocity Ratio. The ratio of the movement of the effort to that of the resistance, in a machine.

Veneered Bricks. Bricks coated, to a depth of $\frac{3}{16}$ or $\frac{1}{4}$ in., with glaze, or superior clay, as the plastic clay leaves the mouthpiece during moulding.

Veneered Construction. Steel-constructed or reinforced-concrete buildings faced with a thin layer, like glass, rustless steel, marble slabs, etc. See *Walls.*

Veneers. Facings of very thin material, superior to the material that it covers. Wood veneers are sawn or sliced from ornamental timber that is generally useless for constructional work, hence the carcass is made from some other cheaper and straight-grained timber. The veneers for plywood are peeled from a revolving log, or flitch. See *Plywood, Peeling,* and *Slicing.*

Venesta. A well-known make of plywood.

Venetian Arch. A combination of camber and semicircular arches. It is structurally weak, and is sometimes supported by mullions at the springing of the semicircular arch. Also a pointed arch with intrados and extrados struck from different centres. See *Arches.*

Venetian Blind, or **Shutter.** One that consists of thin laths, or slats, like louvres, that open and close as required.

Venetian Door. One with side lights formed in the same frame.

Venetian Mosaic. See *Mosaic.*

Venetian Red. A pigment, ferric oxide, used in paint. The rich red colour is due to the iron. It is obtained by igniting ferrous sulphate.

Venetian Window. A window frame with two mullions and three pairs of sashes. One with a large centre light and narrow side lights.

Venice White. Equal parts of white lead and barium sulphate.

Vent. 1. A natural flaw in stone. 2. A small opening or flue.

Vent Duct. A passage between floor and ceiling for ventilating pipes.

VENETIAN WINDOW

VENTILATING BEAD

Ventiduct. An underground air passage.

Ventilating Bead. A wide guard, or inner, bead, at the bottom of a frame for sliding sashes, to enable the bottom sash to be raised a little without causing a draught. See page 347.

Ventilating Flue or **Duct.** A conduit for transmitting fresh air.

Ventilating Fret. Any decorative fret or panel pierced for ventilation.

Ventilating Shutter. A shutter that admits ventilation. See *Venwood*.

Ventilating Trunk. A ventilating shaft. Same as duct, but larger.

Ventilation. Replacing vitiated air or gases by fresh air. The ventilation of large buildings is usually associated with the heating arrangements. Ventilation may take place naturally by convection currents, or it may be forced, as by the Plenum system.

Ventilator. A contrivance for keeping the air fresh in a building, such as an air grid, louvre, fan, jalousie, or registered types such as Tobin, Bax, Archimedian, Sheringham, etc. In large buildings the Plenum or vacuum systems and special air-conditioning plants are used

Ventilock. A patent type of sliding sash window.

Vent Linings. Earthenware pipe linings for flues.

Vent Pipe. Ventilating pipe. See *Fresh-air Inlet* and *Expansion Pipe*.

Vent Stack. Usually applied to a soil pipe above the soil inlet. See *Fresh-air Inlet*.

Venwood. A registered shutter that obstructs light but not air. It is removable and may be hung or sliding.

Verandah. An external platform or gallery to a building, having a sloping roof and supported on slender columns or cantilevers. It is usually enclosed by a balustrade. A covered path against the wall of a house. It may be enclosed by glazed frames.

Verdigris. Acetate of copper used as a green pigment. The rust of copper and brass.

Verditer. A light-blue and green pigment. Neutral acetate of copper.

Verge Board. A barge board, *q.v.*

Verge Mould. A moulding under the projecting slates at a gable.

Verges. The overhanging edges of roof coverings at the gable.

Verge Tile. A tile-and-a-half tile.

Veribest. A bituminous roofing felt. It is water-, acid-, and fume-proof.

Vermiculated. The face of ashlar dressings having a number of irregular sinkings or pockets, suggesting a worm-eaten surface. Also see *Reticulated*.

Vermiculite. A mineral from the Transvaal that expands when heated, and used for light-weight concrete, mechanical plastering, etc.

Vermilion. Sulphide of mercury. A red pigment, *q.v.*

Vernier. A sliding, or auxiliary, scale moving parallel with a fixed scale. The sliding scale is graduated with ten divisions, equal to eleven divisions on the fixed scale, so that each is $\frac{1}{10}$th greater than those on the fixed scale. By this means $\frac{1}{100}$ of the unit can be measured. An alternative vernier has nine divisions against 10.

VERMICULATED

Verre Gravé. A registered glass with deep sandblasted design with shaded and coloured effects. Used for panels, fascias, light fittings, etc.

Versed Sine. A trigonometrical function abbreviated to *versin*.

Versin a = 1 − cos a.

Vertical Plane. See *Planes of Projection*.

Vertical Sliding Sashes. See *Sash and Frame*.

Vertical Tiling. See *Tile Hanging*.

Vesica. A panel in the form of a pointed oval. It consists of two circular arcs.

Vesta. Specially designed and registered moulded glass for fittings, panels, etc.

Vestibule. A small antechamber to an apartment. A small room just inside the entrance to a building, forming a screen or protection to a larger room or hall.

Vestry. A room, near the choir of a church, for the vestments, etc.

V-gutter. A triangular gutter, formed by the intersection of two sloping roofs, as a double lean-to roof or between North-light Trusses. See *Roofs*.

Viaduct. A long bridge over a low level, or valley, carrying a road or railway or waterway.

Vibracork. A proprietary material for damping vibrations.

Vibrated Concrete. Consolidating concrete by vibration to obtain the maximum strength. For pre-cast work the vibration is obtained by a rocking motion, but for *in-situ* work various types of patent electrical or pneumatic machines are used, which are fixed to the formwork.

Vibro Dampers. An adjustable spring-loaded device placed under a machine to absorb vibrations.

Vibro Piles. A patent expanding type of pile for soft ground. The piles are cast *in situ* and ensure dense compressed piles of full cross section.

Vibroplat, Vibropyl. Patent surface vibrators for concrete. They may be provided with teeth for pervibration, *q.v.*

Vicat Needle. An instrument for testing the setting time of cement. A frame carries a flat pointed needle, 1 mm. square in section, which makes an impress on the specimen to be tested.

Vice, or Vise. 1. An adjustable appliance to a bench, for gripping the material being wrought. 2. A solid newel stairs. See *Winding Stairs*.

Victoria. See *Granite*.

Victoria-Simplex. A patent fire- and sound-resisting floor. It is self-centering, and consists of hollow clay tiles resting on filler joists.

Vierandeel Girder. A beam strengthened by struts. The principle is applied to foundations in poor ground. An open frame girder.

Vignette. 1. Carvings to imitate vine leaves. 2. A small ornamental design without border.

Villa. A detached suburban house. A country house of some pretension.

Vinculum. A registered pre-cast floor for large spans.

Vinhatico. An orange-yellow hardwood with variegated streaks from tropical America. Rather light and soft, but strong for its weight. Not difficult to work and polishes well. Used for superior joinery, fittings, flooring, etc. Weight about 36 lb. per c. ft.

Viraŕú. A reddish-grey hardwood with reddish streaks from C. America. It is fairly hard, heavy, durable, and strong. Used for joinery, flooring, construction, etc. Weight about 40 lb. per c. ft.

Virtual Slope. The slope of a plane of saturation or a hydraulic gradient.

Visco. A patent ventilator and air filter.

Viscountess Slates. Roofing slates, 18 in. × 10 in. About 192 per square at 3 in. lap.

Vita Glass. A patent glass that admits the ultra-violet rays.

Vitrea Glass. A proprietary glass having special qualities of whiteness and transparency.

Vitreous. Having the appearance and qualities of glass.

Vitrified Bricks. Impervious bricks. The fusion of the materials fills the pores, making the bricks acid- and damp-proof. Absorption less than 3 per cent.

Vitrines. Display cabinets with daylight background. Daylight cabinets. The background is usually of frosted glass.

Vitrolite. An opaque glass, unaffected by stains, and used for wall linings, surrounds, etc. It is made in 12 colours.

Vitruvian Scroll. A type of scroll named after Vitruvius

V-joint. See *Vee Joint*.

Voids. 1. The spaces between broken, or granular, materials, as in

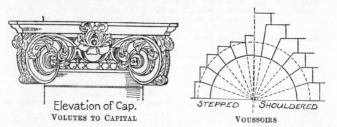

Elevation of Cap.
VOLUTES TO CAPITAL

STEPPED ┊ SHOULDERED

VOUSSOIRS

aggregates for concrete. 2. A term used in *quantities* for openings in walls and floors.

Volet. A wing of a triptych. A panel.

Volspray. A range of registered machines and paint-spraying apparatus.

Volt. The unit of electrical pressure, or electromotive force.

Volute. A scroll in spiral form. See *Ionic Volute*.

Vomitory. A staircase or passage leading from the entrance hall to the middle of the balcony, for convenience of entrance or exit, in a theatre.

Voussoir. A unit of an arch. A wedge-shaped stone or brick for an arch. See *Arch*.

V.P. Abbreviation for *ventilating pipe*.

V-Roof. A double lean-to roof, forming a V-gutter. See *Roof*.

Vronlog. South Wales slates. They are sold in standard sizes, and as randoms and rustics.

Vulcanite. See *Ebonite*.

Vulcanite Roofing. A covering for roofs. It is a hard, bituminous, elastic sheeting.

Vulcanized. Combined with sulphur by heat, as vulcanized rubber.

W

W. Abbreviation for *breaking load*, or *total load*. W_s = safe load. **w.** Abbreviation for *weight per c. ft.*, or *load per ft. run*.

Wabble Saw. A drunken saw. A circular saw running eccentrically for making wide cuts and plough grooves.

Wacke. A greyish-green clay from decomposed volcanic rock.

Wadded. Strutted. A term used by the plasterer when fixing the case for gelatine moulding.

Wafer. A pellet, *q.v.*

Wagon Roof. A trussed rafter roof (2) *q.v.*, with ashler pieces at the feet of the couples, and foot plates connecting the feet of the rafters with the ashler-pieces.

Wagon Vault. A barrel vault, *q.v.*

Wagtail. The parting slip to a sash and frame.

Wainscoting. Wall panelling made of wood, usually of wainscot oak.

Wainscot Oak. Selected oak cut radially. See *Oak*.

Waiting Room. An ante-room, for the convenience of persons waiting.

Walings. Horizontal timbers tying together the poling boards, in timbering for excavations, *q.v.* Also see *Camp Sheeting.*

Walking Line. An imaginary line used to set out the winders for a stairs. It is 18 in. from the centre-line of the handrail.

Walkway. A gangway.

Wallaba. A reddish-brown hardwood with dark streaks, from tropical America. It is very hard and heavy, durable and strong. Used chiefly for structural work. Weight about 54 lb. per c. ft.

Wall-boards. The various types of registered boards used for lining walls and ceilings as a substitute for plastering,

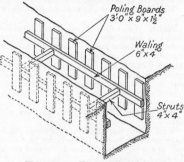

*Poling Boards
3'0" × 9" × 1½"*

*Waling
6" × 4"*

*Struts
4" × 4"*

WALINGS IN TRENCH TIMBERING

and for insulating, etc. The four main types are : wood-fibre, asbestos-cement, plaster, and plywood. Asbestos-cement boards are used for exterior work and fire-resistance. Plaster boards consist of a gypsum core papered on both sides and are extensively used instead of plastering. There is great variation in wood-fibre boards : insulating, medium hard, and hardboard. Similar emulsions are used for all three types. The degree of hardness depends upon heat and pressure in hot plate presses, with a pressure of up to 2,500 tons. The sizes vary but 16 ft. by 4 ft. is a common size. If the width is increased the length is reduced. Insulating boards may be up to $\frac{3}{4}$ in. thick, but hardboards are $\frac{1}{8}$ in. or $\frac{3}{16}$ in. There are many well-known proprietary makes, which are given alphabetically. The boards are used for all purposes: walls, ceilings, floors, joinery, formwork, etc. See *Plywood, Building Boards* and *Strawboards.*

Wall Boxes. Cast-iron boxes built in the wall to carry the end of a timber, such as a binder. They provide for ventilation to the timber.

Waller. One engaged in the erection of stone walls, usually of the rubble type.

Wallfortis. A proprietary liquid waterproofer for walls.

Wall Hangers. Cast-iron or pressed-steel stirrups for carrying the ends of structural timbers not built in the wall. A strong flange to the hanger is built in the wall.

Wall Hold. The bearing of a structural member on the wall.

Wall Hook. A strong nail with right-angled head for fixing anything to a wall. The head is curved for a pipe.

Walling. Stone walls. The various types are : random rubble (coursed, uncoursed, or built to courses), squared or snecked rubble, flint, Kentish rag, block-in-course, ashlar, and composite. See *Walls.*

Wall Paper. Rolls, or pieces, of decorative paper for covering plastered walls. English paper = 11½ yd. × 21 in., French = 9 yd. × 18 in., American = 8 yd. × 18 in. English rolls required = area to be covered in square feet divided by 60, and allow 10 per cent for waste.

Wall Piece, or Plate. A vertical timber receiving the thrust from a shore or strut. It is carried by needles and fixed to the wall by holdfasts or wall hooks. See *Raking Shore.*

Wall Plate. 1. A horizontal timber on a wall to distribute the pressure from joists or spars. See *Eaves, Hearth.* 2. See *Raking Shores.*

Wall Pocket. A small recess in a wall to receive the end of a structural timber, so that it is ventilated. The pocket should be coated with bitumen to keep the timber dry.

Wall Post. A post, or stile, in framing, attached to a wall. See *Pendant Post.*

Walls. Walls may be of brick, terra-cotta, stone, reinforced concrete, blocks of burnt clay or concrete, pisé, etc., or any combination of these materials. Modern construction, for large buildings, is usually of reinforced concrete or structural steelwork, faced with impervious material externally, and absorbent insulating material internally. Glass, marble, and stainless steel facings, with cork or other insulating materials inside, are being extensively used. Some facing materials may act as shuttering for a concrete filling.

Wall String. The string of a stairs placed against a wall.

Wall Ties. Galvanized, or tarred and sanded, iron stays used for tying together the two parts of a cavity wall (*q.v.*). Specially shaped vitrified bricks are used for the same purpose.

Walnut. *Juglans regia.* A rich, dark-brown hard wood with fine even texture. Used for superior joinery and cabinet work. Weight about 41 lb. per c. ft. European timber is artificially burred and sold as Circassian walnut. *Satin walnut* is a light yellowish brown timber used in cabinet-making. *Australian Walnut. Endiandra sp.* A highly decorative wood similar to black walnut and used for similar purposes. *Nigerian Walnut. Lovoa sp.* Golden brown, and more like a hard mahogany. American, *J. Nigra,* is similar to European but darker. There are many other species and they are all excellent woods. Used for superior joinery and fittings.

WALL TIES
AND BRICK

Wandering Heart. Timber in which the pith is not straight, and which causes outcrops of the heart on the face of converted boards. The timber is not so strong as when straight-grained.

Waney Edge. Due to converting timber too economically. The corner of the converted stuff is missing, and the presence of sap makes it unsuitable for certain classes of work.

Wants. The portions in excess of the actual area when an irregular area is measured as rectangular. The term is used in estimating.

Warding File. A flat file for cutting keys.

Wardrobe. A portable closet or cupboard or a small room, for clothes.

Wardrobe Lock. Like a cupboard lock but with a spring latch in addition to the bolt which is operated by a key.

Wards. 1. The metal rings in a lock that are intended to prevent the turning of any but the correct key, or the slots in the key to fit the rings. 2. Rooms for patients in a hospital. 3. See *Bailey.*

Warehouse. A building in which goods are stored.

Ware Pipes. Drain-pipes conforming to the British Standard specification.

Warerite. A proprietary laminated plastic-veneered board for flush doors, panelling, etc. It is moisture resisting and insulating.

Warning Pipe. 1. An overflow pipe, from a lead safe or cistern, discharging into the open, to give warning of a defect in a cistern. 2. A pipe to a W.C. service box to indicate a leakage due to the ball valve.

Warping. Not having a plane surface. Twisting, casting, or *in wind*. The different forms for wood are distinguished as bow, cup, twist, and edge-bend.

Warren Girder. A trussed steel girder. See *Roofs*.

Warren Kiln. A modern type of kiln for brick-making. It is rectangular with about 14 chambers divided by brick partitions. The burning in the various chambers is regulated by dampers.

Warsop. A registered name for power tools: pneumatic drills, etc.

Wash. A thin coat of water-colour paint applied to a drawing. Different colours are used to distinguish different materials: concrete—*Hooker's green*; masonry—*neutral tint or sepia*; brickwork—*scarlet or crimson lake*; timber—*yellow ochre or burnt sienna*; cast iron—*Payne's grey*; wrought iron—*Prussian blue*; steel—*purple*; lead—*indigo*; slates—*neutral tint with crimson lake*; glass—*French blue*.

Washable Distemper. Water paint. There are many proprietary makes prepared on a chemical basis making them fairly permanent for inside work. Usually the pigment is ground in oil and water emulsified with glue or wax saponified with water and linseed oil with an alkali. They are supplied in paste form and diluted with a petrifying liquid. Numerous colours are available and 1 cwt covers about 450 sq. yds.

Washboard. A skirting board.

Wash Bowl. A lavatory basin.

Washdown Closet. The most common type of W.C., in which the deposit enters directly to the trap and is removed by flushing with the water waste preventer. See *Water Closets*.

Washer. 1. A metal disc to distribute the pressure from the head or nut of a bolt. 2. A leather or rubber ring used to make a tight joint. 3. A brass socket for a plug in a lavatory basin.

Washing Gates. The revolving harrows in a wash mill for breaking up the clay to form slurry.

Washita Stone. See *Oilstones*.

Wash Mill. A circular tank with washing gates, used for washing, or churning, the clay and water, to separate the stones and impurities from the resulting slip, or slurry.

Wash-out W.C. An obsolete type of pedestal closet in which the shallow pan is on a higher level than the trap.

Waste-mould Process. A mould, for a plaster cast, made from a clay model. It is broken, or chipped, away from the cast when it is set.

Waste Pipes. The pipes from sanitary fittings conveying the waste to the drains: sinks, baths, and lavatory basins. They are not more than 2 in. diameter, and may be of lead, copper, or cast-iron, and discharge into a gulley trap. The pipes are trapped near the fitting, and a screw access cap is provided for cleaning the pipes.

Wasters. Defective facing bricks, sold as common bricks.

Waste Preventer. A flushing cistern, for a W.C., that only discharges the quantity necessary for flushing the pan. It may have either valve or siphonic action. The latter is compulsory under most authorities. The cistern fills automatically by means of a ball cock, and the flush is released by a pull-down chain attached to a cranked lever.

Wasting. Splitting off surplus stone with a point or a pick, preparatory to squaring and dressing.

Watchman Indicator. A device for indicating a leakage to a flushing valve for a W.C.

Watco. A proprietary preparation for destroying insects that attack wood.

Wate. A prefabricated house.

Water Bar. A strip of oak, galvanized iron, or copper, about 1 in. × ¼ in.

fitted into corresponding grooves in window sill and frame sill, to prevent water from penetrating. It is usually bedded in thick white lead paint. See *Adam's Water Bar* and *Sill*.

Water-carriage System. The term applied to the usual method of conveying sewage by drains and sewers.

Water-cement Ratio. See *Wet Mix*.

Water Checked. Casements with throatings.

Water Closets. The various types are: Valve, Siphonic, Wash-down, Wash-out, and Slop-water. The popular type is the pedestal wash-down. There is great variation of detail in all of the types, according to the different makers, but the essential principles in each type are the same.

Waterex. A proprietary liquid waterproofer.

Water Gas. A gas obtained by passing steam through glowing coke or anthracite coal and adding vaporized mineral oil to the gas. Sometimes used for lighting purposes but requires care because of carbon monoxide.

Water Gate. A gate to control the flow of water.

Water Glass. Solution of silicate of soda. A kind of soluble glass that gives a glossy coating to a surface.

Water Hammer. See *Hammering*.

Water Heating. See *Heating*.

Water-jet Method. 1. A jet of water applied, under pressure, to the foot of a pile to assist the driving of the pile. 2. A method of cleaning and preserving stonework in buildings. It is effective and has no harmful effects on the stone.

Water Joint. See *Saddle Joint*.

Water Lock. The seal of a drainage trap.

Water Logged. Ground saturated with, or full of, water.

Water Main. The main pipe supplying water to a building or town.

Water Mark. A fungus stain in wood that increases the decorative value in some cases.

Water of Cistonage. The amount of water drawn from an underground stratum over the amount absorbed annually, which will eventually exhaust the store.

Water Paint. See *Washable Distemper*.

Waterproofing. Making materials impervious to water. See *Damp-proofing*.

Water-rib Tile. A patent tile with a projecting rib to keep out driving rain or snow.

Water Seal. See *Trap* for drains.

Water Shoot. A pipe or gutter for discharging water.

Watershot. Applied to through stones in thick rubble walls that incline from back to front of wall. The facings and hearting are coursed at the same inclination, which is characteristic of rubble work in the Lake district.

Water Softening. There are numerous methods applied. *Clark's method* is to add 1 oz. of quicklime to every 700 gal. for every degree of hardness. Sodium carbonate and barium carbonate are effective. The *Permutit* process precipitates both *temporary* and *permanent* hardness. Temporary hardness can be removed by boiling. Soft water attacks lead pipes but hard waters *fur* the pipes.

Water Spout. A *Gargoyle*.

Water Stains. Colouring pigments that are mixed with water. These stains raise the grain of the timber more than spirit stain.

Water Table. An offset in a wall, bevelled to throw off the water. Plinth bricks. See *Dripstone*.

Water Tower. A tall structure carrying a tank to give the necessary head to a water-supply system. A high tower for cooling the hot water from a works.

Water Waste Preventer. See *Waste Preventer*.

Waterwite. A registered rolled glass with high light diffusion and obscurity.

Watt. The unit of electrical power. A current of 1 ampere acting through a difference of potential of 1 volt.

Wattle and Dab. Rough timber framework covered with interlaced wicker work and the surfaces plastered.

Waved Tee Bar. A patent bar for reinforcing concrete. The table, or flange, is straight, and the web "S"-shaped.

Wave Moulding. See *Mouldings*.

Wavene. A registered rolled glass, for partitions, etc.

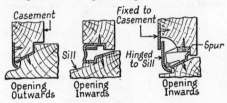

WEATHER BARS

Waving Groins. The intersections of vaults not forming straight lines in plan. See *Vaulting*.

Wavy Grain. Due to the undulation of the elements with the layers crossing each other. The fibres are arranged in short waves or ripples due to wrinkling of the fibres. The figure is common in Sycamore, Birch, Mahogany, Maple, etc.

Wax Moulds. Sometimes used by the plasterer for small casts of intricate mouldings, as egg and dart.

Wax Polishing. Using beeswax for polishing wood. A little turps is put on the rag, mixed with the wax, to make it work freely. The method is often applied to fumed oak, and to hardwood floors.

Wealden. A light-yellow sandstone with brown streaks from Sussex. It is soft with fine grain and used for interior work. Weight 150 lb. per c. ft. Crushing strength 280 tons per sq. ft.

Weather Bar. 1. See *Water Bar*. 2. Patent weather-excluding devices for doors and casements that open inwards.

Weather Board. A piece planted at the bottom of an external door to keep out draughts and driving rain.

Weather Boarding. Boards, or sidings, to cover the external faces of timber buildings. There are numerous types: plain, feather edge, rebated, and novelty. See page 356.

Weather Checks. Throatings, *q.v.*

Weathercock. An ornamental finial that shows the direction of the wind. So called because it is often shaped like a cock.

Weathered Sycamore. Figured sycamore subjected to a steam bath to darken the colour to light-brown for superior joinery.

Weathering. 1. Forming a bevelled or sloping surface to throw off rain-water. 2. The property of a material to resist chemical, physical, or mechanical change due to exposure to the weather. Resisting alternate wet and dry conditions. See *Durability* and *Decay*.

Weather Moulding. 1. See *Dripstone*. 2. A piece fixed to the bottom of a door to exclude draughts and rain.

Weather Resisting. See *Weathering*.

Weather Slating. Slating on a wall.

Weather Struck. Mortar joints pressed inwards at the top edge by the trowel. See *Pointing*.

Web. 1. The vertical connecting piece between the flanges of a B.S.B. or built-up steel girder. 2. The vertical part of a T-iron. 3. Any thin member connecting together two stronger members. 4. Infilling. 5. See *Armoured Floor*.

Webbing. A strong hemp fabric, about 2 in. wide.

Web Covers. Steel plates over the joints of the web of a built-up girder. Two angle-irons on each side of the web are riveted together, through the web and covers, to make the joint rigid.

Wedge. One of the mechanical powers, or simple machines. A tapered

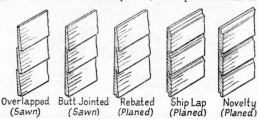

Overlapped (Sawn) Butt Jointed (Sawn) Rebated (Planed) Ship Lap (Planed) Novelty (Planed)

WEATHER BOARDING

piece of wood or metal to give a gradually increasing pressure, or to be used as a fixing. The efficiency of a wedge depends upon the taper, which depends upon the coefficient of friction of the materials in contact.

Wedge Theory. A theory of *earth pressure* for retaining walls. It is assumed that the pressure on the wall is from a triangular prism of earth formed by a *plane of rupture*, the face of the wall, and the top surface of the earth. See *Angle of Repose*.

Weepage. A term applied to water finding its way into undesirable places.

Weep Hole. A hole to allow collected water to escape, as from a condensation groove to a window or from behind an earth retaining wall. See *Casement Window*.

Weighbridge. A platform machine for weighing vehicles.

Weir. A dam or fence of stakes across a river.

Welding. Joining metals together by fusion. The methods are: forging, or by arc, gas, resistance, and thermit welding. Arc and gas welding are used in structural work. See *Oxy-acetylene*.

Well. 1. The space between the outer strings of a stairs with return flights. See *Open Well*. 2. A cylindrical hole sunk to reach a natural source of water.

Well-hole. See *Well* (1) and *Geometrical Stairs*.

Wellinlith. A registered wood-fibre wall-board and roof sheeting.

Welsh Arch. A brick cut wedge-shaped, like a keystone, and usually supported by two projecting stretchers. It is only used for a very small aperture, say 9 in. × 9 in., for a pipe or ventilator, or for a pocket.

Welsh Bricks. Hard impervious bricks having the nature of terra-cotta, from the Ruabon district.

Welsh Groin. The intersection of two cylindrical vaults of unequal heights.

Welt. See *Seam*.

West. A registered sheet pile for heavy structures.

WELSH ARCH

Western Pine. Applied to several important softwoods from Western North America : ponderosa, sugar, and white pine.

Western Red Cedar (*Thuja sp.*). A very durable softwood from Western North America. It is used for carpentry and joinery, shingles, and weather boards.

Westmorland Slates. Varied grey in colour, rather thick and expensive.

Wet Cemented. Applied to the manufacture of plywood in which the drying of the wood depends upon the heat of the plattens. The veneers are not previously seasoned before the adhesive is applied.

Wet Mix. A term applied to concrete having an excess of water in the mixing. The *water-cement ratio* is important, and only sufficient water should be added to make the concrete workable.

Wet Rot. The disintegration of timber, due to exposure to alternate wet and dry and attack by fungi. The term is also applied to a species of dry rot that requires a lot of moisture.

Wetted Perimeter. The amount of cross-sectional circumference of a pipe, or channel, that is covered when liquid is flowing through the pipe not at full bore.

Weymouth Pine. See *Yellow Pine.*

Weyroc. A wood substitute consisting of wood chips and plastic resin. It is made under pressure and heat up to 7 ft. 6 in. by 3 ft. 6 in., in different grades, also as blocks and tiles for flooring. It is very dense, tough, and strong. Also called "man-made" wood. See *Jicwood,*

" W " Glass. A pavement lens with special ingredients to resist wear and retain translucency.

Wh. Abbreviation for *whiten.*

Wharf. A staging from which ships are loaded or unloaded.

Wharf Walls. Quay Walls. They are subjected to the pressure of the earth behind the wall, and the weight of unloaded goods, and the pressure of water in front.

Whatman Paper. A hand-made drawing paper, used for good class coloured drawings.

Wheeling Step. A *winder.*

Wheel Window. A rose, or catherine-wheel, window.

Whetting. Same as sharpening, on an oilstone.

Whinstone. See *Greenstone.*

Whip and Derry. Rope and pulley used in hoisting.

Whipple Murphy Truss. A double intersection N-girder. To solve graphically, it is broken up on the *principle of superposition*; that is, the truss is treated as if consisting of two independent trusses, and the loads divided between them. See *Roofs and Trusses.*

Whippy. Flexible or springy.

Whips. Wire lashings for scaffolding.

Whip Saw. A narrow double-handed saw used by the mason for curved work.

Whitbed. A fine-grained oolitic limestone and the best stratum of Portland stone. One of the most useful building stones in this country, especially in the London area.

Whitby Cement. See *Cement.*

White Ant. An insect very destructive to timber, reducing the heart to powder. Arsenic and creosote are effective in destroying the ants.

White Bombway (*Terminalia sp.*). A brown hardwood from India and the Andaman Islands, similar to chuglum. Polishes well, and is used for interior joinery and fittings.

White Bricks. See *Suffolk Whites.*

White Cement. Imported cement used because of cheapness. It does not conform to the P.C.B.S. specification. See *White Portland Cement.*

White Chuglum. See *Chuglum*.

White Deal, or **Fir.** *Picea abies.* Norwegian spruce. See *Spruce* and *Whitewood*.

White Dhup. Indian white mahogany. Stained as a substitute for mahogany for interior work.

White Lauan. See *Lauan*.

White Lead. A base and pigment for paint, obtained by decomposing lead, in tan, by acetic acid fumes, thus forming basic carbonate of lead. It is one of the best protective materials used in paint, but should not be used for inside work because it is poisonous, causing painters' colic and plumbism. See *Pigments*.

White Metal. An alloy of nickel and brass.

White Pine. See *Yellow Pine*. Other woods marketed as white pine are: Sugar Pine, Idaho White Pine, and New Zealand Kahikatea.

White Portland Cement. Made from specially white materials and used extensively for the facing of buildings. It conforms to the P.C.B.S. specification.

Whitewash, or **Whitening.** A mixture of whiting, size, and water; used for coating ceilings, but now superseded by water paints. See *Lime-whiting*, and *Washable Distemper*.

Whitewood. The standard name for *Picea abies* from N. Europe, originally called White Deal or Fir, *q.v.* It is soft, light, not strong or durable and subject to small hard knots. Used for construction, cheap joinery, flooring, shuttering, temporary timbering, etc. Weight about 30 lb. per c. ft. Several types of wood are called whitewood, but they have other standard names. See *American Whitewood*.

Whiting. Powdered white chalk, used for making putty and whitewash.

Whittle. 1. A long edged tool for dressing the edges of slates on a traverse. 2. Aimless or indiscriminate cutting of wood. Paring or shaping with a knife.

Whole Timbers. Balks.

W.I. Abbreviation for *wrought iron*.

Wibit. A Yorkshire term used in stone slating.

Wicket. A small door, for foot traffic, framed in a large gate. The bottom half of a stable door.

Wild Elevator. A registered machine for hoisting materials on endless belt principle.

Wilderness. A dull-red sandstone from Gloucester. It is used for general building. Weight 141 lb. per c. ft. Crushing strength 695 tons per sq. ft.

Willesden Paper. A tough rot-proof waterproof paper used on roofing boards instead of felt. It is prepared by passing it through a bath of cupramonium, which is obtained from copper turnings and ammonia solution. The paper may be obtained in one-, two-, or three-ply.

Williott Diagrams. Deflection diagrams for framed structures.

Wimble. A kind of augur.

Winch. A hoisting machine based on the *wheel and axle*.

WICKET GATE

Winchester Cut. A term used in tiling where the tiles on a wall intersec the verge of a roof. Two tiles are cut, instead of one, so that each has a nail hole for fixing.

Wind, or **Winding.** To cast or warp. Not forming a plane surface.

Wind Beam. A collar beam.

Wind Bents. A braced frame designed to resist wind stresses.

Wind Brace. A brace in a framed structure to prevent distortion by the wind.

Winders. Radiating steps, to form a change of direction in the stairs. See *Geometrical Stairs*.

Wind Filling. See *Beam-filling*.

Winding Gear. Appliances used for hoisting or moving heavy loads, such as a winch, crab, capstan, or windlass, with the necessary ropes and blocks.

Winding Stairs. See *Spiral* and *Solid Newel Stairs*.

Winding Sticks. Two parallel straight-

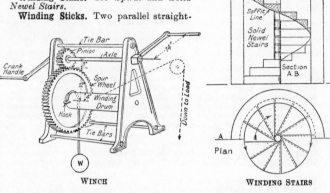

WINCH   WINDING STAIRS

edges used to ascertain, by sighting the top edges, whether a surface is perfectly plane or not.

Windlass. A machine based on the *wheel and axle*. A drum is rotated by levers, and a rope, winding round the drum, hoists the load, either up an inclined plane or vertically.

Window. Used for light, ventilation, and ornamentation. The various types are classified as: fast sheets, sliding sashes, pivoted sashes, casement windows, and Yorkshire lights, and they may be of wood or metal. The frames are divided into two classes: solid, and cased or boxed. Other terms used are: sash and frame, mullion, venetian, bay, french, double or storm, stormproof, hospital or hopper, oriel, bull's-eye, lattice, and dormer windows; skylights, laylights, and lantern lights. Windows may also be named according to their architectural characteristics, and the number of divisions in the frame. In addition, there are many patented forms of steel windows, and combinations of the above types. The area of glass should equal at least one-tenth of the floor space, and at least half must open. The sizes are measured according to the size of the wall opening.

Window Back. 1. The framing forming the back of the enclosure to a shop window. 2. The inside framing, or boarding, between the floor and the window sill of an ordinary window.

Window Blinds. See *Blind*.

Window Board. 1. The inside window bottom, or showboard, to a shop window. 2. A horizontal shelf fixed to the inside of the sill to a window frame.

Window Finishings. The linings, shutters, architraves, etc., to a window

Window Fittings. The necessary fixings for curtains, pelmets, blinds, etc.

Window Frame. A solid or cased frame carrying the sashes. See *Windows*.

Window Guards. A metal grille of an ornamental character, secured in the wall and stone sill, as a protection to a window.

Window Linings. See *Linings*.

Window Sash. See *Sash*.

Window Seat. A fixed seat under a window, especially to a bay window of a living room.

Window Shutters. See *Shutters*.

Wind Portal. See *Portal Bracing*.

Wind Pressure. The formula for vertical plane surfaces, such as walls, is $p = \dfrac{wt^2}{h}$; where w = wt. per c. ft., t = thickness, h = height, p = press. per sq. ft. = $\cdot0032$ Vel.^{2} For tall chimneys the pressure at the base is given by $\dfrac{W}{A} \pm \dfrac{P \times h}{Z}$; when P = projected area $\times p$, h = height to C.G. of projected area. Wind pressure on square chimney = 50 lb., on octagonal = 32 lb., on circular = 25 lb., per sq. ft. of projected area.

Wind Shakes. Cup or ring shakes, and sometimes applied to upsets.

Wind Shelf. A projection in a flue intended to stop down-draught.

Wind Stress. "All buildings shall be designed to resist safely a horizontal wind pressure of not less than 30 lb. per sq. ft. of the upper two-thirds of the surface exposed to wind pressure." (L.C.C. "General Powers" Act, 1909.)

Wind Tie. A diagonal tie or brace.

Wing. A smaller projection from the main part of a larger building.

Wing Compasses. Those in which a set-screw fixes the legs on a quadrant stay.

Winget. A registered concrete mixer.

Wings. The sides of the stage of a theatre. Also see *Slate Ridge*.

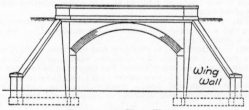

BRIDGE WITH WING WALLS

Wing Wall. A projecting wall to a structure to serve as a retaining wall. They are often curved to bridges, and support the embankment. Also see *Wing*.

Wing Window. A small side window to a doorway or large window.

Winser's Drain Chute. See *Drain Chute*.

Wiped Joint. A method of jointing lead pipes. One pipe is opened with a turnpin and the other pipe is tapered and inserted in the opened end. The pipes are soiled and then shaved clean for the required distance, and rubbed with *touch*. Solder is then poured on the joint and wiped to shape.

Wippets. A wivet. See *Stone Slates*.

Wire Comb. See *Wire Scratcher*.

Wire Cut Bricks. The clay column is cut transversely by wires as it is extruded from the machine.

Wire Gauge. See *Birmingham Gauge.*

Wired Glass. See *Safety Glass.*

Wire Mesh Reinforcement. Galvanized wire netting laid horizontally in the joints of brickwork to provide longitudinal reinforcement.

Wire Netting. Wire woven into the form of a net. It is usually galvanized, and the mesh varies considerably. It is used as a reinforcement to brickwork and concrete, and for fencing enclosures.

Wire Scratcher. A piece of wood with projecting nail points, for scratching the surface of plaster to provide a key for a succeeding coat. See *Scratcher.*

Wireweld. A registered reinforcement for concrete.

Wiring. Running electric wires around a building for an electric installation.

Withdrawing Room. See *Drawing Room.*

Withes, Withy, Withs, or **Wythes.** 1. The divisions in a chimney, separating the flues. 2. Reeds, or osiers, for interlacing thatching.

Wivet or **Wivot.** See *Stone Slates.*

Wolman Salts. A preservative for timber. They may be brushed on, but they are more effective if impregnated by pressure. The treated wood is odourless, clean, and may be painted. The different grades are Wolmanol, Tanolith, Triolith, and Minolith. See *Preservation.*

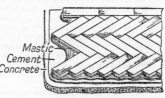

WOOD BLOCKS

Wood. See *Timber.*

Wood Beetles. See *Death Watch Beetle.*

Wood Blocks. Small blocks used for floors. A common size is 9 in. × 3 in. × 1¼ in. They are bedded in a mastic solution.

Wood-fibre Boards. See *Wall-boards.*

Wood Flour or **Meal.** Waste wood broken up into fine sawdust, like meal, and used as a filler for many synthetic materials; jointless flooring, plastics, plastic wood, wallboards, linoleum, etc.

Woodlock. A patent friction casement stay.

Wood Nog. See *Nog.*

Wood Preservation. Treating timber to make it more durable, to resist insect and fungal attack. Mineral or chemical solutions or oils are forced into the cells. The preservatives are either salts or tar oils (creosote), and include salts of arsenic, copper sulphate or chromate, mercuric chloride, tannin, napthenates, sodium fluoride, zinc sulphate or chloride, sugar, and zinc meta-arsenite. Creosote is most commonly used for exterior work, but its odour, appearance, and tendency to creep to other materials often prohibits its use, but several proprietary preparations are less objectionable. The advantage of salt preservatives is that they can be painted and have no effect on adjacent materials.

Many woods resist impregnation and incising is necessary. The naturally durable woods are usually difficult to impregnate. Dry wood and sapwood are usually more absorbent than green wood and heartwood. Pressure treatment is the best, but open-tank dipping is cheaper and commonly used. Surface application is much less effective. Well known proprietary preservatives include Boracure, Carbolineum, Celcure, Cuprinol, Hope's Destroyer, Jodelite, Microleum, Permolite, Peterlineum, Pilcher's Stop Rot, Presotim, Rentokil, Rot-doom, Solignum, Watco, and Zaldecide.

Coatings for preservation include paint, varnish, creosote, tar (coal or Stockholm), oils, charring, cellulose, polish, Also see *Preservation, Fire-resisting, Seasoning, Urea.*

Wood Pulp. Wood reduced to pulp for fibre-boards, etc. The operations are: barking, sawing, chopping, crushing, boiling or digesting, and bleaching. Caustic soda or bisulphate of lime are used in chemical production.

Wood Screw. An ordinary screw used in wood. See *Screws.*

Wood Slips. Pieces, 9 in. × 4½ in. × ¾ in., built into brickwork joints, instead of plugging. They are more satisfactory than nogs because of less shrinkage.

Woodware. Fitments, cupboards, etc. Domestic utensils made of wood.

Wood Wool. Fine shavings from softwoods for which there are many special uses. Also called Excelsior.

Woolaway. A proprietary light-weight concrete used in post and panel system of house construction.

Woolly Butt. A species of Eucalyptus, *q.v.*

Woolton. A dull-red sandstone from Lancashire; used for general building. Weight 160 lb. per c. ft. Crushing strength 400 tons per sq. ft.

Working Chamber. The chamber, under compressed air, in a pneumatic caisson. The air is kept at a slightly greater pressure than that of the water at the cutting edge of the caisson.

Working Drawings. Objects, buildings, etc., shown by plans and elevations. See *Orthographic Projection.*

Working Factor. See *Factor of Safety.*

Working from the Flat. A term used in masonry for double curvature work that is prepared from blocks of just sufficient size, instead of from cubical blocks.

Working Joints. Joints between new concrete and concrete already set.

Working Qualities or Workability. The properties of certain materials, such as stone and wood, through which the labour costs vary. For instance one wood may entail six times as much labour as another wood for the same preparation. The difference may be due to hardness, density, grain structure, seasoning, knots, resin, mineral content, stability, etc.

Working Stress. See *Safe Load.*

Working Up. Preparing a bill of quantities after taking off.

Workshop Acts. See *Factory Acts,* and *Acts.*

Worm Holes. Holes over ⅛ in. diameter, formed by wood-boring larvae.

Worm Roller. A screw roller to a saw bench. A live roller with coarse spiral threads.

Wotton. A patent roof glazing.

Wouldham Hod. A registered, metal light-weight hod for carrying bricks, plaster, etc.

Woven-oak Fencing. Very thin oak boards, like veneer, interlaced to form a close sheet, and used for fencing.

Woven-steel Fabric. The various types of reinforcement for concrete, consisting of wires interlaced and welded at the intersections; such as B.R.C. fabric, Johnson's woven-wire mesh, Lock-woven mesh, etc.

Wrack. A term applied to inferior wood.

Wray's Rule. A rule used in the design of masonry structures consisting of uncemented blocks.

Wreath. A portion of a continuous handrail going round a curved plan.

Wreathed Column. A twisted column. One carved or shaped in helical form.

Wreathed Stairs. Same as geometrical stairs, *q.v.*

Wreath Piece. That part of the string immediately under the wreath. The curved part of a continuous string for a geometrical stairs.

Wrench. A *Spanner*.

Wrot. Abbreviation for *wrought* stuff.

Wrought Iron (W.I.). Almost chemically pure iron. It is fibrous, ductile, and can be welded. It is produced by removing the impurities

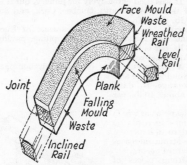

WREATH FOR HANDRAIL

from cast iron by puddling, shingling, or refining. The qualities are: common, best, best best, and treble best. The improved qualities are produced by piling, reheating, and rolling. W.I. cannot be tempered.

Wrought Stuff. Planed-up timber.

W.W.P. Abbreviation for *water waste preventer*. See *Waste Preventer*.

X, Y, Z,

Xestobium. The death watch beetle, *q.v.*

X-tgd. Abbreviation for *cross-tongued*.

Xylonite. A urea-formaldehyde thermo plastic.

y. A symbol, used in structural design, for the distance of the extreme fibres from the neutral axis of the section.

Yacal. A yellowish-brown hardwood from the Philippines. The name is applied to the hard and heavy species of the Lauan family. It is very durable, strong, and resists insects. A good substitute for teak for many purposes. Used for piling, structural work, flooring, etc. Weight about 62 lb. per c. ft.

Yale Lock. A cylindrical lock let into the face of a door and having a very neat appearance. A corrugated key operates a series of pins in the barrel, making the key difficult to duplicate.

Yard. 1. An enclosed space near and belonging to a building. It is usually concrete or paved as distinct from a garden. 2. An obsolete term for a long timber, as a rafter. 3. A measurement of length of 3 ft.

Yard Gulleys. Used for the collection of surface water drainage. They should be specially deep below the trap for the retention of silt, etc., which is removed periodically. See *Gully Trap*.

Yellow Bricks. Made from clays containing free alumina which destroys the red colour of iron oxide.

Yellow Chromes. See *Chromes* and *Pigments*.

Yellow Deal. *Pinus sylvestris*. Baltic Redwood. See *Northern Pine,* and *Redwood*.

Yellow Pine. *Pinus strobus*. The softest, lightest, and most stable of pines, and obtainable in large sizes. Excellent for carving, because of even grain, and as a core for veneer. It has very fine resin ducts, and is pale yellow in colour. It was used extensively for joinery, but it is too expensive now. Also called White, Weymouth, Quebec, Cork, and Soft White Pine. Weight about 27 lb. per c. ft.

Yellow Stringybark. A yellowish-brown species of Eucalyptus. It is excellent wood and used for structural work, piling, paving, flooring, etc. Weight about 62 lb. per c. ft.

Yellowwood. African species of Podo, *q.v.*

Yew. *Taxus.* A coniferous tree, but a hard, compact, fine-grained timber. The sapwood is white and the heartwood orange to dark brown. Weight about 40 lb. per c. ft.

Yield Point. The point at which a material becomes in a plastic state when tested to destruction. At this point the strain increases rapidly, until fracture occurs.

Yokes. 1. Strong pieces of timber bolted round the shuttering, or forms, for columns, to keep the panels in position whilst the concrete is poured. 2. A crossbar to which a bell is hung. 3. The head of a window frame.

Yorcwyte. A proprietary white metal for pipes, etc.

Yorcalbro. A proprietary aluminium brass for pipes, etc.

Yorkshire Bond. See *Flying Bond*.

Yorkshire Light. A solid window frame, one half of which is fitted with a sash sliding horizontally.

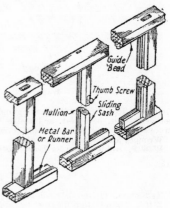

YORKSHIRE LIGHT

York Stone. A general term for a hard, compact sandstone from the Yorkshire area. It is difficult to dress, and used for engineering work, templates, pavements, landings, etc.

Yorky. Applied to a slate with a curved cleavage line.

Young's Modulus. See *Modulus of Elasticity*.

Z. A symbol for *Modulus of Section* (*q.v.*). See page 366.

Zaldecide. An insecticide for wood.

Zariba. A palisaded enclosure (N.E. Africa).

Zat. A proprietary waterproofing emulsion.

Zax. A slater's axe.

Z-bar. A standard steel Z-section.

Z-junction. A right-angled break in a wall.

Z-section. See *British Standard Sections*.

ZIGZAG MOULDING

Zeolites. Used in the Permutit system of water softening.

Zeta. A room over a porch to a church.

Zigzag Bond. Bricks arranged irregularly to form ornamental panels, as herringbone bond.

Zigzag Moulding. A chevron or dancette moulding, *q.v.*

Zigzag Partition. A staggered partition. The staggering of the studs forms two independent partitions which makes it somewhat insulated. Cabot's quilt is often threaded inside the studs to increase the insulation.

Zigzag Riveting. Riveted joints with the rivets in triangular formation.

Zilva. One of the proprietary stainless metals used for shop fronts.

Zinc. A bluish white metal that breaks with a crystalline fracture. It is malleable and ductile at ordinary temperatures, when pure. It is used extensively for coating iron to prevent corrosion, and as a substitute for lead.

Zinc Chrome. A non-poisonous yellow to orange pigment, used largely in distempers. It is used with Chinese blue for making zinc greens.

Zinc Oxide. A non-poisonous pigment for paint. It is a fine white powder, made from metallic zinc, or spelter, and is not affected by sulphurous atmospheres. It is not so good as white lead as a protective agent, but is generally used for inside work. Also called zinc white, Chinese white, and lithopone.

Zinc Templet. See *Templet.*

Zinc White. See *Zinc Oxide.*

Zoccola. The platform, in the form of a step, to a pedestal upon which a column or statue is placed.

Zocle. See *Socle.*

Zone. A band, or belt, running round an object. A horizontal strip of a spherical surface, or dome. The spherical surface between two parallel planes cutting a sphere.

Zoned. Applied to the upper stories of a tall building that are set back from the face of the building.

Zoning. A term applied to town-planning in which a logical arrangement of the different parts and interests is made. The factories, playing fields and parks, artisans' dwellings, etc., each have their own area, or zone.

Zono. A proprietary acoustic plaster.

Zonolite. A black mineral, like biotite mica. It expands to about ten times its original bulk when heated, and weighs about 5 lb. per c. ft. Used as an aggregate for special plasters and wall-boards.

Zophorus. A frieze carved to represent animals.

Zylex. A proprietary reinforced roofing felt.

Case	Shear Stress Diagram	Bending Moment Diagram	Maximum Deflection
Cantilever. W	W	WL	$\dfrac{1}{3}\dfrac{WL^3}{EI}$
w per ft.	wl	$\dfrac{wl^2}{2}$	$\dfrac{1}{8}\dfrac{wl^4}{EI}$
Beam. W	$\dfrac{W}{2}$	$\dfrac{WL}{4}$	$\dfrac{1}{48}\dfrac{WL^3}{EI}$
w per ft.	$\dfrac{wl}{2}$	$\dfrac{wl^2}{8}$	$\dfrac{5}{384}\dfrac{wl^4}{EI}$
Fixed Beam. W	$\dfrac{W}{2}$	$\dfrac{WL}{8}$	$\dfrac{1}{192}\dfrac{WL^3}{EI}$
w per ft.	$\dfrac{wl}{2}$	$\dfrac{wl^2}{24}\quad\dfrac{wl^2}{12}$	$\dfrac{1}{384}\dfrac{wl^4}{EI}$

Rectangle (B, D)	$I_{x.x} = \dfrac{BD^3}{12}$ $I_{y.y} = \dfrac{DB^3}{12}$ $Z_{x.x} = \dfrac{BD^2}{6}$	Hollow rectangle (B, D, b, d)	$I_{x.x} = \dfrac{BD^3 - bd^3}{12}$ $Z = \dfrac{BD^3 - bd^3}{6D}$
I–section	$I_{x.x} = \dfrac{BD^3 - bd^3}{12}$ $I_{y.y} = \dfrac{2TB^3 + dt^3}{12}$ $Z_{x.x} = \dfrac{BD^3 - bd^3}{6D}$	Channel	$I_{x.x} = \dfrac{BD^3 - bd^3}{12}$ $I_{y.y} = \dfrac{2T(B-x)^3 + D_2 x^3 + d(x-t)^3}{...}$ $Z_{x.x} = \dfrac{BD^3 - bd^3}{6D}$
Circle (D)	$I_{x.x} = \dfrac{\pi D^4}{64} = .0491 D^4$ $Z = \dfrac{\pi}{32} D^3$	Hollow circle	$I_{x.x} = \dfrac{\pi(D^4 - d^4)}{64} = .0491(D^4 - d^4)$ $Z = \dfrac{\pi}{32} \times \dfrac{D^4 - d^4}{D}$
Angle	$I_{x.x} = \dfrac{B(x^3 - z^3) + t(y^3 + z^3)}{3}$ $Z_{x.x} = \dfrac{I}{y}$	T–section	$I_{x.x} = \dfrac{B(x^3 - z^3) + t(y^3 - z^3)}{3}$ $Z_{x.x} = \dfrac{I}{y} \ \& \ \dfrac{I}{x}$